S0-ARJ-427

Genome Structure and Function

NATO ASI Series

Advanced Science Institutes Series

A Series presenting the results of activities sponsored by the NATO Science Committee, which aims at the dissemination of advanced scientific and technological knowledge, with a view to strengthening links between scientific communities.

The Series is published by an international board of publishers in conjunction with the NATO Scientific Affairs Division

A	**Life Sciences**	Plenum Publishing Corporation
B	**Physics**	London and New York
C	**Mathematical and Physical Sciences**	Kluwer Academic Publishers
D	**Behavioural and Social Sciences**	Dordrecht, Boston and London
E	**Applied Sciences**	
F	**Computer and Systems Sciences**	Springer-Verlag
G	**Ecological Sciences**	Berlin, Heidelberg, New York, London,
H	**Cell Biology**	Paris and Tokyo
I	**Global Environmental Change**	

PARTNERSHIP SUB-SERIES

1.	**Disarmament Technologies**	Kluwer Academic Publishers
2.	**Environment**	Springer-Verlag / Kluwer Academic Publishers
3.	**High Technology**	Kluwer Academic Publishers
4.	**Science and Technology Policy**	Kluwer Academic Publishers
5.	**Computer Networking**	Kluwer Academic Publishers

The Partnership Sub-Series incorporates activities undertaken in collaboration with NATO's Cooperation Partners, the countries of the CIS and Central and Eastern Europe, in Priority Areas of concern to those countries.

NATO-PCO-DATA BASE

The electronic index to the NATO ASI Series provides full bibliographical references (with keywords and/or abstracts) to more than 50000 contributions from international scientists published in all sections of the NATO ASI Series.
Access to the NATO-PCO-DATA BASE is possible in two ways:

– via online FILE 128 (NATO-PCO-DATA BASE) hosted by ESRIN,
Via Galileo Galilei, I-00044 Frascati, Italy.

– via CD-ROM "NATO-PCO-DATA BASE" with user-friendly retrieval software in English, French and German (© WTV GmbH and DATAWARE Technologies Inc. 1989).

The CD-ROM can be ordered through any member of the Board of Publishers or through NATO-PCO, Overijse, Belgium.

3. High Technology – Vol. 31

TABLE OF CONTENTS

PREFACE

During June 13 - June 23 1996, the 2nd EL.B.A. Foundation course on Genome, a NATO Advanced Study Institute, was held at Marcian Marina, Isle of Elba, Italy, co-sponsored by the North Atlantic Treaty Organization and the EL.B.A. Fundation. The subject of the course was "Genome Structure and Function" with participants selected worldwide from 15 different countries. The purpose of the course and of the resulting book is the study of DNA structure (from the primary to the quinternary) and gene expression in the control of cell function and cell cycle progression; the topics were presented by top experts, covering both structural (down to the atomic resolution) and functional (down to gene level) aspects. The topics were presented by top experts and scientists active in the field, with the goal to give an insight into modern problems of genome study and recent achievements in related fields of molecular and cell biology, genetic engineering, biochemistry and biophysics, oncology and biotechnology. This resulting book is intended to give a broad perspective of the current stand of these fields. The major emphasis is towards a deep understanding of DNA structure and function in interphase and metaphase chromosomes, originating by the parallel biophysical (namely NMR X-Ray and neutron scattering, spectropolarimetry, image analysis, calorimetry) and biochemical study conducted on a wide range of cell systems placing the emphasis on either the higher order DNA structure or gene structure and function. To exemplify the importance of genome studies for practical applications, the book also addresses the industrial utilization of recombinant DNA in agriculture and health, and the novel or non-conventional instrumentation for the study of DNA isolated or in situ (scanning probe microscopy, synchrotron radiation, biosensors, LB trough, silicon chip). The topics cover multiple aspects of genome studies, ranging from gene regulation, transcriptional control of cell cycle, domain organization of the genome, to large scale chromatin structure, structural studies of chromatin up to tumorigenesis and plant gene technology. This book is the result of this Advanced Study Institute and is the thertheente of a series, which began with "Chromatin Structure and Function" (1979, Vol. A21 & B) and continued with "Cell Growth" (1982, Vol. A38), "Chemical Carcinogenesis", (1982 Vol. A52), "Interactions between Electromagnetic Fields and Cells" (1985, Vol. A97), "Modeling and Analysis in Biomedicine" (1984, W.S.P.), "Structure. and Function of the Genetic Apparatus" (1985, Vol. A98), "NMR in the Life Sciences" (1986 Vol. A107), "Cell Biophysics" (1987), "Towards the Biochip" (1988, W.S.P.), "Protein Structure and Engineering" (1989, Vol. A183), Structure and Dynamics of Biopolymers" (1986, Vol. E133, Martinoff) and "Molecular Basis of Human Cancer" (1991, Vol. A209), edited or occasionally coedited by myself as Director and published mostly by Plenum within the NATO ASI Life Science Series and partly by other publishers. This book represents an update along the major trends of the series with its interdisciplinary approach to the genome structure and function.

I wish to express my gratitude to Kostantin Skryabin for his active coleadership in the planning and conduction of the course at Marciana Marina, to Sergey Vakula and Paolo Occhialini for their invaluable and critical cooperation prior, during and after the Institute and publication of this volume and to Fabrizio Nozza for the typing and editorial assistance of these proceedings.

Claudio Nicolini

GENOME STRUCTURE - FUNCTION FROM NUCLEI TO CHROMOSOMES AND NUCLEOSOMES

C. NICOLINI
Institute of Biophysics University of Genoa,
Salita Superiore della Noce 35
16132 Genoa Italy

Over the past several years the evaluation of the interphase and metaphase chromosomes organization down to the submicron resolution has awakened a great interest (see also the chapters by Belmont, Bradbury, Razin in this volume resulting from the proceedings of a NATO-Advanced Study Institute). Recently, several papers have reported the existence of different structural domains both in mitotic chromosomes and in interphase chromatin [1-7]. These domains are usually referred to as "large scale chromatin organization" and each of them can exhibit different levels of chromatin folding, from the 10 nm nucleofilament to the 30-35 nm fiber [3, 6, 8-12] and up to the 120-250 nm one [1, 5]. It is well known that the folding/unfolding of chromatin fiber at different levels of organization is strictly related to the cell functioning (see also the chapters by Stein, Wolffe, Beato, Gilmour, Turner in this volume), but large uncertainty still exists on the higher order chromatin-DNA structures *in situ*.

In the past few decades a large number of studies has been performed by using a variety of techniques which are summarized in this and few following chapters, both at the level of intact nuclei, isolated interphase chromatin and metaphase chromosomes. The fundamental repetitive unit of chromatin is the nucleosome which consists of about 145 base pairs (bp) of DNA wrapped in 1.8 superhelical turns around the histone core. The diameter of this left-handed superhelix is 8.6 nm, 4 times wider than DNA double helix itself [13]. Several papers [14-17] have shown that these superhelical turns are associated with the DNA supercoiling in eukaryotic cells and suggest that the primary role of this supercoiling is to facilitate the formation of a wrapped and functionally active structure - the nucleosome [18] - and to control the gene expression [7, 19]. Several open questions still remain: how does the level of chromatin high order structure alter the accessibility of DNA to intercalating molecules and correlate with cell function, and how is DNA supercoiling influenced by the method of chromatin isolation from intact cells? Several biophysical studies recently carried out down to the atomic resolution [20], here also summarized, confirm the artifact nature of the soluble chromatin prepared by limited nuclease digestion [21], and a close structural comparison with the preparation using hypotonic swelling of nuclei [9, 22, 23] gives new insights on the native high order chromatin-DNA structure and on the role of DNA supercoil in the control of gene expression. Other similarly relevant issues are addressed in the following chapters of this book written by Bradbury, Wolffe, Stein and Beato. Later in this volume the implications of the significant progress in our understanding of chromosome structure and function down to the gene level (see also

1

C. Nicolini (ed.), Genome Structure and Function, 1–37.
© 1997 *Kluwer Academic Publishers. Printed in the Netherlands.*

Gilmour, Harrington in this volume) appear evident by the tremendous advancements in the genetic control of human cancer (Cavenee, Georgiev in this volume) and in the area of plant gene technology (Skryabin in this volume), made also possible through the constant development in DNA sequencing following the Human Genome Project (Cantor, this volume).

1. Samples Preparation

Calf thymus *chromatin* has been extracted with two different procedures.
The first procedure (*Cold Water Method*) gives rise to undigested chromatin [22, 24-25].
Calf thymocytes were prepared as previously reported [25, 26]: small pieces of thymus were homogenized in a homogenization buffer (0.8% NaCl dissolved in 0.1M Tris-HCl pH 7.2) and filtered through a steel mesh. After centrifugation (4 min. at 300 x g) the cellular pellet was washed in PBS and the nuclei were extracted by brief incubation (3 min. at 4° C) with 0.1% Triton X 100. The nuclear pellet was washed with 10 volumes of 0.15 M NaCl - 0.01 M Tris-HCl at pH 8 and resuspended for one hour in cold 2 mM EDTA buffer, pH 8 to inhibit proteolytic enzymes. Finally the nuclei were lysed by homogenization in a Dounce homogenizer (30 strokes) and chromatin was purified by centrifugation through 1.7 M sucrose at 100.000 x g for 80 min., to divide chromatin from the broken membranes. After centrifugation chromatin pellet was gently resuspended in low (0.01 M Tris-HCl, 0.001 M EDTA pH 8) or physiological (0.15 M NaCl in TE) ionic strength buffer and dialyzed overnight versus TE at 4° C.
The second procedure named *Nuclease Digested Method* [27] utilized instead a mild digestion with micrococcal nuclease causing a distribution of the chromatin fragments length. Nuclei (5×10^8 nuclei/ml) prepared as described above, were resuspended in Buffer A (0.3 M Sucrose, 0.05 M Tris-HCl, 0.025 M KCl, 0.005 M $MgCl_2$ 0.001 M, $PhMSO_2F$, 0.001 M $CaCl_2$, pH 8) to a final concentration of 10^8 nuclei/ml, and were digested with 10-15 U/ml micrococcal nuclease (Sigma Chem, Co. St. Louis, Mo., USA) for variable time at 37°C. The reaction was stopped with 10 mM Na-EDTA. After a brief centrifugation (5 min. at 4000 x g) nuclei were lysed as previously described, homogenized and centrifuged at 2000-4000xg; then the supernatant containing soluble chromatin was utilized for the subsequent experiments. Prior to CD and X-ray measurements all the chromatin samples were dialyzed overnight at 4° C versus a suitable buffer. Shearing of the CW samples was done both by sonication (ultrasounds were applied for 8 pulses for 30" with interpulse 15", power 30%, with a SONIC 300 V/T, Imaging Products International, Inc.) or by syringe strokes.
In order to prepare *mononucleosomes*, nuclei (50 OD_{260} units), isolated from CHO-9 (Chinese Hamster Ovary) cells in culture, were extracted in 0.5% NP-40, washed twice and resuspended in a hypotonic buffer A' (Tris HCl 10 mM pH 7.4, $MgCl_2$ 3 mM, NaCl 10 mM), then digested for 2 min. at 37° C with 50 U/ml of micrococcal nuclease. The reaction was stopped by the addition of EDTA (final concentration 10 mM) and the nuclei were centrifuged at 10,000xg for 20 min. This procedure avoids the lysis of nuclei and yields a supernatant fraction comprised mainly of monomeric nucleosomes lacking histone H1 (Allegra et al., 1987).
In order to prepare *metaphase chromosomes,* 3T6 and CHO cells were incubated with PHA (100 µl/ml) and Colcemid (0.05 µl/ml) in humidified atmosphere for 20 hours at

37°C. Then they were collected by using trypsin-EDTA, washed with PBS and treated with hypotonic shock by using 0.075 M KCl for 10 minutes at room temperature.

The sample was treated with ethanol and acetic acid for 30 minutes and then deposited on a cold and wet cover glasses. The sample was air dried and dyed with Giemsa at 5% for 10 minutes.

2. Biophysical Probes

In additional to the spectroscopic methods traditionally employed for structural investigation of chromatin, a new range of complementary biophysical tools has been recently introduced to investigate the native low and high order chromatin-DNA structures.

2.1 GEL ELECTROPHORESIS

Calf thymus DNA was extracted from chromatin samples according to the standard procedure [28] and its molecular weight was determined by an electrophoretic run on 0.8% agarose gel in a Tris-borate buffer (pH 8) containing 1 g/ml of ethidium bromide.

2.2 X RAY SMALL ANGLE SCATTERING

The scattering measurements were carried out on a small-angle X-ray diffractometer AMUR-K with a linear position-sensitive detector. A Philips high voltage generator PW 1830 with Cu-anode X-ray long fine focus tube (35 kV, 35 mA) was employed as a source of X-ray radiation. The diffractometer is supplied with a 3-slit collimation system containing separate edges of the second slit forming the horizontal linear X-ray beam (15 mm length) [29, 30]. The total path of X-ray (the collimator box, the specimen holder box and the space between specimen and detector) can be vacuumed. The linear resolution of detector is 0.3 mm and the angular resolution is 0.4 mrad. The data has been collected from $s = 0.0055$ Å^{-1}, $s = 4\pi \sin\Theta/\lambda$, where 2Θ is the scattering angle and $\lambda = 1.54$ Å is the wavelength (CuK$_\alpha$ radiation). The monochromatization is achieved with a Ni-filtration and an amplitude discriminator.

The experimental data from chromatin samples and buffers were collected during a period of 3000 sec, 5-6 times in succession, with a total measurement time of 15.000-18.000 sec and the determination of the radius of gyration was based on the modified Debye's formula including distortion effects of the second order [31,32]

$$I(s) = I_s(s)\left\{ N + 2\sum_{j>k}\sum \exp\left(-4\pi^2(j-k)\Delta^2 s^2\right) \frac{\sin 2\pi s r_{jk}^a}{2\pi s r_{jk}^a} \right\}$$

where I(s) denotes the scattering intensity of the filament of subunits assumed to be spherical in the small angle region of interest $s < 0.08$ Å^{-1} ($4\pi\sin\theta/\lambda$). $I_s(s)$ is the scattering intensity of the subunit depending on the Radius of gyration of the subunit (R_g), r_{jk}^a is the average distance between the jth and the kth evaluated from the amounts of r_a, θ_a, φ_a. From the slope of the straight line in the cross-sectional Guinier plots (log s I(s) vs. s^2) we determined the radii of gyration of the cross sections (R_c). For all the cross-sectional Guinier plots we have properly chosen the angular regions in which the slope of the curve is constant.

2.3 STM - AFM IMAGING

As substrate for the STM visualization of chromatin structures we used gold-coated mica and highly oriented pyrolitic graphite (HOPG) (AsseZ, Padua). Gold-coated mica samples were prepared by evaporating gold onto freshly cleaved mica sheets keeping the mica sheet at 450°C and obtaining this way substrates with large flat areas. HOPG offers the advantage of yielding automatically flat areas on micrometer scale by simply cleaving it with adhesive tape. Due to its high hydrophobicity which leads to weak adsorption of individual chromatin fragments, HOPG could not appear as the ideal substrate [33]. However, the mobility of these fibers remains conveniently small owing to their high molecular weight being more than 1.5×10^5 Daltons [20, 34 and Table I]. Before deposition the liquid samples were dialyzed overnight against TE buffer in order to minimize the presence of salts in the sample. For each sample a 10 µl drop of the solution was deposited onto the surface of the substrate. The sample was then left undisturbed in the refrigerator at a temperature of 4°C. After drying for $1\frac{1}{2}$ hours the sample was placed on the STM sample holder and dried at room temperature for another 30 minutes. Tunneling conditions were established immediately after the evaporation of the last amounts of water. In this way we avoided dehydration of the object [35].

The instruments used were commercially available ones, STM (Asse-Z, Padua) and STM-AFM (Park Scientific Instruments, USA).

For all acquired *STM* images we used tips mechanically cut from a 0.5 mm PtIr (90/10) wire (Goodfellow, England). In the case of HOPG, before imaging the chromatin fragments the tip was tested on the surface in order to prove the capability of acquiring images at atomic resolution. The various specimens were prepared in about two hours, during which the liquid samples were kept in the freezer. The presence of chromatin structure in the liquid samples was controlled by circular dichroism spectroscopy in the 190-300 nm range before and after STM imaging. All images were acquired in constant current mode with bias voltage in the range 0.1-0.2 V and output current in the range 0.9-1 nA. In this mode the tunneling current was monitored and the tip-sample distance continuously adjusted by the STM electronic feedback in order to maintain the preset tunneling current. The compensation signal of the feedback system was registered and gave the STM image.

Utilizing fleshly cleaved mica substrates, *AFM* study of several chromatin samples was carried out in parallel. A commercially available AFM Microscope (Park Scientific Instruments, California, USA) was used in the study. Samples were prepared as in the case of STM imaging and were scanned in air with standard Si_3N_4 cantilever (tip radius < 400 Å) in contact mode, with typical forces in the range 1-3 nN. Because of the finite dimensions of AFM tips, the lateral dimensions of the biological structures were overestimated in AFM measurements. Image deconvolution was obtainable considering geometry and size of the utilized AFM probes. In particular, the width of the chromatin fibers was calculated according to a tip-sample interaction model which sketch is shown in the following figure. Supposing that chromatin fibers were not affected by the pressure of the tip, the proposed model gives a correct width by the following formula:

$$w_D = \frac{w_{app}^{\ 2}}{8R}$$

[1]

where w_D is for deconvoluted width, w_{app} is for apparent width, R is the radius of a spherical tip. To calculate the tip radius, all SFM probes have been characterized by imaging polystyrene latex spheres (15 nm in diameter) deposited onto a mica sheet. The height of the spheres and their apparent width could be related to the dimension of the tip by still using the formula [1].

According to this, the apparent width of the fiber (w_{app}) is given by the following formula:

$$w_{app} = 4\frac{h}{2}\sqrt{R}$$

[2]

where R is the radius of the tip and h the height of the fiber.

2.4 CIRCULAR DICHROISM AND CIRCULAR INTENSITY DIFFERENTIAL SCATTERING

Molar ellipticity measurements of chromatin and DNA samples were made on a Jasco 710 spectropolarimeter interfaced to a personal computer supplied with operative software for data acquisition and elaboration [9, 36]. Circular dichroism measurements were made in a nitrogen atmosphere at 25°C in a 1 cm path-length quartz cell, within a wavelength range between 360 and 220 nm. In order to reduce random error and noise, each acquired spectrum was the average of five different measurements (standard deviation < 5.5%). The following parameters were used in our experiments: time constant 4 sec, scanning speed 10-20 nm/min., band width 0.2 nm, sensitivity 10-20 mdegree, step resolution 0.5 nm and PMT voltage below 400 V [25,37]. The acquired signal is usually expressed in terms of molar ellipticity (degrees $\times$ cm^2 dmol^{-1}). The CIDS, corresponding to the differential scattering of left and right circularly polarized light, was determined outside the absorption band using an "in-house built" attachment to the standard Jasco spectropolarimeter [38]. For all the measurements the DNA concentration of samples was kept constant (7.5×10^{-5} M), and was spectrophotometrically determined by using an extinction coefficient for nucleosomal DNA at 260 nm of 21,000 cm^2/g in 0.2% sodium dodecyl sulfate buffer [39]. In three series of *ethidium bromide titration* the dye concentration was 1 mg/ml as determined using a molar extinction coefficient of 5600 at 480 nm [40].

The PTC-343 Jasco system was used with the temperature control: the temperature of the quartz cell was increased with a predetermined rate (100 K/hour in our experiments) by using a Peltier thermostatic cell holder.

2.5 HIGH RESOLUTION FLUORESCENCE IMAGE ANALYSIS BY CCD

A Zeiss Axioplan light microscope properly modified for computer controlled data acquisition, and a high-scientific grade Charge Coupled Device (CCD) camera (Photometrics, Tucson) have been used as previously described [41]. The images of about 50 cells for each sample were acquired through both a 100x/NA=1.3 oil immersion lens (depth-of-field of 1.23 m) and a 40x/NA=0.75 lens (depth-of-field of 5.17 m). Optical sections at 1 micron focal intervals were collected through the entire nucleus. In order to obtain quantitative information shading correction, dark image subtraction and nearest-neighbor algorithm were applied to the images [41]. For each image the Integrated Fluorescence Intensity (IFI) was calculated as the sum of the intensity of each pixel composing the acquired 2D nuclear image and it can be related to the amount of dye contained in the corresponding nuclear slice [26, 41]. The quantitative analysis was carried out on the restored images. Since the higher is the lens numerical aperture (N.A.) the lower is the depth of focus, the nuclear cytometric measurement of the DAPI fluorescence intensity is related to the distribution of DNA within the thickness of each optical section. The fluorescence intensity of each pixel represents the amount of DNA-bound dye contained per unit picture point, related to the amount of DNA and to its higher order folding in native chromatin.

The model which permits us to obtain a better interpretation of the fluorescence data follows the equation

$$I = \frac{N\alpha}{(d+\beta)} \ p.r.$$

where d is the dimension of the fibers of chromatin in nm; $p.r.$ is the packing ratio of the fiber; N is the number of fiber filling the physical pixels.

The values of the numerical parameters α and β ($\alpha = 13.958$ and $\beta = 18.737$) were determined following the hypothesis that peaks I and II, present both in unfixed and in ethanol/acetic acid treated samples, were due to free DNA and nucleosomal fiber, respectively. Since the positions of the peaks were not strictly identical in the Gaussian decompositions from unfixed, glutaraldehyde and ethanol/acetic fixed cells, we take as reference the weighted average positions, namely 44 gray level for peak I and 68 gray level for peak II. Earlier differential scanning calorimetry [26] confirms the validity of this assumption, considering that after ethanol/acetic acid fixation the enthalpy profiles shift towards protein-free DNA and nucleosomal DNA having melting transitions at around 353 K and 362 K respectively.

2.6 DIFFERENTIAL SCANNING CALORIMETRY

DSC experiments were performed on a Perkin Elmer DSC2 (Perkin-Elmer Corporation, Norwalk, CT) using a temperature range between 310 K and 410 K and 75 µl aluminum capsules [42, 43].

A computerized system interfaced to the calorimeter allowed us to obtain a good reproducibility and sensitivity of the signal and also offered great potential for signal acquisition, background subtraction and data display [43]. During thermal scanning the following parameters were used: low thermal rate (5 K/min.), high sensitivity (0.1 mcal/s), high sample size (60 mg per capsule).

After each measurement the corresponding baseline was acquired by performing a new thermal scanning of the denatured sample: the subtraction of these baselines from the raw data provides us with correct thermograms.

Deconvolution of the heat capacity profiles into Gaussian components was carried out by least square fitting of the acquired data.

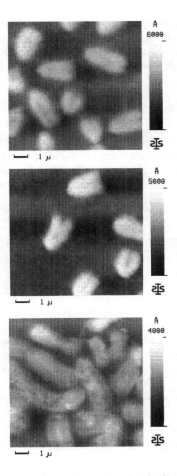

Figure 1. SFM images of (top, middle) 3T6 chromosomes acquired in air (top) and in water (middle), and of (bottom) CHO cells in air.

3. Metaphase Chromosomes

3T6 and CHO metaphase chromosomes have been studied in air and in water by using a scanning force microscope (SFM). We worked in air and in water by using standard pyramidal (low aspect ratio) and Carbon modified tips (high aspect ratio). According to their optical images (not shown) most of 3T6 chromosomes are acrocentric and only in limited cases metacentric while CHO chromosomes appear metacentric. A second strong difference consists in their dimensions with 3T6 chromosomes being smaller than the CHO ones. SFM images of 3T6 cells in air and in water and of CHO in air are shown in Figure 1. Chromatids features do not appear strongly affected by the environment: the increase in volume observed when chromosomes are in water is much less than that observed by other authors [44]. The best resolution was obtained by using the high aspect ratio carbon modified tips. High resolution images in particular pointed out the presence of a basic unit of 0.5 μm on metaphase chromatids. A similar longitudinal periodicity has already been seen in human chromosomes elsewhere [45], pointing to a strict relationship between this longitudinal pattern and the G banding.

4. Interphase chromosomes

Distinct features of the high order structure of isolated metaphase chromosomes appear present also in the native intranuclear organization of interphase chromosomes, both in the native nuclei and in native chromatin. The size and molecular weight of DNA in various chromatin preparations are summarized in Tables I and II.

TABLE I Percentage of DNA with given molecular weight, in number of base pairs (bp), as obtained for the four different chromatin preparations. DNA preparation by gel electrophoresis (1bp = 628 Daltons).

	188 ± 10	332 ± 20	491 ± 20	654 ± 30	845 ± 15	> 20kbp
cold water	0.0	0.0	0.0	0.0	0.0	100.0
1 min. nuclease digestion	29	28	23	14	8	0
5 min. nuclease digestion	48	26	14	12	0	0
30 min. nuclease digestion	100.0	0.0	0.0	0.0	0.0	0.0

TABLE II Percentage of fiber with given length (Å), as obtained by scanning tunneling microscopy for the indicated four different chromatin preparations.

Sample	120±60	240±60	360 ±60	480±60	600±60	1080±60	> 26000
Cold Water	0.0	0.0	0.0	0.0	0.0	0.0	100.0
1 min. nuclease digestion	7.0	14	28	21	21	7.0	0.0
5 min. nuclease digestion	44	29	17	7	3	0	0
30 min. nuclease digestion	100	0	0	0	0	0	0

4.1 NATIVE NUCLEI

The calorimetric profiles (Figure 2) of thymocytes and CHO cells under physiological conditions show four thermal transitions centered at about 338 K (0 Transition), 347 K (I Transition), 362 K (II Transition) and 375 K (III Transition). These transitions have been previously assigned [26, 42] as follows: Transition 0 to the melting of membranes and debris, Transition I to nuclear proteins, Transition II to nucleosome organized in 10 nm filament, Transition III to nucleosome organized in higher order structures (see Table III).

TABLE III Comparison between the relative areas (%) of the three main thermal transitions observed in thermograms of CHO nuclei and thymocytes. The values of standard deviation are also reported for each transition

Sample	I transition	II transition	III transition
CHO	22±3.9	43±11	27±2
Thymocytes	8.61±0.11	12.06±3.7	79.33±14.2

Relative area (%) of the three main thermal transitions observed in thermograms of CHO nuclei as function of the treatment with Na-butyrate. The values of standard deviation are also reported.

After the Gaussian decomposition Transition III resulted constituted by two components centered at 372 and 379 K that we referred to [see later and 46] as III_a and III_b respectively: these components are probably related to the melting of chromatin higher order structures that exhibit a different level of condensation or different DNA writhing. CHO cells show an amount of chromatin-DNA organized in the 10 nm fiber more consistent than the thymocytes having chromatin organized in the higher order structures. As reported in Table III, in terms of relative peak area, the second transition is about the 43% of the total area in CHO thermograms, while in calf thymocytes is only 12%. On the other hand the relative area of the III transition is 27% in CHO and threefold higher in thymocytes. These data can be correlated with the different metabolic activity of the cells analyzed. CHO, in fact, are characterized by an intense mitotic and synthetic activity in such a way that their chromatin has to be accessible for DNA transcription and duplication, therefore in a more relaxed conformation with respect to cells showing reduced metabolic activity such as calf thymocytes.

a

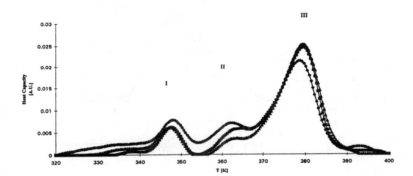

b

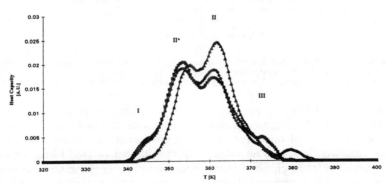

Figure 2. Profiles of heat capacity (Arbitrary Units) versus temperature (K) for three different samples of native calf thymocytes (A) and isolated chromatin (B) prepared as described in Table I.

4.1.1 Effect of permeabilization and ionic strength

Figure 3 reports the effects of ionic strength (upper panel) and of permeabilization (lower panels) on thermograms of thymocytes. Namely, when thymocytes are resuspended in Buffer A instead of in PBS:

10

- a minor shift of Transition I towards lower temperatures becomes apparent (both mercaptoethanol and trivalent ions are in fact known to stabilize proteins);
- an increase in Transition III_b enthalpy explainable by an increase of chromatin condensation when the ionic environment contains polycations such as spermine and spermidine: this effect of polycations matches the predictions of the polyelectrolyte theory [47, 48].

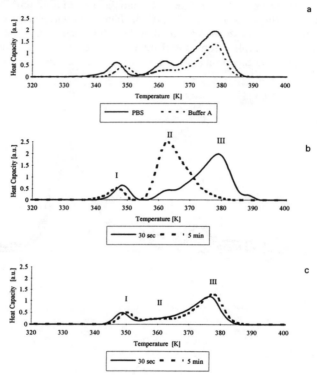

Figure 3. Profiles of heat capacity (arbitrary units) versus temperature (K) for native thymocytes resuspended in Buffer A and in PBS **(a)**; for thymocytes in PBS before and after permeabilization for 30 sec and for 5 min. **(b)**; for thymocytes in Buffer A before and after permeabilization for 30 sec and for 5 min. **(c)**.

Similarly, the effect of permeabilization is lacking in Buffer A, but is rather pronounced with PBS, leading to complete disappearance of Transition III after 5 min. (Figure 3 below). Parallel measurements by fluorescence microscopy on the same thymocytes consistently shows an increase in the number of DAPI binding sites in comparison with Buffer A, also compatible with a decrease in chromatin condensation. In agreement with other authors [2], Buffer A is needed to preserve the truly native chromatin-DNA structure during *in situ* characterization.

4.1.2 *Effect of fixation*

The calf thymocytes, either unfixed or fixed, display a multimodal distribution of their fluorescence intensity (Figure 4), where each histogram has been calculated by averaging the histograms of 10 cells and 5 stacked optical sections per each section

[41]. The interphase chromatin is indeed known to be organized in domains, folding and unfolding in different states of condensation which in turn correspond to different high order chromatin structures as it was recently confirmed for these very same thymocytes by means of differential scanning calorimetry (Figure 5 and [26]).

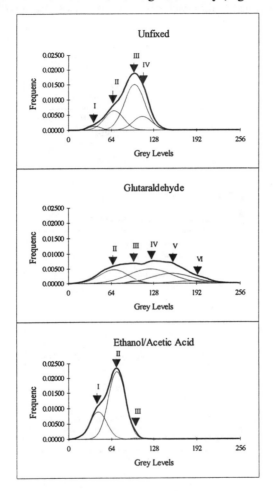

Figure 4. Fluorescence intensity distributions on the equatorial plane of nuclear images for unfixed (top) and fixed nuclei either by glutaraldehyde (middle) and ethanol/acetic acid (bottom),containing also decompositions into Gaussian components, enable us to assign the states of chromatin condensation.

The corresponding fractions of chromatin-DNA in the various conformational states clearly emerge from enthalpy profiles versus temperature shown in Figure 5 for the unfixed and fixed thymocytes; the thermogram of protein-free DNA is given as reference in the same Figure 5. Even if physical dimensions of native fibers are below optical resolution, we can attempt to explain, as shown in previous pages, the experimental fluorescence frequency distribution in terms of various levels of

12

chromatin-DNA structures and intranuclear distribution. In a simple model the frequency distribution of the nuclear fluorescence intensity (gray level histogram) can be considered as the sum of intensity distribution peaks, each related to different DNA condensation state, reflecting the number, the packing ratio and the diameter of chromatin fibers contained within each pixel, computed as indicated in Materials and Methods.

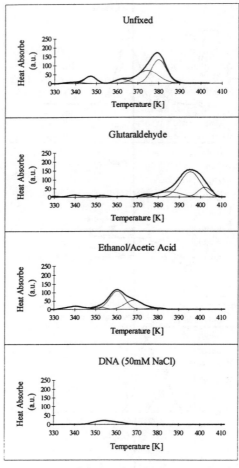

Figure 5. Enthalpy profile upon temperature for unfixed (top) and fixed nuclei either by glutaraldehyde (first middle) and ethanol/acetic acid (second middle) containing also their Gaussian decomposition summarized in Table II [26]. The figure shows also the enthalpy profile of the free DNA.

To determine objectively the number of components and the related parameters, a Gaussian decomposition method has been applied to both the fluorescence intensity distribution and the enthalpy profiles. In unfixed whole thymocytes the best fitting is achieved for four distinct levels of DNA condensation and for higher number of components the fitting optimization returns only the same four components shown in the upper panel of Figure 4. Strikingly similar results are obtained by the Gaussian decomposition of the enthalpy profiles of the same samples (Figure 5), where the number of thermal transitions closely correspond to the number of fluorescence intensity levels for each sample - with the transitions at 345 °K and at 352 °K being clearly associated to the melting of proteins and of free DNA (secondary structure) respectively.

4.1.3 Effect of ethidium bromide intercalation

When ethidium bromide is added to calf thymocytes a reduction of Transition III can be observed for low dye concentrations (Figure 6). For all the reported data we indicate the ethidium bromide concentration as a ratio (R) between dye and DNA (namely phosphate residues) concentration [25, 40, 49-52]. The reduction of Transition III is not clearly evident at very low dye concentrations (until R=0.08) and becomes more evident at increasing dye concentrations (from R=0.08 to R=0.12). When ethidium bromide concentration further increases Transition III still decreases until it completely disappears at a critical dye concentration (between R=0.15 and R=0.17). A further addition of ethidium bromide (R=0.2) causes a reappearance of Transition III and a disappearance of Transition II. For higher EB concentration (R=0.4) a broadening of this last peak becomes apparent together with a shift of Transition III_b towards higher temperature values (from 380 to 386 K). For each main endotherm exhibited by thymocytes the relative peak area is reported in figure 6 as function of EB intercalation; the corresponding relative melting enthalpy is reported in Table IV.

TABLE IV. Relative melting enthalpy (Kcal/mol) of the main thermal transitions appearing in thymocytes thermograms at different ethidium bromide concentrations: the I Transition is assigned to the melting of nuclear proteins, the II Transition to nucleosome organized in 10 nm filament, the III Transition to nucleosome organized in higher order structures. Transition III has been deconvoluted into two Gaussian components that in native thymocytes are centered at 372 and 380 K (here referred to as III_a and III_b).

R (EB/DNA)	Thermal Transitions			
	I (341-348 K)	II (362-365 K)	III_a (370-376 K)	III_b (380-386 K)
0.000	5.55	3.36	11.13	28.70
0.010	7.44	18.25	27.35	28.68
0.050	6.76	14.50	17.55	55.44
0.080	8.73	10.09	21.15	49.64
0.120	2.76	25.65	26.51	19.54
0.150	0.05	26.18	19.91	18.66
0.175	4.48	2.50	1.60	50.12
0.200	2.27	2.18	0.00	51.77
0.400	2.15	1.33	3.56	36.78

The following standard deviations have been calculated for the three main transitions: 1.3% (I Transition), 31.8% (II Transition) and 18.4% (III Transition).

While the enthalpy of Transition I does not change significantly following ethidium bromide intercalation (Table IV), it is interesting to point out that for this transition a shift towards lower temperatures (from 347 to 341 K) becomes evident with higher ethidium bromide concentrations. On the contrary, under analogous conditions, the II and the III transitions show significant and opposite alterations of their enthalpy; while for low dye concentrations (R between 0.01 and 0.1) the enthalpies change

14

slightly, for higher R values (R above 0.1) the EB effects are quite evident. Finally for R=0.15 the II transition shows an increase of its relative melting enthalpy, while the III one exhibits a drastic reduction. Further observations about ethidium bromide binding can be pointed out at higher dye concentration (R=0.2 and R=0.4): whereby the II transition completely disappears bringing about the reappearance of the III one.

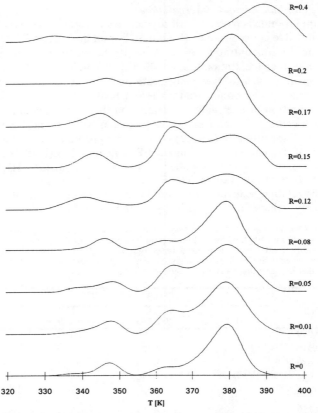

Figure 6. Profiles of heat capacity (Arbitrary Units) versus temperature (K) for calf thymocytes at increasing ethidium bromide concentrations: R=0; R=0.01; R=0.05; R=0.08; R=0.12; R=0.15; R=0.17; R=0.2; R=0.4.

4.2 ISOLATED POLYNUCLEOSOMAL CHROMATIN

The two different procedures most commonly utilized for chromatin preparation gave rise to quite different electrophoretic profiles. As shown in Figure 7, DNA fragments extracted from the cold water chromatin were longer than 25 kbp (1.65×10^7 Daltons), while the DNA corresponding to the nuclease digested chromatin presented a defined nucleosomal ladder. In particular, in the sample digested for 1 minute were evident fragments whose size ranged from the mononucleosome (about 190 bp, 1.2×10^5 Daltons) to polynucleosomes (up to seven nucleosomes). In the sample digested for 5 minutes, DNA corresponding to mono- and dinucleosomes were prevailing but it was possible to see also fragments from tri- and tetranucleosomes.

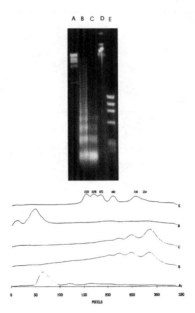

Figure 7. (Left panels) Electrophoretic profiles of chromatin isolated by the "cold water" procedure (D) and by "limited nuclease digestion" for 1 (B) and 5 (C) minutes. For molecular weight calibration l DNA of known size (bp) were also separately run in the gel electrophoresis at the same voltage and time [A and E].

Circular dichroism spectra between 200 and 330 nm, both within and outside the absorption band (Figure 8), have been acquired to confirm the two distinct structural features typical of the two chromatin preparations, whereby only the cold water chromatin displays the native feature of a circular intensity differential scattering even at low ionic strength [11, 24, 53], while mononucleosomes and nuclease digested chromatin do not (figure 8). CD spectra have been obtained at low ionic strength [19, 37]. Both chromatin preparations show at low ionic strength the typical positive band with two different maxima, one at 272 nm and one at 284 nm; this positive bi-ellipticity is characteristic of chromatin CD spectrum, with respect to the free DNA one, and brings about the appearance of the well-known signal decrease in the positive band associated with the presence of chromosomal proteins.

Figure 9 reports the molar ellipticity values at 272 nm and 224 nm acquired for native "cold water" chromatin at increasing temperatures. It is well-known that circular dichroism measurements of chromatin give a characteristic ellipticity spectrum with the positive region (above 258 nm) related to the DNA component and the negative one (below 230 nm) mostly related to protein components of chromatin. Therefore the molar ellipticity values at 272 nm and 224 nm (corresponding to the absorption band of DNA and proteins respectively) can be considered indicative parameters for investigating structural changes of these two chromatin components as function of temperature; this allows to interpret the differential scanning calorimetry data on isolated chromatin preparations (Figure 2 B), which confirm earlier CIDS and CD data (Figures 8 and 9).

16

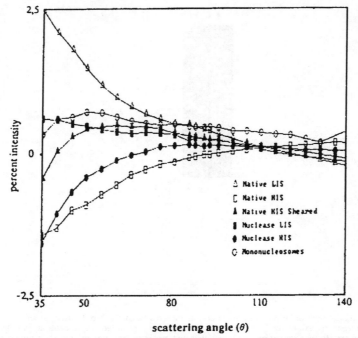

Figure 8. Circular intensity differential scattering of chromatin isolated from calf thymus by limited nuclease digestion and by cold water.

Conversely, in a recent X-ray scattering study (Maccioni et al, 1996) the major difference between the two chromatins appeared due to the band at $s=0.03Å^{-1}$ which was more pronounced in the nuclease digested sample (Figure 10). This band in the scattering curve can be considered a specific feature of the uncondensed chromatin at low ionic strength. In fact it has been reported that when the ionic strength increases it disappears, probably related to the transition of chromatin fiber from the low order structure (100Å) to the higher order one (300Å) with a progressive increase in the pitch [54]. The higher level of condensation could be possibly due to the folding on itself of the long nucleofilament resulting from truly native chromatin fibers. With shearing cold water chromatin decreases its mean Rc value from 74 to 42 Å, quite similar to the ND value.

As shown earlier (Table II), STM images of the very same samples yield results in terms of fiber length and width quite compatible with the independent molecular weight determination by gel electrophoresis (Figure 7). Fig. 11 shows an STM image of the cold water sample on gold-coated mica, which can be described as two parallel fibers made of thinner ones which appear to superfold. The size of these combined fibers was estimated to be over 20,000 Å in length and about 800 Å in width (Table II). While with the exception of one fiber of about 110 Å all fibers appeared to have a bimodal width distribution, centered respectively at about 700 Å±35 Å and 363.4±40Å; the average height was 80.2±22 Å and the length ranged between 2000 and 11,000 Å, quite compatible with AFM determination on the same chromatin preparation

[20]. Table II summarizes all geometric features of various chromatin preparations, quite compatible with the gel electrophoretic data on the same samples (Table I).

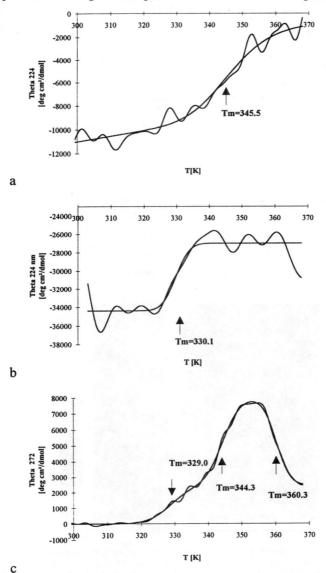

a

b

c

Figure 9. Plots of molar ellipticity values at 224 nm (θ_{224}) acquired for native cold water chromatin at increasing temperatures (from 300 to 370 K). (A) isolated chromatin without ethidium bromide; (B) isolated chromatin with ethidium bromide (R=0.4). Plots of molar ellipticity values at 272 nm (θ_{272}) acquired for native cold water chromatin without ethidium bromide (C). Samples have been analyzed in TE buffer pH 8 under standard acquisition conditions (standard deviation of the measurement was · 5.5%)

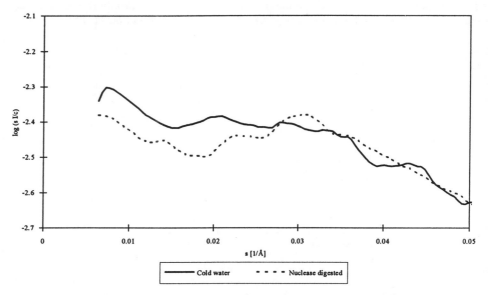

Figure 10. X-ray scattering data of chromatin isolated by the cold water method and by the limited nuclease digestion.

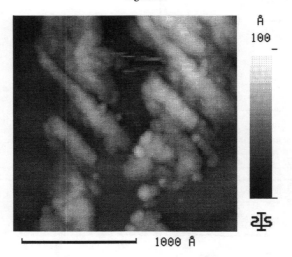

Figure 11. Two dimensional view of an STM scan of the cold water sample, voltage 0.2 V, current 0.9 nA. The SPM study here reported appears to confirm previously reported data [34, 53] on the effect of nuclease digestion and confirms that the true quaternary and quinternary chromatin-DNA structures under native conditions would seem to consist of fibers 300-350 Å and 700-800 Å wide, respectively.

4.2.1 Effect of Ethidium Bromide Intercalation

In order to better understand the nature of the structural changes observed on *in situ* chromatin after ethidium bromide intercalation, analogous measurements have been carried out on the physiologically isolated chromatin (Figure 12 and Table V).

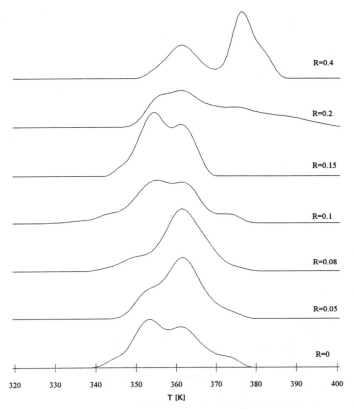

Figure 12. Profiles of heat capacity (Arbitrary Units) versus temperature (K) for isolated chromatin at increasing ethidium bromide concentrations: R=0; R=0.05; R=0.08; R=0.1; R=0.15; R=0.2; R=0.4.

TABLE V. Relative melting enthalpy (Kcal/mol) of the main thermal transitions appearing in chromatin thermograms at different ethidium bromide concentrations. In comparison with thymocytes chromatin shows an additional thermal transition at 354 K which is probably related to the melting of linker DNA, while the Transition III_b is absent and appears only at R=0.4.

R (EB/DNA)	Thermal Transitions				
	I (341-348 K)	II* (350-355 K)	II (362-365 K)	IIIa (370-375 K)	IIIb (380-386 K)
0.00	1.61	17.90	23.53	6.91	0.00
0.05	1.25	6.71	29.14	3.91	0.00
0.08	2.04	0.00	23.91	4.35	0.00
0.10	0.00	8.18	8.66	1.59	0.00
0.150	0.00	10.92	4.14	1.42	0.00
0.200	0.00	2.30	7.94	8.90	2.06
0.400	0.00	0.06	8.27	10.01	2.41

We can point out some differences between isolated chromatin and intact thymocytes at the level of their thermograms. First of all, the procedure of chromatin isolation provides a partial loss of high order chromatin structure which can be revealed by an enthalpy reduction of Transition III that is only evident as a shoulder in the last peak of the thermogram. It is also interesting to underline that this shoulder corresponds to Transition III_a, Transition III_b being completely lost after chromatin extraction.

Second, a reduction of Transition I is apparent probably because of the loss of non-chromosomal proteins following chromatin extraction. Third, Transition II which in native thymocytes was centered at 362 K, in chromatin results split into two different transitions centered respectively at 354 K (Transition II*) and 362 K (Transition II). Transition II* could be ascribed to the melting of the linker DNA. After ethidium bromide intercalation chromatin changes its calorimetric profile: at low dye concentrations (R ranging from 0 to 0.125) Transition III shows no significant alteration; by increasing ethidium bromide concentrations an enthalpy redistribution between Transitions II and III becomes evident. In particular, for R value of 0.15 Transition III completely disappears: at this point when EB concentration is still increasing (R=0.2 and R=0.4) a further reduction of Transition II appears together with a corresponding increase of the III one. These results are consistent with the data previously obtained for *in situ* chromatin: the only difference is a less evident change of Transition III after EB intercalation in isolated chromatin because this transition is already poor when the dye is absent. Similar experiments have also been carried out on digested chromatin and calf thymus DNA in order to clarify the kind of structural changes appearing in chromatin (isolated and *in situ*) when EB intercalates DNA. No significant enthalpy reduction is observed both for free DNA (R=0.05 and R=0.2) and for digested chromatin (R=0 and R=0.4) at increasing EB concentrations. It is interesting to point out that in physiologically isolated chromatin a pronounced enthalpy redistribution becomes apparent between Transitions II* and II: at low EB concentrations the enthalpy of Transition II* (centered at 354 K) decreases until it disappears at R values ranging from 0.05 and 0.1. For higher EB concentrations this transition starts to increase until it becomes more pronounced than Transition II (centered at 362 K). At last these two peaks merge in a single broad transition (R=0.2 and R=0.4).

In order to verify the peak assignment proposed here we carried out a series of selective digestions by using micrococcal nuclease, DNase I and proteinase K. While the digestion with micrococcal nuclease produces a disappearance of Transition III without changing Transition I, the digestion with DNase I induces both a loss of Transition III and a minor shift of Transitions II and I to lower temperatures (the melting temperatures of these two transitions move respectively to 357 and 343 K). When thymocytes are treated with proteinase K their thermograms exhibit a disappearance of Transition I (which confirms its assignment to protein melting) together with loss of Transition III.

4.3 MONONUCLEOSOMES

The length of DNA extracted from mononucleosome has been checked by electrophoresis and measured less than 200 bp and therefore lacked the H1 histone. As previously shown [55] electrophoretic profiles of proteins extracted from our mononucleosome preparation point to a lack of histone H1 (Figure 7). The comparison between mononucleosome and chromatin shows an interesting difference at the level of their CD spectra: while chromatin spectrum shows, as seen previously, two different maxima (one at 272 nm and one at 284 nm) in the positive band, the mononucleosome sample is instead characterized by only one maximum at 272 nm, resembling free DNA in its linear form. Furthermore it shows a lack of any differential light scattering outside the absorption band above the 310 nm and a lack of ethidium bromide intercalation

[25]: the mononucleosomes appear unable to bind this dye at any concentration, as shown by the zero CD signal at 308 nm for the EB-mononucleosomes complexes and by the invariance of the CD spectra in the 250-300 nm region (pointing to the lack of any DNA unfolding which had been earlier reported upon EB intercalation for other mononucleosome preparations). At the same time free-DNA and chromatin display a strong CD signal corresponding to the EB absorption band in the 300-320 nm range.

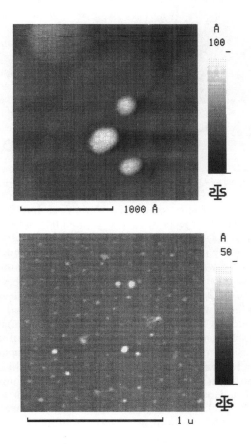

Figure 13. STM (above) and AFM (below) scan of mononucleosomes obtained by extended nuclease digestion up to 30 minutes.

New enlightening geometric details down to a very high resolution have been recently obtained by STM and AFM (Figure 13) for the mononucleosomal subunit obtained by extended nuclease digestion up to 30 minutes [20]. This result confirms the possibility of STM to image large adsorbates [20]. The mechanisms of image formation involve very likely the measurements of ionic currents which would appear because of the relative humidity conditions and the presence of salt ions from the physiological buffer in which chromatin is suspended. The 3D image of a single mononucleosome deposited onto HOPG (Figure 14), in spite of the problems which can arise using this substrate, shows features in complete agreement with the known atomic resolution structure of this basic genome unit by X-ray crystallography [56] and neutron low angle scattering [57].

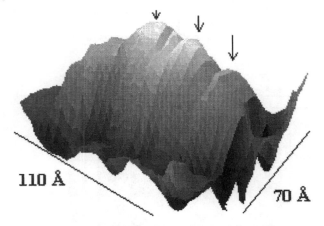

Figure 14. STM scan of mononucleosome on HOPG.

SPM images of the fourth sample at 30 minutes nuclease digestion show images displaying isolated objects which size matches well with that of the mononucleosome (Figure 13). Fig. 14, measured by STM on HOPG, provides unique high resolution details of the geometric feature of a single mononucleosome, an oblate like shape quite compatible with independent earlier X-ray and neutron crystallography [13]. Striking is the clear appearance of 2.5 turns of DNA fiber 20 Å wide, superfolding around the octamer core 70 Å wide, 110 Å long and 70 Å high; the nucleosomal DNA superhelix appears 4.5 times wider than the DNA double-helix itself.

Worthy of notice is that these biological adsorbates, imaged on HOPG by STM, often display depressions relative to constant current level of uncovered substrate surface. Similar depression of 20-30 Å depth is frequently apparent in free DNA samples, in mononucleosome samples (100 Å long and 70 Å wide) and in polynucleosomes of different length which appeared as depressions 100-120 Å deep. On these images the nucleosome rod appears as a depression accompanied by a protrusion, due to mechanical removal of the biopolymer and subsequent "hole" effect. The detailed explanation of this phenomenon, however, is given in Nevernov et al. [58].

5. Alternatives models of overall genome organization in situ

At the limit of optical resolution, the dimensions of the elementary unit have been considered to be equal to 200x200 nm². Whenever each pixel is filled with only one type of fiber, characterized by specific size and DNA packing ratio (p.r.) with N being the number of identical fibers for each pixel, the DAPI fluorescence distributions, resulting from the various models and experimental observations reported in the literature, appear unimodal and centered at distinct fluorescence intensity (Figure 15).

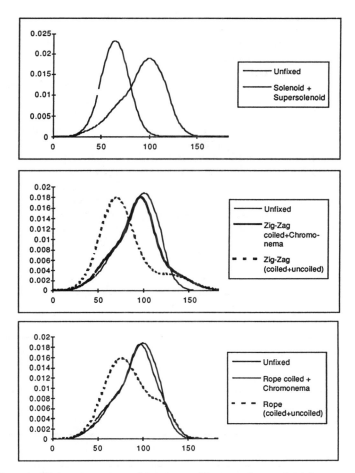

Figure 15. Theoretical fitting of experimental fluorescence histograms from unfixed thymocytes based on the assumption that physical pixels of the corresponding nuclei are homogeneously filled by the free DNA (3%), the nucleofilament (23%), the 130 nm 'chromonema' superfiber (58%) and the 28-33 nm fiber (16%), either rope-like (bottom) or ribbon-like (center), all having known chromatin-DNA packing ratio (p), diameter (d) and number of fibers per pixel (N), as reported in Table VIII For comparison (above) is given the best theoretical fit obtainable with the well-known solenoid (16%) and supersolenoid (58%) fibers, which parameters are also given in Table VIII

Only the combination of four differently packed chromatin-DNA fibers properly distributed within the overall nuclear volume can optimally match the experimental data, with each pixel containing only fibers of the same diameter and packing. DNA (2.2 nm wide) is folded within the interphase chromosome through interaction with histone and non-histone proteins into multiple levels of organization with increasing linear condensation [10, 59, 60]. The lowest and best characterized level of packing, up to atomic resolution [61, 62], is the 10 nm nucleofilament accounting for a 7:1 packing ratio and consisting of a string of nucleosomes having 210 bp of DNA wrapped around the exterior of a histone octamer [56, 63]. For this reason we have determined the value of the α and β coefficients by best fitting with the above equation the first two Gaussian distributions present in all three experimented fluorescence histograms of fixed and unfixed thymocytes, whereby the nucleus is assumed filled by two DNA fibers with known and universally accepted packing ratios (1 and 7) and diameters (2.2 and 10 nm).

We are then in a position to simulate the unimodal fluorescence distributions expected from models having been proposed in the last two decades for both the quaternary [6, 10-11, 62, 64-67] and quinternary [1-2, 20, 60, 68-71] chromatin-DNA structures, either prior or after glutaraldehyde fixation. Most models proposed so far in the literature do not appear compatible with the experimental DNA fluorescence distribution, with the exception of two models which instead appear to simulate with different but similarly high accuracy the experimental data from unfixed thymocytes. Both models, the helical rope-like [10, 67, 72] and the helical ribbon-like [66, 71], are based on a 30-35 nm fiber resulting from the strong intertwisting among helically arranged 10 nm nucleofilaments and appear uniquely capable to explain with their coiled forms the IV experimental peak, while the 130 nm superfiber identified by Belmont et al [2] in native interphase nuclei optimally fits the III fluorescence peak.

Table VI. Fiber width (D) and pitch (P) and DNA packing ratio (P.R.) experimentally determined for the chromatin DNA structures being discussed so far world-wide.

Structure	Reference	D(nm)	P.R.*	N	P (nm)
Secondary	[59]	2	1	100	
	[20]	2.2	1	91	
	[60]	2.2	1	91	3.4
Tertiary	[56]				
	[73]	10	7	20	11
	[71]	10	7	20	11
	[20]	10.5	8	19	
	[10]	11	7	18.2	11.7
Quaternary	[64]				
	[65]				
	[66]	30	42	6.6	11
	[11]				
	[10]	33 (coiled)	77 *	6.0	32
	[67]	23 (relaxed uncoiled)	30	8.8	32
	[6]				
	[62]	28 (coiled)	66	7.0	11
		30 (uncoiled)	35	6.6	11
Quinternary	[73]	40	11	5	
	[74]	60	7	3.3	
	[75]	33	77*	6.0	32
	[2]	130	690#	1.5	
	[76]	300	462*	0.66	
	[1]	200	250	1	
	[77]	200	824	1	
	[78]	200	1000	1	
Glutaraldehyde-fixed High order	[71]	23.5	69	8.0	11
	[20]	140	1309	1.4	
	[79]	130 (coiled chromonema)	1000	1.5	
	[79]	70 (uncoiled chromonema)	160	3.3	

* This quantity has been estimated from the equation $\frac{2\pi}{P}\sqrt{(D/2)^2+(P/2\pi)^2}$ coherently with the model proposed by the authors.

This quantity has been estimated taking into account that this fiber could explain the third peak of the Gaussian decomposition.

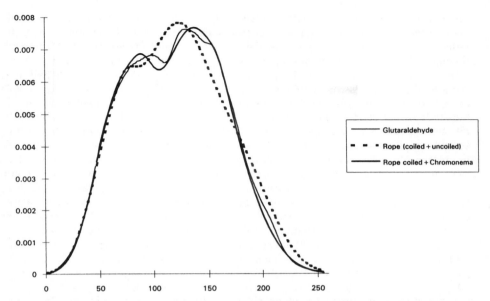

Figure 16. Theoretical fitting of experimental fluorescence histograms from glutaraldehyde-fixed thymocytes considering the rope model for higher structures of chromatin.

The other closest alternative superstructure for the III level of fluorescence, the quaternary uncoiled 23-30 nm fiber (either rope- or ribbon-like), fits to a lower degree the experimental distribution of the unfixed thymocytes appearing furthermore unlike such unfolding at the physiological ionic strength of the native nuclei. The effect of gluteraldehyde fixation further corroborates this conclusion (Figure 16), whereby the two higher fluorescence peaks (V and VI) can only be explained respectively by the 130-140 nm coiled chromonema super-fiber (as observed by Nicolini et al, 1996 [20] and Belmont and Bruce, 1994 [79] in glutaraldehyde-fixed samples), and by the zig-zag ribbon shrank to about 23 nm (as observed recently by Woodcock, 1994 [71] in fixed nuclei). The results of the fitting to the experimental mean fluorescence histograms (with and without the higher order superfolding) are given on figure 16 for glutaraldehyde-fixed thymocytes, even in presence of uncoiled 23-30 nm fiber. Finally, from the cross-section of an unfixed nuclei we can identify regions of the highest pixel intensity (state IV) forming a ring ~1.0 microns thick near the nuclear border, compatibly with earlier findings [4]; "chromatin-DNA bodies" with distinct level of intensity III, II, and I are also clearly recognizable with the expected Gaussian local distribution, which are consistently preserved after gluteraldehyde fixation, compatibly with earlier findings [8]. In summary, the parallel acquisition of DAPI-stained nuclei from optical sections of both unfixed and fixed thymocytes, properly reconstructed by the modeling procedure previously outlined, allow to infer new quantitative insights on the *in situ* high order chromatin-DNA structure to a degree previously unmatched by earlier low resolution (0.2 micron) light microscopy. This is made possible only by the simultaneous consideration of the highly reproducible effect of both the ethanol/acetic acid mixture and the glutaraldehyde on the native intranuclear DNA distribution resulting from the DAPI-stained thymocyte images. A close comparison with differential scanning calorimetry consistently points to the presence of disperse nucleofilaments after both ethanol-acetic acid and glutaraldehyde treatment, with the latter inducing cross-linking within the most condensed 30 nm fibers. Moreover, recently gluteraldehyde fixation appeared to produce no significant changes in the ultrastructure or diameter of chromatin fibers observed *in situ* in frozen hydrated sections [71]. The possibility to objectively identify chromatin fibers in unfixed whole cells, here and earlier [62, 10, 75] demonstrated, establishes an unequivocal baseline for the study of native chromatin and chromosome architecture. Without such a baseline, all previous experimental approaches to these questions, based on conventional thin sectioning of nuclei and on studies of isolated chromatin, neither of which proved able to preserve the native state of chromatin organization [7, 9, 11, 12, 73, 80] can only lead to a further increase of the existing large ambiguity for the highest levels of chromatin DNA folding. It does not escape our notice that in unfixed thymocytes the four levels of fluorescence intensity can be strictly explained in terms of the four known levels of chromatin-DNA structures, namely: state I = free DNA fibers (2.2 nm, secondary structure), state II = nucleofilament (10 nm, tertiary structure), state III (relaxed or uncoiled ribbon- or rope-like 30 nm structure) and state IV (coiled ribbon- or rope-like 30 nm structure, as independently identified long time ago by Woodcock et al, 1984 [62] and Nicolini et al, 1983 [10]), the latest being the single universal structure adopted by inactive genes *in vivo* and *in vitro* which may assume a wide range of conformations with distinct packing ratio and diameter in the 70-35 and 25-35 nm range. Large-scale chromatin domains consisting of the 30 nm fiber packed into a 130 nm or larger solenoidal superfiber named "chromonema", which appears predominant in metaphase chromosomes [1, 2] and interphase mainly after gluteraldehyde fixation [1, 20, 79], do also fit the III level of fluorescence intensity and appear thereby a possible DNA conformation during interphase, an alternative to the uncoiled ribbon - or rope-like 23-30 nm fiber.

6. Structure-Function Relationship

Transcriptional deregulation of genes occurs through chromosoma rearrangement at the level of the higher order genome structures described earlier, which involves up to 300 kbp of the genome and appears a prerequisite of gene translocation. Furthermore, genes frequently are expressed by clusters (Stein, this volume). The role of chromatin - DNA structures in the control of gene expression and cell function is confirmed by the fact that gene translocation, namely of *abk*, *bcl*, *bcm* and *tell* genes, between chromosomes and subsequent gene fusion appears related to acute leukemia and leukemic cell lymphoma. More than 40 genes have been identified as capable of being activated through the mechanism leading eventually to cancer (Cavenee, this volume).

6.1 CELL CYCLE ALTERATIONS

Time ago [81], by monitoring the effect of NaCl on the H-NMR spectra of nucleosomes in S-phase versus M-phase, we noticed a monotonous decrease of the line width with increasing NaCl pointing to a change in the degree of motion of DNA phosphates. Conversely the resonance peak area, being a rough estimate of the number of free phosphate of the molecule, appears strongly affected by the binding of histones to DNA that "freeze out" a two base-pairs long DNA region adjacent to the binding site, increasing with protein removal due to the higher number of free phosphates in the nucleosomal DNA backbone. Going from 0 to 2 M NaCl a loss of sensitivity was apparent, likely due to the worsening of the field homogeneity with increasing the ionic molarity of the solution.

The DNA phosphate resonance areas were therefore normalized with respect to the reference TMP resonance area. If we make a conservative assumption that complete chromosomal proteins removal occurs at 2.0 M NaCl the fraction F_M of bound phosphates defined as $F = 1-(A_R/A_{2.0})$ can be plotted against the NaCl molar concentration for both mono- and oligonucloesomes, where $A_{2.0}$ is the resonance area at 2.0 M NaCl. The maximum value of F_M reached at 10 mM Tris-KCl is proportional to the fraction of ^{31}P resonance area which can be lost in the native complex by chromosomal proteins binding. From Table VII it is also possible to understand which is the effect of different ionic strengths on the overall nucleosome stability. When we consider mononucleosomes in S-phase, we can see a noticeable effect at 0.35 M and 0.6 M NaCl, little effect at 1.2 M and again significant changes at 2.0 M NaCl. On the other hand, oligonucleosomes still behave in the same way but the effect at 0.35 M NaCl disappeared. When compared to S-phase, M-phase mononucleosomes appear to be similar in line-widths. Such similarities end when we consider oligonucleosome salt-induced line-widths. F_M values for mononucleosomes and oligonucleosomes in M-phase point out a very little effect at 0.35 M and 0.6 M NaCl when compared to S-phase nucleosomes.

TABLE VII. ^{31}P NMR resonance area of oligonucleosomes in S- and M-phase versus salt concentration at 10 mM Tris-HCl, pH 7.0. Normalized area was obtained dividing the nucleosome area by the TMP area, calculated by integrating the peaks at each salt concentration from Fig. 3. The numbers are expressed in arbitrary units.

[NaCl]	S-phase		M-phase	
	Monomers	Oligomers	Monomers	Oligomers
0 mM	3.6	3.9	1.9	2.4
350 mM	4.7	3.8	1.9	2.5
600 mM	5.4	4.7	1.8	2.6
1.2 M	5.9	5.4	2.1	2.9
2.0 M	8.0	11.5	2.7	3.3

Moreover, the fraction of bound phosphate for NaCl concentration lower than 1.2 M seems to be much smaller in M-phase than in S-phase. From Table VII it is evident how the change in F_M is much smaller in M-phase than in S-phase. Such an effect is even worse if we consider either F_M or resonance area versus HWHH being in the latter quite sharper for S than M-phase.

In summary we have shown that DNA internal motion is highly and selectively reduced within nucleosomes by the binding of chromosomal proteins and is function of cell cycle progression (M versus S, S versus log-phase). Differential scanning calorimetry of the native chromatin-DNA fiber display similar cell-cycle alteration (Figure 17).

The lower fraction of bound phosphates in M-phase, when compared to S-phase, points out to a smaller affinity of core histones in the binding of the phosphate DNA backbone during mitosis, sometimes associated with the observed H4 acetylation during the mitosis [82]. Indeed the overall ^{31}P spectra ionic strength dependence is compatible with earlier suggestions on the differential role of the various chromosomal subfractions, namely of H1 histones in the sealing of mononucleosomes and in the stability of polynucleosomes [81].

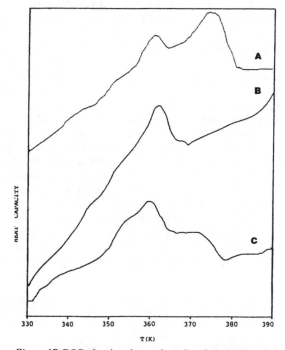

Figure 17. DSC of native chromatin as function of cell cycle.

6.2 ENZYMATIC MODIFICATIONS

Among the multiple factors that can affect the higher order chromatin structure and function, the physiological role of histone acetylation is a subject of interest. Our experiments have been therefore devoted to the study of the bulk chromatin in CHO cells, either untreated or treated with Na-butyrate. Our results (Table VIII) are compatible with modest but significant alterations in the various levels of chromatin organization, as a result of the charge neutralization of some lysine residues within the N-terminal region of the histonic octamer.

Table VIII DSC of native nuclei upon histones acetylation.

CHO Sample	I transition	II transition	III transition
Untreated	22±3.9	43±11	27±2
Na-butyrate treated	25 ±2	54±16	18±12

The calorimetric data of native nuclei (Table VIII) show indeed that the ratio between the III and the II peak of the thermograms is quite lower in the Na-butyrate treated nuclei (0.33) with respect to untreated (0.62) samples, suggesting that histone acetylations tend to relax chromatin-DNA. Namely, large statistically significant differences do exist in the heat capacity thermograms of native nuclei (Table VIII), where the unfolding of the highly packed native chromatin superfiber into a relaxed nucleofilament appears associated with histone acetylation.

At the same time circular dichroism, ethidium bromide and affinity chromatography point to quite a modest but consistent increase in the compactness of isolated nucleosomes and polynucleosomes which has been frequently associated with a change in the degree of supercoil [6].

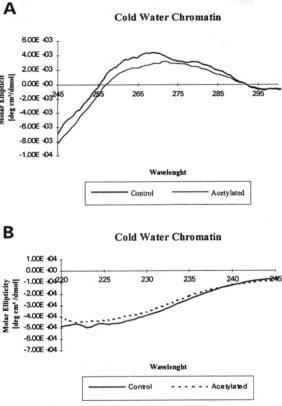

Figure 18. Circular dichroism spectra of control and acetylated cold water chromatin for either the spectral region relative to DNA (A) or the region relative to proteins (B).

The CD spectrum of the hyperacetylated chromatin in fact presents (Figure 18) the loss of the bi-ellipticity in the DNA region, earlier correlated to the DNA supercoil [25]. In the case of linear lambda DNA the CD spectrum is characterized by the main peak at 272 nm, while in circularized DNA, i.e. supercoiled, the peak was shifted around the 280 nm. Following this interpretation, the disappearance of the peak around the 270 nm in the cold water acetylated chromatin with respect to untreated one could be ascribed to a minor increase in the degree of supercoil which is furthermore associated with a minor increase in the EB affinity at low concentration [83]. This can be confirmed from the nuclease digested chromatin and from mononucleosomes [83] that lack, either in the untreated and in the Na-butyrate treated samples, the bi-ellipticity of the DNA peak being instead centered at 265-270 nm; in both these cases in fact the constraints are disrupted by the enzymatic digestion.

The distribution of active and inactive mononucleosomes in untreated and Na-butyrate treated CHO, as results after separation on a Hg^{II}-chromatography, is compatible with the above findings on isolated chromatin suggesting that acetylation induces in CHO cells an increase in the fraction of folded nucleosomal particle without exposure of the sulphydril group of histone H3.

In summary, it can be inferred that *in vivo* treatment with Na-butyrate induces a significant release on the overall higher order superfolding of the native chromatin fiber into relaxed nucleofilament [83], while at the level of the single nucleosome appears to induce a surprising but consistent increase in the compactness and supercoil of nucleosomal DNA, which then would appear confirmed as the paradoxical prerequisite of gene activation [6].

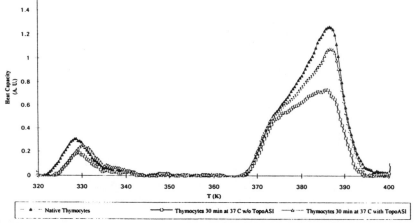

Figure 19. Profiles of heat capacity (arbitrary units) versus temperature (K) for calf thymocytes at saturating concentration of EB (R=0.2) but differently treated.
(Reference): thymocytes measured without incubation at 37°C;
(Control): thymocytes incubated at 37°C for 30 min. without enzyme;
(Sample): thymocytes incubated at 37°C for 30 min. with 500 U/ml of topoisomerase I.

6.3 DNA SUPERCOIL

Few considerations are necessary for a more precise definition of supercoiling [84-86]: in a closed DNA molecule the Linking Number defines the number of times the two strands are interwound. The difference in the Linking Number between a topologically closed DNA molecule and the corresponding relaxed one determines the degree of supercoiling. Two kinds of processes contribute to the Linking Number: the first one (*twist*) describes the intertwining of the double helix strands around their common axis, while the second one (*writhe*) describes the path of the helical axis itself. In a relaxed DNA molecule *twist* is equal to that expected for the B form of DNA and *writhe* is close to zero. In the negatively supercoiled DNA the molecule is instead under tension, therefore the processes reducing the difference in either *twist* or *writhe* will be favored (for example DNA wrapping into a higher order structure). It is generally considered that DNA supercoiling has strong effects on the interactions between DNA and other molecules [16, 87], namely the intercalating molecules.

When native thymocytes are treated with topoisomerase I, as described in detail in Vergani et al, 1996 [46], their thermogram shows significant and reproducible alterations that can be related to changes in geometrical or topological parameters of DNA. Figure 19 reports three thermograms of native thymocytes: all of them are at saturating concentration of EB (R=0.2) but differently treated. While *reference* thymocytes have been measured without incubation at 37°C, *control* thymocytes have been incubated at 37°C for 30 min. without enzyme and *sample* thymocytes have been incubated at 37°C for 30 min. with 500 U/ml of topoisomerase I. As previously described *reference* thymocytes at EB saturation (R=0.2) exhibit the increase of Transition III$_b$ together with the disappearance of Transition II; when the same sample is incubated at 37°C for 30 min. a significant reduction of Transition III$_b$ can be observed as a consequence of the activity of endogenous nucleases.

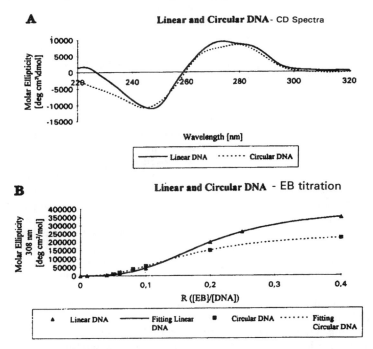

Figure 20. (A) Circular dichroism spectra of linear versus circular DNA: the acquired spectra are normalized for the corresponding DNA concentration (7.5 x 10^{-5} M) calculated by spectrophotometric methods (as described in the text). Both samples are analyzed in TE buffer pH 8 under standard acquisition conditions (standard deviation of the measurement is < 5.5%). (B) Molar ellipticity plots of molar ellipticity values at 308 nm (θ_{308}) versus R (ratio between total ethidium bromide concentration and DNA concentration) for linear and circular DNA. The markers represent the experimental values and the lines represent the corresponding curves fitting these data.

When the sample is incubated at 37°C for 30 min. with topoisomerase I we can observe a less evident decrease in Transition III$_b$: this enzyme gives rise to an increase of the most condensed chromatin structure and topoisomerase I is therefore able to balance the loss of this structure produced by the activity of endogenous nucleases, with EB intercalation. Topoisomerase I is able to relax the negative and positive supercoiling by changing the Linking Number (Lk) of a topologically closed DNA molecule. Because of the relationship [16, 88, 89] between this topological parameter (Lk) and the geometrical parameters *twist* (Tw) and writhe (Wr) (in fact Lk = Tw + Wr) when Lk is modified this change has to be balanced by related modifications of the other two parameters. When EB intercalates a negatively supercoiled DNA this binding produces a progressive reduction of *twist* and an opposite increase of *writhe* from its negative value (Lk in fact is not modified) until it becomes nul at a critical dye concentration. When EB concentration further increases *writhe* starts to assume positive values while *twist* decreases progressively.

A comparison between the effects of EB intercalation and topoisomerase activity shows that both of them produce an increase of the melting enthalpy of Transition III_b: therefore we could assign it to the denaturation of chromatin fiber with high *writhe* value (this parameter being increased by both of these treatments). This result is very interesting because it allows to differentiate chromatin fibers with different negative supercoiling on the basis of their melting temperatures directly inside the nucleus: in the future this possibility would be used to investigate the DNA supercoiling of chromatin in relation to its role in regulating the gene transcription by using chromatin fibers *in situ*.

6.3.1 Linear versus circular Phage DNA

With regard to the above problems it is illuminating the study, here summarized, of native chromatin isolated by these two procedures in correlation with the analysis of phage DNA in two different conformations, the linear form and the circular one, to underline the role [88, 90] of a topologically closed system (like circular DNA). The typical CD spectrum of DNA (Figure 20) shows a positive band with a maximum at 272 nm and a symmetric negative band with a maximum at 246 nm. At the level of the CD spectrum the comparison between linear DNA and the corresponding circular form shows an interesting difference: in the circularized sample the maximum of positive band moves from 272 nm to 280 nm displaying a red shift with respect to the linear DNA spectrum. The appearance of this new maximum following DNA circularization could be associated to a change in DNA *twist* and/or *writhe*. A further difference between linear and circular DNA spectrum is the position of the cross-over that occurs at the 257 nm for the linear DNA and shifts to 260 nm for the circularized one.

The plot of θ_{308} versus R (ratio between ethidium bromide and DNA concentration) reassuringly shows that circular DNA has a reduced number of primary binding sites and a significantly larger affinity, for low dye concentrations, with respect to the linear form: this last result is compatible with the well-known presence of negative supercoiling in a closed DNA molecule. When both chromatin samples are exposed to increasing ethidium bromide concentrations the resulting plots of θ_{308} versus R show that at low ethidium bromide concentrations both cold water (partly) and nuclease digested chromatins appear accompanied by a higher affinity for the dye with respect to the free-DNA.

The existence of a hidden negative band, in the 285-292 nm region, showing an identical amplitude with respect to the positive band at 308-315 nm was previously deduced from the CD spectra of the DNA-EB complex (Parodi et al., 1975). It is indeed only after the correction of this hidden band that the true DNA structure can be estimated as a function of EB intercalation: when the θ_{282} values are plotted versus R, after the subtraction of the corresponding molar ellipticity values at 318 nm [51], the linear region of the curve represents a measure of the direct proportionality between the normalized EB concentration and the corresponding ellipticity signal. The parameters which have been obtained by a linear regression fitting of these data give quantitative information about the characteristics of the binding: in particular the R value corresponding to the loss of the linearity can be related to the number of primary binding sites in DNA.

TABLE IX

This table reports the parameters values of the sigmoidal equation $y = \dfrac{ax^c}{b+x^c}$ which has been employed to best fit (by minimizing the roo mean square error) the experimental data (θ_{308}) recorded at increasing ethidium bromide concentrations (R).
"r^2" is the correlation coefficient that represents a measure of the goodness of the fit and "N" is the number of the experimental data points.

Source of DNA	a	b	c	r^2	N
Linear DNA	24828	0.0086	3.121	1.00	9
Circular DNA	15351	0.0117	2.448	0.96	10
Cold Water Chromatin	13223	0.0829	1.901	0.99	7
Nuclease Digested Chromatin	11415	0.0590	1.571	1.00	8

Number of binding sites per nucleotide "n" for different DNA and chromatin-DNA samples: these values have been evaluated by plotting the θ_{282} values (properly corrected for the hidden band) versus R, as described in details in the test.

Source of DNA	n Number of binding sites per nucleotide
Linear DNA	0.20-0.25
Circular DNA	0.10-0.13
Cold Water Chromatin	0.08-0.10
Nuclease Digested Chromatin	0.09-0.11

The differences between CD spectra of linear and circular DNA could really be explainable in terms of DNA supercoiling: following DNA circularization, supercoiling becomes indeed apparent together with a change in *twist* and *writhe* parameters. The appearance of a maximum over the 280 nm in the circular DNA spectrum can probably be due to these changes. It is interesting that the maximum at 284 nm is also present in chromatin but not in the mononucleosomes spectra: we can therefore deduce that both circularization and nucleosomal superhelical organization cause similar modifications in the DNA structure. In fact the prolonged nuclease digestion employed for preparing the mononucleosomal sample is responsible for the lack of H1 histone and the consequent loss of the negative DNA supercoiling at this level of the nucleosomal organization. These results are in accordance with previously published data that pointed to the nucleosomal arrangement as the principal responsible for eukariotic DNA supercoiling [15, 17, 18].

The similarity between circularized free-DNA and nucleosomal DNA results also evident when the number of primary binding sites is evaluated: the number of binding sites per nucleotide "n" is between 0.2 and 0.25 for the linear DNA and decreases to 0.1-0.12 for the circular one; a similar "n" value has been found for the nuclease digested and for the cold water chromatin. This result points out that circularization and the nucleosomal arrangement of DNA molecule produce the same effect in topological terms. The disappearance of the maximum at 284 nm in mononucleosomes deprived of H1 histone can therefore be interpreted as a loss of DNA supercoiling.

The above CD data and the difference in ethidium bromide affinity between chromatin prepared by the cold water method and that prepared by the nuclease digested one point to a difference in their higher order structure (as earlier reported by [9] and [24] but not in the degree of DNA supercoiling. While the first procedure partially preserves the chromatin higher order structure the second does not.

7. Conclusions

In summary, the structural characterization down to the Angström level of the mammalian genome carried out by several and complementary biophysical methods gives a coherent picture of the DNA organization into successive foldings and superfoldings maintained in negative superhelical turns by the topological constraints present *in situ* and even in chromatin whenever isolated by the non-invasive cold water swelling. This supercoiled DNA higher order structure appears to regularly modulate through fiber folding and unfolding during gene expression, enzymatic modifications, cell cycle and any change in the cell function.

This chapter represents the structural framework, from native nuclei down to the single chromosomes and nucleosomes, for all the contributions to genome science and technology contained in the following pages. This book indeed aims to a concise and comprehensive description of the state of the art of on-going research in the structure and function of genes (Stein, Wolffe, Gilmour), chromosomes (Razin, Belmont, Bradbury), and their constituents (Turner, Harrington, Beato), and in the development of increasingly powerful applications of gene technology (Cantor) to human cancers (Cavenee, Georgiev) and plants (Skryabin).

Acknowledgments

This work was supported by research contract from CNR (89.02845.CT04/115 22744 and 90.03107.CT04/115.22734) by MURST (DP 40% pf 1992, 1993) and by a collaborative research grant from NATO Scientific Affairs Division.

References

1. Belmont, A.S., Sedat, J.W. and Agard, D.A. (1987) A Three Dimensional Approach to Mitotic Chromosome Structure: Evidence for a Complex Hierarchical Organisation, *J. Cell Biol.* **105**, 77-92.
2. Belmont, A.S., Braunfeld, M.B., Sedat, J.W. and Agard, D. A. (1989) Large-scale chromatin structural domains within mitotic and interphase chromosomes in vivo and in vitro, *Chromosoma* **98**, 129-143.
3. Kendall, F.M., Swenson, R., Borun, Y., Rowinski, J. and Nicolini, C. (1977) Nuclear Morphometry During the Cell Cycle, *Science* **196**, 1106-1109.
4. Kendall, F.M., Beltrame, F., Zietz, S., Belmont, A. S. and Nicolini, C. (1980) The Quinternary Chromatin-DNA Structure: Three Dimensional Reconstruction and Functional Significance, *Cell Biophys.* **2**, 373-404.
5. Manuelidis, L. and Chen, T.L. (1990) A Unified Model of Eukaryotic Chromosomes, *Cytometry* **11**, 8-25.
6. Nicolini, C. (1986) *Biophysics and Cancer,* Plenum Press, New York, 1-455.
7. Nicolini, C. (1991) The Beginning of a Molecular Description of Human Cancer: Chromatin 3D-Structure, DNA Supercoil, H1 Histone, Microtubules and the Modulation of Unbalanced Gene Expression. *In Molecular Basis of Human Cancer.* N.Y.: Plenum Press.
8. Belmont, A.S., Kendall, F.M. and Nicolini, C. (1984) Three-Dimensional Intranuclear DNA Organization in situ: Three States of Condensation and Their Distribution As a Function of Nuclear Size Near the G_1-S Border in Hela S-3, Cells, *J. Cell Sci.* **65**, 123-138.
9. Diaspro, A., Bertolotto, M., Vergani, L. and Nicolini, C. (1991) Polarized light scattering of nucleosomes and polynucleosomes: in situ and in vitro studies, *IEEE Transactions on Biomedical Engineering BME* **38**, 670-678.
10. Nicolini, C., Cavazza, B., Trefiletti, V., Pioli, F., Beltrame, F., Brambilla, G., Maraldi, N. and Patrone, E. (1983) Higher-Order Structure of Chromatin from Resting Cells: High-Resolution Computer Analysis of Native Chromatin Fibers and Freeze-Etching of Nuclei from Rat Liver Cells, *J. Cell Sci.* **62**, 103-115.
11. Nicolini, C. and Kendall, F. (1977) Differential light scattering in native chromatin: corrections and inferences combining melting and dye-binding studies. A two-order superhelical model, *Physiol. Chem. Phys* **9**, 13, 265-283.
12. Nicolini, C., Carlo, P., Martelli, A., Finollo, R., Patrone, E., Trefiletti, V. and Brambilla, G. (1982) Viscoelastic properties of native DNA from intact nuclei of mammalian cells. Higher-Order DNA Packing and Cell Function, *J. Mol. Biol.* **161**, 155-175.
13. Travers, A.A., Klug, A. (1987) The bending of DNA in nucleosomes and its wider implications, *Philos. Trans. Royal Soc. London Ser.B-Biol. Sci.* **317**, 537-561.
14. Germond, J.E., Hirt, B., Oudet, P., Grossbelllard, M. and Chambon, P. (1975) Supercoiling of sv40-dna and chromatin structure, *Experentia*, **31**, 739.
15. Richet E., Abcarien P. and Nash H.A. (1986) The interaction of recombination proteins with supercoiled dna - defining the role of supercoiling in lambda integrative recombination,*Cell* **46**, 1011-1021.
16. Giaever, G.N. and Wang, J.C. (1988) Supercoiling of intracellular DNA can occur in eukariotic cells, *Cell* **55**, 849-856.
17. Clark, D.J. and Felsenfeld, G. (1991) Formation of nucleosomes on positively supercoiled dna, *EMBO J.* **10(2)**, 387-395.
18. Ohba, R., Tabuchi, H. and Hirose, S. (1992) Dna supercoiling facilitates the assembly of transcriptionally active chromatin on the adenovirus major late promoter, *Biochem Biophys Res Comm* **186**, 2, 963-969.
19. Nicolini, C., Diaspro, A., Bertolotto, M., Facci, P. and Vergani, L. (1991) Changes in DNA superhelical density monitored by polarized light scattering, *Biochem Biophys Res Commun* **177**, 3, 1313-1318.
20. Nicolini, C., Facci, P. and Alliata, D. (1996) Scanning probe microscopy of nucleosomes and polynucleosomes, Nanobiology, in press.
21. Noll, M. (1974) Subunit structure of chromatin, *Nature* **251**, 249-251.
22. Nicolini, C. and Baserga, R. (1975) Role of nonhistone chromosomal proteins in determining circular dichroism spectra of chromatin, *Arch. Biochem. Biophys.* **169**, 678-685.
23. Nicolini, C., Catasti, P., Nizzari, M. and Carrara, E. (1989) in *"Protein Structure and Engineering (Jardetsky, O., Ed.)* pp 221-242, Plenum Press, New York.
24. Nicolini, C., Vergani, L., Diaspro, A., Di Maria, E. (1989) Native chromatin-DNA structure and cell cycle: DSC and gel electrophoresis, *Termochimica Acta* **252**, 307-327.
25. Vergani, L., Gavazzo, P., Mascetti, G. and Nicolini, C. (1994) Ethidium bromide intercalation and chromatin structure. A spectropolarimetric analysis, *Biochemistry* **33**, 21, 6578-6585.
26. Vergani, L., Mascetti, G., Gavazzo, P. and Nicolini, C. (1992) Effects of Fixatives on High Order Chromatin Structure: a Calorimetric Study. *Thermochimica Acta* **206**, 175-179.
27. Noll, M., Thomas, J.O. and Kornberg, R.D.(1976) Preparation of native chromatin and damage caused by shearing, *Science* **187**, 1203.

28. Sambrook, J., Fritsch, E.F. and Maniatis, T. (1989) Molecular Cloning, Cold Spring Harbor Laboratory Press Cold Spring Harbor, N.Y.
29. Feigin, L.A. and Svergun D.I. (1987) *Structure Analysis by Small-Angle and Neutron Scattering*, Plenum Press, New York and London.
30. Maccioni, E., Dembo, A. and Nicolini, C. (1996) X-ray low-angle scatering of polynucleosomes, *Molecular Biology Reports.* submitted.
31. Fujiwara S.,(1992) Interpretation of the x-ray-scattering profiles of chromatin at various nacl concentrations by a simple chain model,*Biophys. Chem.* **43**, 81-87.
32. Debye, P. (1915)*Ann. Phys* **46** 809.
33. Travaglini, G., Rohrer, H., Amrein, M., Gross, H. (1987) Surface studies by scanning tunneling microscopy, *Surf. Science* **181**, 380.
34. Nicolini, C., Baserga, M. and Kendall, F. (1976) DNA structure in sheared and unsheared chromatin, *Science* **192**, 796-798.
35. Beebe, T.P., Wilson, T.E., Ogletree, D.F., Katz, J.E., Balhorn, R., Salmeron, M.B., Siekhaus, W.J. (1989) Direct observation of native DNA structures with scanning tunnelling microscopy, *Science* **243**, 370.
36. Diaspro, A., Scelza, P., Nicolini, C. (1990) MUCIDS: an operative C-environment for acquisition and processing of polarized-light scattered from biological specimens, *Computer Application in Biosciences* **6**, 229-236.
37. Nicolini, C. and Baserga, R. (1975) Circular dichroism and ethidium bromide binding of 5-deoxybromouridine-substituted chromatin, *Biochemical Biophysical Research Communication* **64**, 189-195.
38. Diaspro, A., Radicchi, G. and Nicolini, C. (1995) Polarized light scattering: a biophysical method for studying bacterial cells, *IEEE Transactions on Biomedical Engineering*, **42**, 1038-1043.
39. Augenlicht, L., Nicolini, C. and Baserga, R. (1974) Circular dichroism and thermal denaturation studies of chromatin and DNA by BrdU-treated mouse fibroblasts, *Biochem. Biophys. Res. Commun.*, **59**, 920-924.
40. Bailly, C., Waring M.J., Travers, A.A. (1995) Effects of base substitutions on the binding of a dna-bending protein, Journal of Molecular Biology **253**, 1-7.
41. Mascetti, G., Vergani, L., Diaspro, A., Carrara S., Radicchi, G. and Nicolini, C. (1996) Effective of fixatives on calf thymocytes chromatin: 3D hight resolution fluorescence microscopy,*Cytometry* **23**, 110-119.
42. Nicolini, C., Trefiletti, V., Cavazza, B., Cuniberti, C., Patrone, E., Carlo, P. and Brambilla, G. (1983) Quaternary and Quinternary Structures of Native Chromatin-DNA in Liver Nuclei: Differential Scanning Calorimetry, *Science* **219**, 176-178.
43. Nicolini, C., Diaspro, A., Vergani, L. and Cittadini, G. (1988) In situ thermodynamics characterization of chromatin and other macromolecules during the cell cycle, *Int. J. Biol. Macrom.* **10**, 137-142.
44. Degrooth, B.G. and Putman, C.A.J. (1992) High-resolution imaging of chromosome-related structures by atomic force microscopy, *J. Microscopy* **168**, 239-247.
45. Musio, A., Mariani, T., Frediani, C., Sbrana, I., Ascoli, C. (1996) Longitudinal patterns similar to G-banding in untreated human chromosomes: evidence by atomic force microscope, Chromosoma, in press.
46. Vergani, L., Mascetti, G. and Nicolini, C. (1996) Ethidium Bromide intercalation and chromatin structure: a thermal analysis. *Thermochimica Acta* , submitted.
47. Manning, G.S. (1978) Molecular theory of polyelectrolyte solutions with applications to electrostatic properties of polynucleotides, *Quart. Rev. Biophys.* **11**, 179-246.
48. Belmont, A. and Nicolini, C. (1981) Polyelectrolyte theory and chromatin-DNA quaternary structure: role of ionic strenght and H1 histone, *Journal of Theoretical Biology* **90**, 169-179.
49. Le Pecq, J.B. and Paoletti, C. (1967) A fluorescent complex between ethidium bromide and nucleic acids, *Journal of Molecular Biology*, **27**, 87-98.
50. Williams R.E., Lurquin P.F. and Seligy V.L. 1972 Circular dichroism of chromatin and Ethidium Bromide, *European Journal of Biochemistry* **29**, 3, 427-432.
51. Parodi, S., Kendall, F. and Nicolini, C. (1975) A clarification of the complex spectrum observed with UV circular dichroism of EB bound to DNA, *Nucleic Acis Research* **2**, 471-477
52. Nicolini, C. and Baserga, R. (1975) Circular dichroism and EB binding studies of chromatin from Wi38 fibroblasts stimulated to proliferate, *Chemical Biological Interaction* **11**, 101-115.
53. Diaspro, A., Nicolini, C. (1986) Circular intensity differential scattering and chromatin DNA structure, *Cell Biophysics*, **10**, 45-60.
54. Fujiwara, S., Inoko, Y., Ueki, T. (1989) Synchrotron x-ray-scattering study of chromatin condensation induced by mono-valent salt - analysis of the small-angle scattering data, *J. Biochem* **106**, 119-125.
55. Allegra, P., Sterner, R., Clayton, F. and Allfrey, V. (1987) Affinity chromatographic purification of nucleosomes containing transcriptionally active DNA sequences, *J.Mol. Biol.* **196**, 379-388.
56. Richmond; T.J., Finch J.T., Rushton B., Rhodes D. and Klug A. (1984) Structure of the nucleosome core particle at 7 A resolution, *Nature* **311**, 532-537.
57. Suau, P., Bradbury, E.M. and Baldwin, J.P. (1984) Higher-order structures of chromatin in solution, *Eur. J. Biochem.* **97**, 593-602.

58. Nevernov, I., Kurnikov I., Nicolini, C. (1995) Mechanical interactions in STM imaging of large insulating adsorbtes, *Ultramicroscopy* **58**, 269-274.
59. Dupraw, E. (1965) Macromolecular organization of nuclei and chromosomes, *Nature* **206**, 338-340.
60. Dickerson, R. E., Drew, H. R., Conner, B., Fratins, A. and Kopka, M.L. (1982) The anatomy of A-, B- and Z- DNA, *Science* **216**, 475-478.
61. McGhee, J.D. and Felsenfeld, G., (1980) Nucleosome structure, *Annu Rev. Biochem*, **49**, 1115-1128.
62. Woodcock C.L., Frado L. and Rattner J. (1984) The higher-order structure of chromatin. Evidence for a helical ribbon arrangement, *J. Cell Biol.* **99**, 42-52.
63. Nicolini, C., Carrara, S. and Mascetti, G. (1996) High order DNA structure by fluorescence analysis and thermal profile of thymocytes, *Molecular Biology Reports* submitted.
64. Finch, J.T. and Klug, A. (1976) Solenoidal model for superstructure in chromatin. *Proc. Nat. Acad. Sci.* **73**, 1897-1901.
65. Widom J. and Klug A. (1985). Structure of the 300 Å chromatin filament: X-ray diffraction from oriented samples. *Cell* **43**, 207-213.
66. Williams, S.P., Athey, B.D., Muglia, L.J., Schosse, R.S., Gough, A.H. and Langmore, J.P. (1986) Chromatin fibers are last handed double helixes with dameters and mass per unit length that depend on linker lent, *Biophys. J.* **49**, 233-248.
67. Nicolini, C. (1983) Chromatin Structure: from Nuclei to genes, *Anticancer Research* **3**, 63-86.
68. Marsden, M P.F. Laemmli U.K. (1979) Metaphase chromosome structure: evidence for a radial loop model, *Cell*, **17**, 849-858.
69. Adolph, K.W. (1980) Organization of chromosomes in mitotic Hela cells. *Exp. Cell Res* **125**, 95-103.
70. Rattner J.B. and Lin C.C. (1985) Radial loops and helical coils coexist in metaphase chromosomes, *Cell* **42**, 291-296.
71. Woodcock, C.L. (1994) Chromatin fibers observed in-situ in frozen-hydrated sections - native fiber diameter is not correlated with nucleosome repeat length, *J. Cell Biol.* **125**, 11-19.
72. Cavazza, B., Trefiletti, V., Pioli, F., Ricci, E., and Patrone E., (1983) Higher-order structure of chromatin from resting cells .1. electron-microscopy of chromatin from calf thymus, *J. Cell Sci.* **62** 81-102.
73. Langmore, J.P. and Paulson, J.R. (1983) Low angle X-ray diffraction studies of chromatin structure in vivo and in isolated nuclei and metaphase chromosomes, *J. Cell Biol.* **96**, 1120-1131.
74. Paulson, J.R. and Laemmli, U.K. (1977) Structure of histone-depleted metaphase chromosomes, *Cell* **12**, 817-828.
75. Nicolini, C., Vernazza, B., Chiabrera, A., Maraldi, I.N., Capitani, S. (1984) Nuclear Pores and Interphase Chromatin: High-Resolution Image Analysis and Freeze Etching, *J. Cell Sci.* **72**, 75-87.
76. Agard, D.A., Sedat, J.W. (1980) Three dimensional analysis of biological specimens utilizing image processing techniques, *Spie.* **264**, 110-117.
77. Sedat, J. and Manuelidis, L. (1978) A direct approach to the structure of eukaryotic chromosomes, *Cold Spring Harbour Symp. Biol.* **42**, 331-345.
78. Nagl, W. (1977) In: Rostl T.L. and Gifford EM (eds) *Mechanism of cell division*, Dowden, Hutchinson and Ross, Stroudsburg, Pa.
79. Belmont, A.S. and Bruce, K. (1994) Visualization of G1 Chromosomes: a folded, twisted, supercoiled Chromonema model of interphase chromatid structure, *J. Cell Biol.* **127**, 287-302.
80. Giammasca, P.J., Horowitz, R.A. and Woodcock, C.L. (1993) Transitions between in situ and isolated chromatin, *J. Cell Sci.* **105**, 551-561.
81. Nicolini, C., Catasti, P., Yao, P., Szylagy, L., (1993) DNA internal motion within nucleosomes during the Cell cycle and as a function of ionic strenght, *Biochemistry* **32**, 6465-6469.
82. Alfrey, V., Faulkner, R. and Mirsky, A. (1964) Acetylation and methylation of histones and their possible role in RNA synthesis regulation, *Proc. Natl. Acad. Sci.USA* **51**, 786.
83. Gavazzo, P., Vergani, L, Mascetti, G. and Nicolini, C. (1996) Effects of histone acetylations on higher order chromatin structure and function. *Journal Cell Biochemistry* in press.
84. Nash, H.A. (1990) Bending and supercoiling of dna at the attachment site of bacteriophage-lambda, *Trends in Biochemical Sciences* **15**, 222-227.
85. Brotherton, T.W., Jagannadham, M.V. and Ginder, G.D. (1989) Heparin binds to intact mononucleosomes and induces a novel unfolded structure, *Biochemistry* **28**, 3518-3525.
86. Van Holde (1988) Chromatin. Springer Verlag ed..
87. Sen, D. and Crothers, D.M. (1986) Influence of dna-binding drugs on chromatin condensation, *Biochemistry* **25**, 1503-1509.
88. Bauer, W.R., Crick, P.H.C. and White, J.H. (1980) DNA supercoling, *Scientific American* **2**, 45-50.
89. Travers, A.A., and Klug, A. (1987) DNA wrapping and writhing, *Nature* **327**, 280-282.
90. Wang J.C., Caron, P.R., Kim, R.A. (1990) The role of dna topoisomerases in recombination and genome stability - a double-edged-sword *Cell*, **62**, 403-406.

CHROMOSOMAL DNA LOOPS AND DOMAIN ORGANIZATION OF THE EUKARYOTIC GENOME

S.V. RAZIN
Laboratory of Structural and Functional Organization of Chromosomes, Institute of Gene Biology, Vavilov St. 34/5, 117334 Moscow, Russia.

Over twenty years ago it was found that chromosomal DNA in eukaryotic cell nuclei is organized into large loops by periodic attachment to the high salt-insoluble proteinous nuclear (chromosomal) matrix. The specificity of genomic DNA organization into loops has been intensively studied with the aim to find out whether they may constitute quasi-independent structural-functional units of the genome. These studies have resulted in conflicting findings and consequently in conflicting conclusions. In this paper different experimental approaches used to analyse the above problem will be critically reviewed. The discussion will be followed by the presentation of a novel approach for mapping DNA loop anchorage sites which has been developed in our laboratory. Based on the excision of whole DNA loops by topoisomerase II - mediated DNA cleavage at matrix attachment sites this approach seems to constitute a unique tool for analyzing the topological organization of chromosomal DNA in living cells.

After analysis of methodological problems we shall return to the problem of structural-functional organization of the eukaryotic genome. A relationship between DNA loops and replicons, as well as a possible correlation between DNA loops and transcriptionally active/repressed genomic domains will be discussed.

1. Periodical attachment of chromosomal DNA to the nuclear matrix organizes this DNA in large closed loops

The genome of typical eukaryotic cells consists of an approximately 2 metres long DNA molecule (actually a set of shorter DNA molecules of individual chromosomes) which is packed in a certain manner within a nucleus about 10 microns in diameter. It has long been recognized that the above compactization of DNA is accomplished in several steps. Interaction of DNA with histones to form the so-called chromatin is essential for roper DNA packaging within nuclei. However, the organization of DNA into nucleosomes followed by packaging of the nucleosomal thread into a 30 nm chromatin fiber accounts for only about 40 fold level of DNA compactization. Recent data indicate that further packaging of chromosomal DNA occurs through the organization of the 30 nm chromatin fiber into loops. Observations suggesting that chromosomal DNA is organized into large loops were first made in the middle seventies by Cook and collaborators [1-5]. It was demonstrated that after disruption of nucleosomes by 2M NaCl extraction the nuclear DNA was not released into solution but remained organized into constrained domains. These domains possessed the typical

C. Nicolini (ed.), Genome Structure and Function, 39–56.
© 1997 *Kluwer Academic Publishers. Printed in the Netherlands.*

properties of closed circular DNA, as followed from the analysis of sedimentation properties of high salt- extracted nuclei (in further discussion referred to as "nucleoids", as proposed by Cook and collaborators [1-3]). In particular, a biphasic dependence of the sedimentation coefficient of nucleoids on the concentration of intercalating agents was observed (Fig. 1) To explain these observations, chromosomal DNA was proposed to be organized into loops bound to high-salt-insoluble nuclear remnants. The latter were called "nuclear ghost" or "nuclear cage" but presently they are more known as "nuclear matrix" [6-8]. After extraction of histones, DNA loops may be expected to become negatively supercoiled. By titration with an increasing concentration of ethidium bromide it is possible to eliminate all negative supercoils and, furthermore, to induce positive supercoiling. Changes in the superhelical density influence sedimentation properties of nucleoids (Fig. 1). The conclusions made by Cook and collaborators were soon supported by direct visualization of large DNA loops in high salt extracted metaphase chromosomes and interphase nuclei [9-11]. Analysis of the size distribution of DNA fragments released from eukaryotic cell nuclei treated for increasing time intervals with nucleases also indicated that DNA was periodically attached to some internal nuclear structure [12,13].

The size of chromosomal DNA loops can be estimated by different experimental approaches. The simplest approach is the measurement of the contour length of DNA loops on electron microscopic photos of spread nucleoids. Naturally, for correct measurement the loops should be relaxed. In many cases the nicks unintentionally introduced into DNA in the course of isolation of nuclei and metaphase chromosomes are sufficient for complete relaxation of DNA loops [9,11]. Otherwise, the spreads should be treated with topoisomerase I.

In many works the so-called nucleoid relaxation analysis was used for estimation of an average size of chromosomal DNA loops. There are two variants of this approach. In the first one nucleoids are treated with nucleases in order to introduce precisely so many nicks as necessary for the complete relaxation of loop DNA. Then an average distance between the nicks is determined experimentally by DNA sedimentation in an alkaline sucrose gradient [14-16]. It is easy to see that this distance is equal to a double average length of DNA loops. The second variant of the nucleoid relaxation analysis is technically more simple. DNA loops in nucleoids are relaxed either by limited treatment with DNase I or by treatment with topoisomerase I. Then the diameter of the DNA loop halo is measured under a fluorescence microscope. In simplest case, i. e. when there is no discontinuity in the distribution of loops according to their sizes, the diameter of the loop halo should be equal to the size of an average DNA loop.

Treatment of nucleoids with nucleases producing double stranded DNA breaks causes gradual cleavage of distal parts of DNA loops. Within a 2 to 50 kb area the size of DNA fragments remaining bound to the nuclear matrix after limited DNA cleavage correlates linearly with the percentage of DNA recovered in the nuclear matrix [13,19]. On the basis of this correlation it is also possible to estimate an average size of DNA loops by approximation to the starting point of digestion.

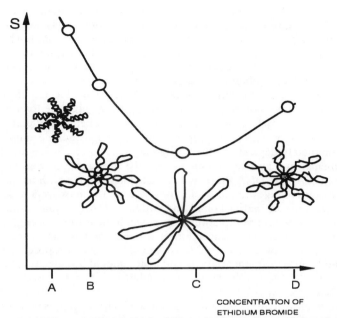

Figure 1. A scheme illustrating the biphasic dependence of sedimentation coefficient of nucleoids on concentration of the ethidium bromide in the medium. With an increase of the ethidium bromide concentration the negatively supercoiled DNA loops are gradually unwound (A-C) until a full loss of supercoils (C). Thereafter positive supercoils are introduced (D). Unwinding of DNA loops leads to an increase of the diameter of the loop halo (A-C). Correspondingly, the sedimentation coefficient of nucleoids decreases due to the "parachute" effect. Positive supercoiling of DNA loops introduced at high concentrations of ethidium bromide decreases the diameter of nucleoid and increases its sedimentation coefficient.

Each of the above described approaches for estimation of an average size of chromosomal DNA loops has certain limitations and under some conditions they may give incorrect results. One can find a more detailed discussion of the problem in several recently published review articles [19,20]. Nevertheless it is important to stress here that most researchers estimated an average size of DNA loops to be within the range of 50 to 100 kb [9, 11-15, 18, reviewed in ref. 20,21].

2. Characterization of specificity of DNA organization into loops

2.1. INTRODUCTION TO THE PROBLEM

It is well known that the eukaryotic genome is subdivided into quasi-independent functional units such as replicons and transcriptons. Since the discovery of chromosomal DNA loops the question has been discussed about a possible relationship between the organization of DNA into loops and the organization of the genome into functional units. There were two rationales for the above supposition. First, the average size of chromosomal DNA loops is comparable with the average sizes of replicons and transcriptons. Second, the transcriptional status of large genomic domains seems to be regulated at the level of DNA packaging in chromatin. Indeed, this can be the only reason for preferential nuclease sensitivity (and hence, preferential accessibility) of active genes [22]. DNase I-sensitive regions usually have relatively sharp borders. This suggests that there should be some structural units of chromatin within which the mode of DNA packaging may be changed without affecting the neighbouring regions. In order to find out whether DNA loops constitute such structural units it was necessary to find a way to map the loop borders within characterized genomic areas. At first sight the task seems to be easy. One has to isolate DNA fragments located close to the loop basements or even at the basements themselves. It is then possible to choose a certain genomic area and to study by hybridization experiments whether any part of this area is overrepresented in tloop basement preparations. Although looking quite simple, the experiment completely depends on the availability of a correct procedure for isolation of loop basements. In a number of experiments these basements were identified with nuclear matrix DNA, i. e. DNA that can be isolated in complex with the nuclear matrix after mild or extensive digestion of nuclei (or nucleoids) with exogenous nucleases and subsequent separation of matrix-bound and soluble DNA fractions. It is known that large pieces of chromatin are not soluble in physiological ionic strength. Hence, the extraction with a concentrated salt solution was commonly used for disruption of chromatin and elution of cleaved-off DNA [23-28]. It was proposed later that this extraction might cause different artefacts and hence should be substituted by extraction with a weak ionic detergent, lithium diiodosalicylate (LIS) [29]. However, more recent data suggest that this extraction destroys all or almost all preexisting interactions of DNA with the nuclear matrix (for a discussion see [20]). Complexes of LIS-extracted nuclei with a specific subset of DNA fragments are likely to originate as a result of *in-vitro* reconstitution taking place in the course of treatment of LIS-extracted nuclei with restriction enzymes [20].

In order to avoid the high-salt extraction it was also proposed to separate cleaved-off pieces of DNA organized in chromatin by electroelution in physiological ionic strength [30,31]. As shall be discussed below, the results obtained with this procedure of isolation of matrix-bound DNA fragments fully support the conclusions made in experiments with high salt extraction.

Two new approaches for mapping DNA loop anchorage sites have been developed recently. One of them is based on the excision of whole loops and their oligomers by specific DNA cleavage at matrix attachment sites with nuclear matrix high salt-insoluble topoisomerase II [32-34]. The other approach implies the technique of multicolour *in situ* hybridization for determination of relative positions of different DNA probes within the halo of DNA loops in nucleoids and high salt-extracted metaphase chromosomes [35,36].

Below we shall discuss the results of mapping of DNA loop anchorage sites by all of the above mentioned approaches.

2.2. CHARACTERIZATION OF THE SPECIFICITY OF DNA LOOP ORGANIZATION IN HIGH SALT EXTRACTED NUCLEI

2.2.1. *Association of transcriptionally active genes with the nuclear matrix*
A general procedure for determining approximate positions of different genes within chromosomal DNA loops was developed by Cook and Brazell [23] who reasoned that the estimation of the relative portion of a DNA sequence under study in matrix-bound DNA fractions (operationally defined as "nuclear matrix DNA") of different sizes

should show at what distance from the loop anchorage site the DNA sequence under study is located. Indeed, according to the simplest model of DNA organization in more or less uniform loops any given DNA sequence must be found preferentially in nuclear matrix DNA as long as the average sizes of the nuclear matrix DNA fragments exceed more than twice the distance between this DNA sequence and the loop anchorage site. Then (when the average size of the nuclear matrix DNA fragments decreases beyond the above point) the same DNA sequence may be expected to accumulate in the cleaved-off DNA fractions. It is clear that for correct estimation of a distance between a sequence of interest and a loop anchorage site the whole DNA loop should be equally accessible to a nuclease used for the detachment of DNA from the nuclear matrix. Furthermore, the above nuclease should cut all DNA sequences with an equal probability.

Although in most subsequent experiments an equal accessibility of different DNA sequences was assured by histone removal, the second of the above conditions was usually disregarded, as sequence-specific restriction endonucleases were used for cleavage of distal parts of DNA loops. Consequently, the resolution of the approach was decreased so that it was possible only to discriminate DNA sequences located at or close to the matrix attachment sites from DNA sequences located somewhere in the loops. In experiments of the above type a dependence between the transcriptional status of a gene and its attachment to the nuclear matrix. was observed. In other words, active genes were generally found in nuclear matrix DNA, while inactive ones - in cleaved-off DNA fractions [24-28,37]. Especially convincing results were, perhaps, obtained in experiments with hormone-induced premature cell differentiation. Cell type-specific genes starting their expression in differentiated cells were bound to the nuclear matrix in these cells and were not bound to the nuclear matrix in maternal (non-differentiated) cells [26,37].

Obviously, these results were far from what was expected. Nowonder that many scientists met them with a certain suspicion. It was proposed that the apparent association of transcriptionally active DNA with the nuclear matrix was an artefact resulting from high-salt extraction stimulating precipitation of the transcriptional machinery [29,38]. However, an apparent association of active genes with the nuclear matrix could also be observed in experiments with electroelution of cleaved-off DNA at physiological ionic strength [30,39,40]. Furthermore, it was demonstrated in different types of experiments that continuous transcription was not necessary for the attachment of potentially active genes to the nuclear matrix [28,37]. This association is likely to be a condition of transcriptional activation, rather than a consequence of transcription. At the basis of recent findings [41-43] one may suggest that the association of active genes with the nuclear matrix reflects their targeting to the transcription *foci*.

2.2.2. *Permanent sites of DNA attachment to the nuclear matrix*

The relationship between the transcriptional status of a genomic region and positioning of this region close to the basement of a DNA loop could be hardly explained by the simple model of DNA organization into relatively uniform loops anchored to the nuclear matrix. On the other hand, the assumption that all specific DNA interactions with the nuclear matrix are transcription-related [27,44] also does not explain some experimental observations, for example a correlation between an average size of DNA loops and an average size of replicons in cells of different species [17]. To solve the problem, it was proposed that attachment sites of a special type delimited the borders of basic DNA loops in nucleoids [45,46]. This supposition was corroborated by the results of experiments on mapping matrix attachment sites within the domain of α-globin genes in chicken erythroblast, mature erythrocyte and mature sperm nuclei [45-48]. While the whole transcriptionally active area was bound to the nuclear matrix in erythroblast nuclei, in erythrocyte and sperm nuclei only two major attachment sites framing the domain from upstream and downstream were detected. The same DNA fragments were also attached to the nuclear matrix in cultured chicken fibroblasts [48]. Hence, at least within the domain of chicken α -globin genes, it was possible to discriminate between permanent (preserved in inactive nuclei) and transcription-dependent interactions of DNA with the nuclear matrix. In order to find out whether permanent sites of DNA attachment to the nuclear matrix are present also in other genomic areas, we compared the sequence complexities of nuclear matrix DNA samples prepared from erythroblasts and mature erythrocytes [49]. Trace amounts of either total DNA or nuclear matrix DNA from erythroblasts or erythrocytes were reannealed in reactions driven by a vast excess of either erythrocyte or erythroblast nuclear matrix DNA. The most important observation made in these experiments was that in both renaturation reactions with C_0t values sufficient for complete renaturation of homologous nuclear matrix DNA probes the total DNA was reannealed only partially (25% of unique sequences in the reaction driven by erythroblast nuclear matrix DNA and 10% of unique sequences in the reaction driven by erythrocyte nuclear matrix DNA). This means that both nuclear matrix DNA fractions contain only subsets of unique DNA sequences present in total DNA. Consequently, one may conclude that matrix attachment sites have specific positions in the genome. Otherwise all unique sequences present in total DNA would be found in the vicinity of matrix attachment sites (i. e. in nuclear matrix DNA). Similar considerations made it possible to conclude that erythrocyte nuclear matrix DNA is a subfraction of erythroblast nuclear matrix DNA. Indeed, the latter was not reannealed completely in the reaction driven by erythrocyte nuclear matrix DNA while both nuclear matrix DNA samples were reannealed completely in the reaction driven by erythroblast nuclear matrix DNA. Hence, the erythroblast nuclear matrix DNA is composed of two subfractions: DNA fragments permanently attached to nuclear matrix (i.e. erythrocyte nuclear matrix DNA) and DNA fragments interacting with the nuclear matrix in accordance with the functional processes.

2.3. MAPPING OF DNA LOOP ANCHORAGE SITES USING DNA LOOP EXCISION MEDIATED BY HIGH SALT-INSOLUBLE TOPOISOMERASE II OF THE NUCLEAR MATRIX

Recently we have developed a new procedure for mapping the DNA loop anchorage sites in large genomic areas. The procedure is based on previous observations indicating that DNA-topoisomerase II is an integral part of the nuclear matrix [49] and the scaffold of metaphase chromosomes [50]. Topoisomerase II is known to introduce double-stranded scissions into DNA in the course of relaxation reaction catalysed by this enzyme [51]. Normally, these scissions are religated after relaxation of DNA. However, intermediate complexes of topoisomerase II with cleaved DNA may be accumulated due to the inhibition of the ligation step of the reaction with a variety of drugs including VM-26, VP-16 or m-AMSA [52,53]. After denaturation of the enzyme with SDS followed by digestion of proteins with proteinase K it is possible to obtain samples of DNA cleaved at the sites of preferential action of topoisomerase II. This approach is widely used for mapping preferential sites of topoisomerase II action on DNA [54-57]. The enzyme demonstrates a limited specificity of cleavage of free DNA [54,55]. In living cells the positions of preferential cleavage sites are determined by the chromatin structure. In particular, DNase I hypersensitive sites constitute preferential targets for DNA cleavage by topoisomerase II [56,57]. Although topoisomerase II was found in the nuclear matrix and in the scaffold of metaphase chromosomes, most of the enzyme is soluble in a medium-salt solution [58-61]. It was reasonable to assume soluble topoisomerase II of the nucleoplasm to have preferential access to DNase I hypersensitive sites in chromatin. We reasoned that the extraction of nuclei with a concentrated salt solution (which does not affect the nuclear matrix (chromosomal scaffold) integrity and does not release DNA loops from the nuclear matrix) must remove soluble topoisomerase II capable of interacting with DNA loops outside the matrix attachment regions. Furthermore, the preferential accessibility of DNase I hypersensitive regions for soluble topoisomerase II must be lost in any case with the disruption of nucleosomes by 2M NaCl extraction. Topoisomerase II is not inactivated by incubation in a concentrated salt solution. It was shown previously that high-salt-insoluble topoisomerase II of the nuclear matrix was able to introduce scissions in DNA [62]. As follows from the above discussion, in high salt extracted nuclei only the DNA loop anchorage sites (i. e. the parts of the loops which are in contact with the nuclear matrix) should be accessible for cleavage with high salt insoluble topoisomerase II. Consequently, incubation of high salt extracted nuclei (nucleoids) with topoisomerase II-specific drugs should result in the excision of chromosomal DNA loops (Fig. 2). On the basis of these considerations we have developed the following procedure for excision of chromosomal DNA loops and mapping loop anchorage sites in characterized genomic areas [33,34]:

Cells embedded into agarose blocks (to prevent DNA fragmentation by mechanical shearing) were extracted with a buffer solution containing 2M NaCl and a non-ionic detergent (NP40 or Triton X100). This extraction was followed by washing off 2M NaCl and an incubation under conditions favouring topoisomerase II-mediated DNA cleavage (in a low salt magnesium-containing buffer supplemented with VM-26). After termination of the incubation by addition of SDS and digestion of proteins with proteinase K, the excised loops and their oligomers were separated by Pulsed Field Gel Electrophoresis (PFGE) and subjected to Southern analysis.

For mapping the positions of the loop ends (loop anchorage sites) versus known sites of DNA cleavage by rear-cutting restriction enzymes (Not I, Sfi I) an indirect end labelling approach was used. Correspondingly, the DNA samples were additionally treated with the above enzymes before separation by PFGE.

The above protocol for the excision of chromosomal DNA loops was first tested in experiments with mammalian ribosomal genes [33]. These genes were found to be organized into loops equal in size to an individual repeated unit and separated by the attachment sites in nontranscribed spacers. When the products of DNA cleavage by endogenous topoisomerase II in high salt-extracted nuclei were separated by PFGE and subjected to Southern analysis with a ribosomal DNA (rDNA) probe, multiple bands with the sizes divisible by the size of an individual ribosomal gene repeat were observed. At high VM-26 concentrations virtually all nucleolar DNA was cut to fragments having the size of an individual rDNA repeat. This fragmentation pattern was not altered in mitotic cells indicating that the specificity of DNA organization into loops is the same in interphase and metaphase chromosomes. Interestingly, the same pattern of bands, although with a higher non-specific background, was observed when living cells were treated with VM-26 [33]. Apparently, the matrix attachment sites constitute preferential targets for topoisomerase II-mediated DNA cleavage even in non-disturbed nuclei. That is why the regular pattern could still be seen in the above experiment. The high background observed in experiments on topoisomerase II - mediated DNA loop excision *in vivo* can be easily explained by a less specific cleavage of DNA with soluble topoisomerase II reported to occur at all accessible places in chromatin such as DNase I HS and even internucleosomal spacers. In full accordance with the proposed model (Fig. 2), the high salt extraction eliminated or at least reduced significantly the above background without affecting the regular pattern of the "loop-sized" bands. It is therefore highly unlikely that the high salt treatment causes disintegration, redistribution or creation *de novo* of matrix attachment sites organized with the participation of topoisomerase II.

A high-resolution analysis performed using indirect end-labelling of topoisomerase II cleavage products additionally cut by restriction enzymes made it possible to conclude that both in living cells and in high salt-extracted nuclei topoisomerase II-mediated cleavages occurred in a non-transcribed spacer within an about 4 kb area located just upstream to the start of the 45S rRNA precursor [33]. Hence, at least in this case it would be appropriate to speak about a DNA loop anchorage region (LAR) rather then a DNA loop anchorage site.

The most important conclusion following from experiments with a ribosomal gene cluster is that an experimental protocol based on a particular model of topological organization of DNA within eukaryotic cell nuclei has permitted us to obtain exactly those results that were predicted by the model. Clearly, this not only justifies the suggested experimental approach but proves as well the initial model.

One specific problem concerning observations made in experiments with a ribosomal gene cluster is that these observations may or may not be reproduced in the rest part of the genome. Indeed, it is well known that ribosomal genes occupy a special compartment in eukaryotic cell nuclei (nucleoli). One can argue that the spatial organization of non-nucleolar genes may differ from those of nucleolar genes. In order to find out whether this is the case, the DNA loop organization in an amplified human *c-myc* gene locus was studied using the topoisomerase II - mediated DNA loop excision protocol [34]. In these experiments the separation of excised loops of chromosomal DNA was combined with indirect end-labelling of large fragments of individual loops obtained after additional cleavage of DNA with rear-cutting restriction nucleases. This made it possible to localize the approximate positions of loop ends with respect to the sites of DNA cleavage by rear-cutting restriction enzymes. As a result, the borders of several DNA loops were mapped at the level of PFGE-resolution. Again, the preferential targets for topoisomerase II-mediated DNA cleavage generating specific long-range DNA fragmentation were found to be the same in living cells and in high salt extracted nuclei. Furthermore, the choice of a drug (VM-26, VP-16, m-AMSA) to be used for entrapment of topoisomerase II cleavage complexes did not influence the results of the mapping experiments [34]. One of the loop basements mapped in the *c-myc* gene locus roughly colocalized with the *c-myc* gene itself. High-resolution studies of the loop anchorage region in the vicinity of the *c-myc* gene permitted the identification of multiple sites of DNA contact with nuclear matrix topoisomerase II scattered along a 5 kb area including the first two exons of the *c-myc* gene and the upstream sequences [34].

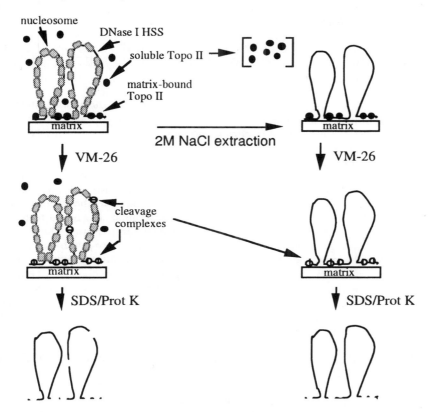

Figure 2. A scheme illustrating the principle of DNA loop excision by topoisomerase II-mediated DNA cleavage at matrix attachment sites. In living cells the 30 nm chromatin fibril is arranged into loops by periodic attachment to the nuclear matrix. The latter contains topoisomerase II which can cut DNA at loop basements. In addition a soluble topoisomerase II can interact with DNA at a number of accessible sites creating an apparently random cleavage pattern (as far as the low-resolution analysis by PFGE is considered). The resulting cleavage pattern will represent the superimposing of this random cleavage pattern over the regular pattern of cleavages at loop basements. With the soluble topoisomerase II removed by high-salt extraction the cleavages can occur only within matrix attachment sites. Consequently, inhibition of the topoisomerase II activity results in excision of chromosomal DNA loops.
For a simplicity the increase of DNA loop contour length and supercoiling of loop DNA after histone removal by 2M NaCl extraction are not shown in the scheme.

Topoisomerase II-mediated DNA loop excision was also successfully used for characterization of the DNA loop organization within a non-amplified 500 kb long region of the first chromosome of Drosophila. The whole region was previously cloned by Miassod and collaborators [63] and the positions of DNA cleavage by rear-cutting restriction enzymes have been determined. Hence it was easy to apply the mapping strategy developed in experiments with an amplified *c-myc* gene locus (i. e. a combination of topoisomerase II-mediated DNA loop excision with indirect end-labelling of large fragments of individual loops). As a result of realization of this project eleven loop anchorage regions (LARs) delimiting ten DNA loops ranging in size from 20 to 90 kb were mapped [64]. Interestingly, one of the LARs was located within the active transcription unit. Hence, the transcription machinery can pass in some way the loop anchorage sites.

2.4. ANALYSIS OF THE SPECIFICITY OF DNA ORGANIZATION INTO LOOPS BY HIGH RESOLUTION *IN SITU* HYBRIDIZATION OF SPECIFIC PROBES WITH NUCLEAR HALOS

Development of a technique for multicolour fluorescence *in situ* hybridization has opened the possibility of determining approximate positions of different genes within DNA loops by direct hybridization of corresponding probes to the halos of DNA loops in high salt-extracted nuclei and metaphase chromosomes [35,36]. Although the resolution of the approach is presently not very high, the approach has an important advantage, as it excludes the possibility of artificial reorganization of attachment sites in the course of isolation of nuclear matrix DNA. Two different research teams have used the fluorescence *in situ* hybridization (FISH) to analyse the specificity of chromosomal DNA organization into loops. This organization was found to be non-random [35,36]. In experiments with interphase nuclei the previous observations indicating that active genes are bound to the nuclear matrix were confirmed [35]. In metaphase chromosomes a preferential association of replication origins with the nuclear matrix was observed. In particular, in a long stretch of unique DNA on human chromosome 11 the most early replicating region was found in close proximity to the axis of a metaphase chromosome [37]. In the ribosomal gene cluster the part of the non-transcribed spacer containing preferential sites of initiation of DNA replication was located close to the loop basements [37].

2.5. RECONSTITUTION *IN VITRO* OF THE COMPLEXES OF DNA WITH THE NUCLEAR MATRIX

It has long been known that nuclear matrices contain a number of DNA-binding proteins [65]. Cockerill and Garrard have demonstrated that isolated nuclear matrices contain specific affinity sites that can detain a subset of sequences present in total DNA [66-68]. The number of these affinity sites was estimated as 50'000 per nucleus. The DNA sequence elements detained selectively (in the presence of a non-specific competitor) by isolated nuclear matrices were called Matrix Associated Regions or MARs [66]. Binding of these DNA sequences to the nuclear matrices was not tissue- or even species-specific [67]. Although most, if not all, DNA fragments found in complexes with nuclear matrices were relatively rich in AT - pairs, some other DNA fragments with a similar nucleotide composition were not detained by the nuclear matrices [69]. Hence, it was logical to assume that DNA fragments capable to interact with affinity sites on the nuclear matrix possessed some common property determined at the level of nucleotide sequence. Comparison of nucleotide sequences of several cloned DNA fragments containing MARs did not reveal any prominent homology [70]. Recent evidence suggests that an essential feature of MARs is the ability to melt under relatively mild conditions [71]. This ability is probably determined by the presence of $(AT)_n$ motifs [71] and ATATTT motifs [66]. Another characteristic feature of MARs determined by the presence of numerous short homopolymeric runs of A and T is a narrow DNA minor groove [72]. The distribution of MARs in the genome has been intensively studied. MARs were foud to colocalize frequently with regulatory sequences (such as enhancers), recombination hot-spots and the borders of DNase I sensitive genomic domains [66,68,73-76]. In yeast cells MARs seem to constitute an essential part of autonomously replicating sequences [77]. At the same time some MARs were mapped within transcriptional units and even within structural genes [78-81].

For the present discussion the most important question is whether MARs participate (possibly along with other genomic elements) in the formation of DNA loop anchorage sites. In the previous section experiments on mapping LARs in a 500 kb region of the first chromosome of Drosophila [64] have been described. The distribution of MARs in the same region was studied previously by Miassod and collaborators [63,82]. Comparison of the two sets of data made it possible to conclude that a fraction of MARs does participate in the organization of loop anchorage sites. Indeed, with a single exception (the loop anchorage site at map position 470) the loop anchorage sites mapped by the topoisomerase II - mediated DNA loop excision protocol colocalized with or at least partially overlapped fragments bearing MARs. At the same time, a number of other MARs were located in loop DNA. Hence, it is clear that at best a MAR could constitute a part of a loop anchorage site but the presence of a MAR is not sufficient for creation of high salt resistant DNA loop anchorage at the nuclear matrix. It is important that no correlation between the "strength" of MARs, as determined by the *in vitro* binding procedure [63,82], and the location of these MARs at or close to actual DNA loop anchorage sites was observed [64].

3. Is there any relationship between the functional organization of the genome and the chromosomal DNA organization into loops?

3.1. REPLICATION ORIGINS ARE LOCATED AT THE BASES OF DNA LOOPS

Initial observations relevant to the problem of spatial organization of replicons in eukaryotic cell nuclei were made by Buongiorno-Nardelli and collaborators [17]. It was found that the size of supercoiled loop-domains in the genome of different eukaryotic cells directly correlated with the average size of replicons in these cells. This observation suggested that a specific element of each replicon (and there are actually two such elements, namely a replication origin and a termination site) was attached to the nuclear matrix. Several lines of experimental evidence indicates that the above element is a replication origin. Observations indicating that replication origins are attached to the nuclear matrix were first made by Wanka and collaborators in pulse-chase experiments performed with synchronized cells [83-85]. It was found that a pulse label incorporated into DNA in the middle of the S-phase was first concentrated in the nuclear matrix DNA and then, after a chase, moved to the cleaved-off (loop DNA) fraction. This was exactly the result that one would expect knowing that replication forks are attached to the nuclear matrix [7,86]. A different distribution of the pulse label between the matrix-bound and cleaved-off DNA fractions was, however, observed when the pulse was given at the very beginning of the S-phase (i. e., when the label was incorporated close to the replication origins). In this case the label could not be chased from the matrix-bound DNA fraction, indicating that the replication origins were located close to the matrix throughout the whole cell cycle [83-85]. Similar results were later reported by other authors [87,88]. The supposition that replication origins are permanently bound to the nuclear matrix was further supported by the results of comparison (in corenaturation experiments) of unique DNA sequences surrounding the replication origins and permanent sites of DNA attachment to the nuclear matrix [89]. It is also noteworthy that most of the presently mapped mammalian and avian replication origins do colocolize with MARs or loop anchorage regions mapped by other approaches. The list includes the replication origin of the domain of chicken α-globin genes [89], the replication origin located in the upstream area of the human *c-myc* gene [34], the replication origin located in proximity to the immunoglobulin heavy chain enhancer [90] and replication origins located in non-transcribed spacers of ribosomal genes [33,91,92].

In yeast cells certain elements of genomic DNA, known as ARS sequences, are able to support the autonomous replication of plasmids. At least some of these elements were also shown to function as chromosomal replication origins (for a review see [93]). From the point of view of the present discussion the most remarkable observation is that MARs constitute a part of full-sized ARS [94] and that the deletion of MARs drastically suppreses the ability of ARS to support the autonomous replication of plasmids [77]. Finally, it is worth mentioning that MARs and known replication origins of both lower and higher eukaryotes share common DNA sequence motifs including AT and TAT repeats [Boulikas, 1992]

Summarizing all the above data, one may conclude that the organization of chromosomal DNA into looped domains is directly linked to the organization of replication units due to the location of replication origins at or close to the DNA loop anchorage sites.

3.2. DO THE DNA LOOP ANCHORAGE SITES DEFINE THE BORDERS OF TRANSCRIPTION UNITS OR TRANSCRIPTIONALLY ACTIVE GENOMIC REGIONS?

There are two approaches to answer the question posed in the title of this section. The first approach can be referred to as a topographical one. Indeed, it is possible simply to map the loop anchorage sites in different genomic areas and to see whether they frame the transcription units. The results will, of course, depend on the method used for mapping anchorage sites. In spite of the fact that MARs constitute the sites of *in vitro* DNA binding to the nuclear matrix, they are still frequently considered as the DNA loop borders. As have been mentioned before (section 2.5), MARs were found at the borders of some DNaseI-sensitive (i. e. transcriptionally active) genomic domains [74-76]. However, MARs can be located also within genes [78-81]. Consequently, the statement that MARs delimit the borders of thranscriptionally active genomic domains can not be correct as long as all MARs are taken into consideration. However, recent experiments on mapping the loop anchorage sites in a 500 kb region of the first chromosome of Drosophila have demonstrated that only a fraction of MARs may actually be involved in the anchorage of DNA loops on the nuclear matrix [64]. As far as LARs mapped by topoisomerase II-mediated DNA loop excision are concerned, there is not enough of mapping data to make a final conclusion. Yet, at least in one case a LAR was mapped within a gene [64]. It is also clear that some genes of higher eukaryotes are too long to be organized in one loop (for example the human Dystriphin gene is more than 2,000 kb long). Our preliminary data suggest that these genes are organized into several loops (Razin et al., unpublished). Hence, it is likely that transcriptional complexes can pass in some way the loop anchorage sites.

The second approach which can be used in order to find out whether DNA loop anchorage sites define the borders of active genomic domains may be referred to as a functional one. It is based on the construction of artificial mini-domains insulating the reporter genes from position variegation effects in the genome of transgenic animals. It is well known that in differentiated mammalian cells most of the genomic domains are not active. When exogenous DNA is microinjected into the pronuclei of zygotes, its integration into any genomic domain seems to be equally probable. In differentiated cells of an adult organism, only genes that happen to integrate into domains active in the given cell type will evidently be expressed (see the scheme in Fig. 3). The logical way to solve the problem of position effects is to create artificial mini-domains including a foreign gene and a block of regulatory sequences. The active domains in chromatin which are characterized by preferential sensitivity to DNaseI usually have relatively sharp borders. This suggests that there should be some special genomic elements defining the above borders. So, the question is: what are these elements and whether they can be found among the DNA sequences located at the basements of DNA loops? A direct experimental approach to answer the above questions consists in the creation of artificial genomic domains containing a reporter gene surrounded by putative domain border elements (Fig. 3). In the positive case the above elements are expected to protect the reporter gene from position variegation effects. In other words, the expression of this gene in different tissues of transgenic animals should be position-independent and copy number-dependent. Up to now only MARs have been tested for the ability to insulate transgenes from suppressing effects of host chromatin. In several experiments made by different authors it was observed that various MARs, when placed upstream and downstream to the transgene, suppressed (in some cases only partially) the position variegation effects [96-99]. Other authors have found that MARs flanking a reporter gene stimulate the expression of this gene but do not dampen the position variegation effects [100]. One problem, which was not recognized in the initial experiments with mini-domains, arises because of multiple integration of transgenes in the same genomic position. Recent observations suggest that tandem arrays of multicopy transfectants may become transcriptionally inactive in spite of the presence of all necessary regulatory elements [99,101]. Hence, further experiments with a single copy integration or multiple integration in different positions are necessary to find out whether MARs may indeed serve as functional borders of genomic domains.

In conclusion we may say that the relationship between chromosomal DNA loops and transcriptional units or active genomic domains remains unclear. The mapping data indicate that both MARs and LARs can be found within transcription units. So, at least some DNA loop anchorage regions do not colocalize with the borders of active genomic domains. Yet certain MARs were found to possess functional properties of domain border sequences, i. e. the ability to insulate a transgene from suppressing effects of the host chromatin domain. As a matter of fact, the present data do not exclude the possibility that all borders of chromatin domains colocalize with the loop anchorage sites, although an opposite statement would not be correct.

52

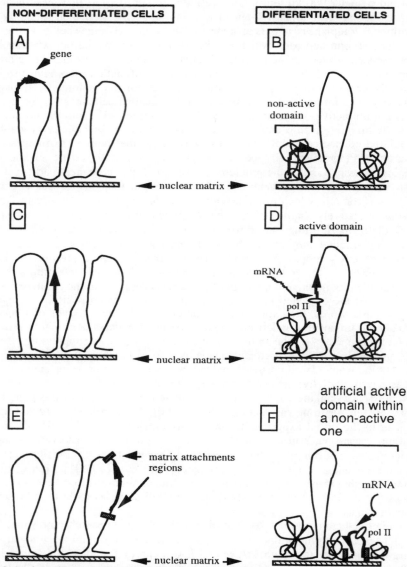

Figure 3. A scheme illustrating the strategy of dampening position effects by creation of mini-domains in the genome of transgenic animals.

In embryonic (non-differentiated) cells (A, C, E) all domains are open. The probability of integration of a transgene into each domain is supposed to be similar or even equal. In differentiated cells the majority of genomic domains becomes non-active (condensed loops in sections B, D, F). In a non-active domain a transgene is not expressed (B). The transgene happened to reside in an active domain is expressed (D). In order to insulate a transgene from the influence of the host-cell chromatin the transgene is places between matrix attachment regions (E). It is supposed that interaction of the latter with the nuclear matrix should create an independent domain (F). Within the above domain expression of the transgene will be determined only by regulatory elements placed into this domain along with the transgene. In a result expression of the transgene will not depend on the site of integration.

Acknowledgements

The work of S.V. Razin is supported by grant 96-04-49120 from Russian Foundation for Support of Fundamental Science, grant # 097 from Russian State Program "Frontiers in Genetics" and ICGEB grant CRP/RUS93-06.

References

1. Cook, P.R. and Brazell, I.A. (1975) Supercoils in human DNA, *J. Cell Sci.* **19**, 261-279.
2. Cook, P.R. and Brazell, I.A. (1976) Conformational constrains in human DNA, *J. Cell Sci.* **22**, 287-302.
3. Cook, P.R., Brazell, I.A., and Jost, E. (1976) Characterization of nuclear structures containing superhelical DNA, *J. Cell. Sci.* **22**, 303-324.
4. Cook, P. and Brazell, I.A. (1978) Spectro fluorometric measurement of the binding of ethidium bromide to superhelical DNA from cell nuclei, *Eur. J. Biochem.* **84**, 464-477.
5. Warren, A.C. and Cook, P.R. (1978) Supercoiling of DNA and nuclear conformation during the cell cycle, *J. Cell Sci.* **30**, 211-226.
6. Berezney, R. and Coffey D. S. (1974) Identification of a nuclear protein matrix, *Biochem. Biophys. Res. Commun.* **60**, 1410-1417.
7. Berezney, R. and Coffey, D. S. (1975) Nuclear protein matrix: association with newly synthesized DNA, *Science* **189**, 291-292.
8. Berezney, R. and Coffey, D.S., (1977) Nuclear matrix: Isolation and characterization of a framework structure from rat liver nuclei, *J. Cell. Biol.* **73**, 616-637.
9. Paulson, J.R. and Laemmli, U.K. (1977) The structure of histone-depleted metaphase chromosomes, *Cell* **12**, 815-828.
10. McCready, S.J., Akrigg, A., and Cook, P.R. (1979) Electron microscopy of intact nuclear DNA from human cells, *J. Cell. Sci.* **39**, 53-62.
11. Hancock, R. and Hughes, M.E. (1982) Organization of DNA in the eukaryotic nucleus. *Biol. Cell.* **44**, 201-212.
12. Igo-Kemennes, T. and Zachau, H.G. (1977) Domains in chromatin structure, *Cold Spring Harbor Symp. Quant. Biol,* **42**, 109-118.
13. Razin, S.V., Mantieva, V.L., and Georgiev, G.P. (1979) The similarity of DNA sequences remaining bound to scaffold upon nuclease treatment of interphase nuclei and metaphase chromosomes, *Nucl. Acids Res.* **17**, 1713-1735.
14. Benyajati, C. and Worcel, A. (1976) Isolation, characterization and structure of the folded interphase genome of Drosophila melanogaster, *Cell* **9**, 393-407.
15. Hartwig, M. (1982) Organization of mammalian chromosomal DNA: supercoiled and folded circular DNA subunits from interphase cell nuclei, *Acta Biol. Med. Germ.* **37**, 421-432.
16. Nakane, M., Ide, T., Anzai, K., Ohara, S., and Andoh, T. (1978) Supercoiled DNA folded by nonhistone proteins in cultured mouse carcinoma cells, *J. Biochem.* **84**, 145-157.
17. Buongiorno-Nardelli, M., Gioacchino, M., Carri, M.T., and Marilley, M. (1982) A relationship between replicon size and supercoiled loop domains in the eukaryotic genome, *Nature* **298**, 100-102.
18. Vogelstein, B., Small, D., Robinson, S., and Nelkin, B. (1985) The nuclear matrix and the organization of nuclear DNA, In Bekhor I (ed), *Progress in nonhistone protein research,* CRC Press, Inc., Boka Raton, Florida, Vol. 2, pp. 115-129.
19. Jackson, D. A., Dickinson, P., and Cook, P. R. (1990) The size of chromatin loops in HeLa cells, *EMBO J.* **9**, 567-571.
20. Razin, S.V., Gromova, I.I., and Iarovaia, O.V. (1995) Specificity and functional significance of DNA interaction with the nuclear matrix: New approaches to clarify the old questions, *Int Rev Cytol* **162B**, 405-448.
21. Hancock, R (1982) Topological organization of interphase DNA: the nuclear matrix and other skeletal structures, *Biol. Cell.* **46**, 105-122.
22. Weintraub, H. and Groudine, M. (1976) Chromosomal subunits in active genes have an altered conformation, *Science* **73**, 848-856.
23. Cook, P.R. and Brazell, I.A. (1980) Mapping sequences in loops of nuclear DNA by their progressive detachment from the nuclear cage, *Nucl. Acids Res.* **8**, 2895-2906.
24. Cook, P.R., Lang, J., Hayday, A., Lania. L., Fried, M., Chiswell, D.J., and Wyke, A.(1982) Active viral genes in transformed cells lie close to the nuclear cage, *EMBO J.* **1**, 447-452.
25. Robinson, S.I., Nelkin, B.D., and Vogelstein, B. (1982) The ovalbumin gene is associated with the nuclear matrix of chicken oviduct cells, *Cell* **28**, 99-106.
26. Robinson, S.I., Small, D., Idzerda, R., McKnight, G.S. and Vogelstein, B. (1983) The association of active genes with the nuclear matrix of the chicken oviduct, *Nucl. Acids Res.* **15**, 5113-5130.
27. Small, D. and Vogelstein, B. (1985) The anatomy of supercoiled loops in the *Drosophila* 7F locus, *Nucl. Acids Res.* **13**, 7703-7713.

54

28. .Small, D., Nelkin, B. and Vogelstein, B. (1985) The association of transcribed genes with the nuclear matrix of *Drosophila* cells during heat shock, *Nucl. Acids Res.* **13**, 2413-2431.
29. Mirkovich, J., Mirault, E., and Laemmli, U.K. (1984) Organization of the higher order chromatin loop: specific DNA attachment sites on nuclear scaffold, *Cell* **39**, 323-332.
30. Jackson, D.A. and Cook, P.R. (1985) Transcription occurs at a nucleoskeleton, *EMBO J.* **4**, 919-925.
31. Jackson, D.A. and Cook, P.R. (1986) Replication occurs at a nucleoskeleton, *EMBO J.* **5**, 1403-1410.
32. Razin, S.V., Petrov, P., and Hancock, R. (1991). Precise localization of the α-globin gene cluster within one of 20 to 300 kb fragments released by cleavage of chicken chromosomal DNA at topoisomerase II sites in vivo: evidence that the fragments are DNA loops or domains, *Proc. Natl. Acad. Sci. USA* **88**, 8515-8519.
33. Razin, S.V., Hancock, R., Iarovaia, O., Westergaard, O., Gromova, I., and Georgiev, G.P. (1993) Structural-functional organization of chromosomal DNA domains, *Cold Spring Harbor Symp. Quant. Biol.* **58**, 25-35.
34. Gromova, I.I., Thomsen, B., and Razin, S.V. (1995) Different topoisomerase II antitumour drugs direct similar specific long-range fragmentation of an amplified *c-myc* gene locus in living cells and in high salt-extracted nuclei, *Proc. Natl. Acad. Sci. USA* **92**, 102-106.
35. Gerdes, M.G., Carter, K.C., Moen, Jr., P.T., and Lawrence, J.B. (1994) Dynamic changes in the higher-level chromatin organization of specific sequences revealed by *in situ* hybridization to nuclear halos, *J. Cell. Biol.* **126**, 289-304.
36. Bickmore, W.A., Oghene, K. (1996) Visualizing the spatial relationships between defined DNA sequences and the axial region of extracted metaphase chromosomes, *Cell* **84**, 95-104.
37. Jost, J.-P. and Seldran, M. (1984) Association of transcriptionally active vitellogenin II gene with the nuclear matrix of chicken live,. *EMBO J.* **3**, 2205-2208.
38. Kirov, N., Djondjurov, L., and Tsanev, R. (1984) Nuclear matrix and transcriptional activity of the mouse α-globin gene, *J. Mol. Biol.* **180**, 601-614.
39. Thorburn, A., Moore, R., Knowland, J. (1988) Attachment of transcriptionally active DNA sequences to the nucleoskeleton under isotonic conditions, *Nucl. Acids Res.* **16**, 7183.
40. Thorburn, A. and Knowland, J. (1993) Attachment of vitellogenin genes to the nucleoskeleton accompanies their activation, *Biochem. Biophys. Res. Commun.* **191**, 308-313.
41. Jackson, D.A., Hassan, A.B., Errington, R.J., and Cook, P.R. (1993). Visualization of focal sites of transcription within human nuclei, *EMBO J.* **12**, 1059-1065.
42. Xing, Y., Johnson, C.V., Dobner, P.R., and Lawrence, J.B. (1993) Higher level organization of individual gene transcription and RNA splicing, *Science* **259**, 1326-1330.
43. Xing, Y., Johnson, C.V., Moen, Jr., P.T., McNeil J.A., and Lawrence J.B. (1995) Nonrandom gene organization: Structural arrangements of specific pre-mRNA transcription and splicing with SC-35 domains, *J. Cell. Biol.* **131**, 1635-1647.
44. Cook, P.R. (1994) RNA polymerase: structural determinant of the chromatin loop and chromosome, *Bioessays* **16**, 425-430.
45. Razin, S.V., Rzeszowska-Wolny, J., Moreau, J., and Scherrer, K. (1985) Localization of regions of DNA attachment to the nuclear skeleton within chicken alpha-globin genes in functionally active and functionally inactive nuclei, *Mol. Biol. (Moscow)* **19**, 456-466.
46. Razin, S.V. (1987) DNA interactions with the nuclear matrix and spatial organization of replication and transcription, *Bioessays* **6**, 19-23.
47. Farache, G., Razin, S.V., Rzeszowska-Wolny. J., Moreau. J., Recillas-Targa, F., and Scherrer, K. (1990) Mapping of structural and transcription-related matrix attachment sites in the α-globin gene domain of avian erythroblasts and erythrocytes, *Mol. Cell. Biol.* **10**, 5349-5358.
48. Kalandadze, A.G., Bushara, S.A., Vassetzky, Y.S., and Razin, S.V. (1990) Characterization of DNA pattern in the site of permanent attachment to the nuclear matrix located in the vicinity of replication origin, *Biochem. Biophys. Res. Commun.* **168**, 9-15.
49. Berrious, M., Osheroff, N., and Fisher, P. (1985) In situ localization of DNA topoisomerase II, a major polypeptide of Drosophila nuclear matrix, *Proc. Natl. Acad. Sci. USA* **82**, 4142-4146.
50. Earnshaw, W. C., Halligan, B., Cooke, C. A., Heck, M. M. S., and Liu, L.F. (1985) Topoisomerase II is a structural component of mitotic chromosome scaffolds, *J. Cell. Biol.* **100**, 1706-1715.
51. Wang, J.C. (1985) DNA topoisomerases, *Annu Rev. Biochem.* **60**, 513-552.
52. Chen, G.L., Yang, L., Rowe. T.C., Halligan, B.D., Tewey, K.M. and Liu, L. (1984) Nonintercalative antitumor drugs interfere with the breakage-reunion reaction of mammalian topoisomerase II, *J. Biol. Chem.* **259**, 13560-13566.
53. Robinson, M.J. and Osheroff, N. (1990) Stabilization of the topoisomerase II-DNA cleavage complex by antineoplastic drugs: Inhibition of the enzyme-mediated DNA religation by 4'-(9-acridinylamino)-methanesulfon-*m*-anisidide, *Biochemistry* **29**, 2511-2515.
54. Sander, M. and Hsieh, T.S. (1985) Drosophila topoisomerase II double-strand DNA cleavage: analysis of DNA sequence homology at the cleavage site, *Nucl. Acids Res.* **13**, 1057-1072.
55. Udvardy, A., Schedl, P., Sander, M., and Hsieh, T. (1985) Novel partitioning of DNA cleavage sites for Drosophila topoisomerase II, *Cell* **40**, 933-941.
56. Reitman, M. and Felsenfeld, G. (1990) Developmental regulation of topoisomerase II sites and DNase I - hypersensitive sites in the chicken b-globin locus, *Mol. Cell. Biol.* **10**, 2779-2786.

57. Udvardy, A. and Schedl, P. (1991) Chromatin structure, not DNA sequence specificity, is the primary determinant of topoisomerase II sites of action in vivo, *Mol. Cell. Biol.* **11**, 4973-4984.
58. Fernandes, D.J., Danks, M.K., and Beck, W.T. (1990) Decreased nuclear matrix DNA topoisomerase II in human leukemia cells resistant to VM-26 and m-AMSA, *Biochemistry* **29**, 4235-4241.
59. Kaufmann, S.H. and Shapper, J.H. (1991) Association of topoisomerase II with the hepatoma cell nuclear matrix: the role of intermolecular disulfide bond formation, *Exp. Cell. Res.* **192**, 511-523.
60. Danks, M.K., Qiu, J., Catapano, C.V., Schmidt, C.A., Beck, W.T., and Fernandes, D.J. (1994) Subcellular distribution of the a and b topoisomerase II-DNA complexes stabilized by VM-26, *Biochem Pharmacol.* **48**, 1785-1795.
61. Zini, N., Santi, S., Ognibene, A., Bavelloni, A., Neri, L.M., Valmori, A., Mariani, E., Negri, C., Astaldi-Ricotti, G.C., and Maraldi, N.M. (1994) Discrete localization of different DNA topoisomerases in HeLa and K562 cell nuclei and subnuclear fractions, *Exp. Cell. Res.* **210**, 336-348.
62. Vassetzky, Y.S., Razin, S.V., and Georgiev, G.P. (1989) DNA fragments which specifically bind to isolated nuclear matrix in vitro interact with matrix-associated DNA topoisomerase II, *Biochem. Biophys. Res. Commun.* **159**, 1263-1268.
63. Surdej, P., Got, C., Rosset, R., and Miassod, R. (1990) Supragenic loop organization: mapping in Drosophila embryos, of scaffold-associated regions on a 800 kilobase continuum cloned from the 14B-15B first chromosome region, *Nucleic Acids Res.* **18**, 3713-3722.
64. Iarovaia, O.V., Hancock. R., Lagarkova, M.A., Miassod, R., and Razin, S.V. (1996) Mapping of genomic DNA loop organization in a 500-kilobase region of the Drosophila X chromosome using the topoisomerase II-mediated DNA loop excision protocol, *Mol. Cell. Biol.* **16**, 302-308.
65. Comings, D.E. and Wallack A.S. (1978) DNA-binding properties of nuclear matrix proteins, *J. Cell. Sci.* **34**, 233-246.
66. Cockerill, P.N. and Garrard, W.T. (1986) Chromosomal loop anchorage of the kappa immunoglobulin gene occurs next to the enhancer in a region containing topoisomerase II sites, *Cell* **44**, 273-282.
67. Cockerill, P.N. and Garrard, W.T. (1986) Chromosomal loop anchorage sites appear to be evolutionary conserved, *FEBS Lett.* **204**, 5-7.
68. Cockerill, P.N., Yuen, M.-H., and Garrard, W.T. (1987) The enhancer of the immunoglobulin heavy chain locus is flanked by presumptive chromosomal loop anchorage elements, *J. Biol. Chem.* **262**, 5394-5397.
69. Das, A.T., Luderus, M.E., and Lamers, W.H. (1993) Identification and analysis of a matrix-attachment region 5' of the rat glutamate-dehydrogenase-encoding gene, *Eur. J. Biochem.* **215**, 777-785.
70. Mielke, C., Kohwi, I., Kohwi-Shigematsu, T., and Bode J. (1990) Hierarchical binding of DNA fragments derived from scaffold-attached regions: correlation of properties in vitro and function in vivo, *Biochemistry* **29**, 7475-7485.
71. Bode, J., Kohwi, I., Dickinson, L., Joh, T., Klehr, D., Mielke, C., and Kohwi-Shigematsu, T. (1992) Biological significance of unwinding capability of nuclear matrix-associated DNAs, *Science* **255**, 195-197.
72. Kas, E., Izaurralde, E., and Laemmli, U.K. (1989) Specific inhibition of DNA binding to nuclear scaffolds and histone H1 by distamycin: The role of oligo(dA)·oligo(dT) tracts, *J. Mol. Biol.* **210**, 587-599.
73. Dijkwel, P.A. and Hamlin, J.L. (1988) Matrix attachment regions are positioned near replication initiation sites, genes, and an interamplicon junction in the amplified dihidrofolate reductase domain of Chinese hamster ovary cells, *Mol. Cell. Biol.* **8**, 5398-53409.
74. Phi-Van, L. and Stratling, W.H. (1988) The matrix attachment regions of the chicken lysozyme gene comap with the boundaries of the chromatin domain, *EMBO J.* **7**, 655-664.
75. Bode, J. and Maass, K. (1988) Chromatin domain surrounding the human interferon-b gene as defined by scaffold attached regions, *Biochemistry* **27**, 4706-4711.
76. Levy-Wilson, B. and Fortier, C. (1989) The limits of the DNase I - sensitive domain in the human apolipoprotein B gene coincide with the locations of chromosomal anchorage loops and define the 5' and 3' boundaries of the gene, *J. Biol. Chem.* **264**, 21196-21204.
77. Amati, B. and Gasser, S.M. (1990) Drosophila scaffold-attached regions bind nuclear scaffolds and can function as ARS elements in both budding and fission yeasts, *Mol. Cell. Biol.* **10**, 5442-5454.
78. Ito, T. and Sakaki, Y. (1987) Nuclear matrix association regions of rat alpha 2-macroglobulin gene, *Biochem. Biophys Res Commun* **149**, 449-454.
79. Kas, E. and Chasin, L.A. (1987) Anchorage of the Chinese hamster dihydrofolate reductase gene to the nuclear scaffold occurs in an intragenic region, *J. Mol. Biol.***198**, 577-692.
80. Beggs, A.H. and Migeon, B.R. (1989) Chromatin loop structure of the human X chromosome: Relevance to X inactivation and CpG clusters, *Mol. Cell. Biol.* **9**, 2322-2331.
81. Romig, H., Ruff, J., Fackelmayer, F.O., Patil, M.S., and Richter, A. (1994) Characterization of two intronic nuclear-matrix-attachment regions in the human DNA topoisomerase I gene, *Eur. J. Biochem.* **221**, 411-419.
82. Brun, C., Dang Q., and Miassod, R. (1990) Studies of an 800-kilobase DNA stretch of the Drosophila X chromosome: comparing of a subclass of scaffold-attached regions with sequences able to replicate autonomously in Saccharomyces cerevisiae, *Mol. Cell. Biol.* **10**, 5455-5463.
83. Aelen, J.M., Opstelsten, R.J. and Wanka, F. (1983) Organization of DNA replication in Physarum polycephalum. Attachment of origins of replication and replication forks to the nuclear matrix, *Nucl. Acids Res.* **11**, 1181-1195.

56

84. Van der Velden, H.M., Poot, M., and Wanka, F.(1984) In vitro DNA replication in association with the nuclear matrix of permeable mammalian cells, *Biochim. Biophys. Acta.* **782**, 429-436.
85. Van der Velden, H.M.V. and Wanka, F. (1987) The nuclear matrix - its role in the spatial organization and replication of eukaryotic DNA, *Mol. Biol. Rep.* **12**, 69-77.
86. Pardoll, D.M., Vogelstein, B, and Coffey, D.S. (1980) A fixed site of DNA replication in eukaryotic cells, *Cell* **19**, 527-536.
87. Carri, M.T., Micheli, G., Graziano, E., Pace, T., and Buongiorno-Nardelli, M. (1986) The relationship between chromosomal origins of replication and the nuclear matrix during the cell cycle, *Exp. Cell. Res.* **164**, 426-436.
88. Dijkwel, P. A., Wenink, P. W., and Poddighe, J. (1986) Permanent attachment of replication origins to the nuclear matrix in BHK cells, *Nucl. Acids Res.* **14**, 3241-3249.
89. Razin, S.V., Kekelidze, M.G., Lukanidin, E.M., Scherrer, K., and Georgiev, G.P. (1986) Replication origins are attached to the nuclear skeleton, *Nucl. Acids Res.* **14**, 8189-8207.
90. Ariizumi, K., Wang, Z., and Tucker, P.W. (1993) Immunoglobulin heavy chain enhancer is located near or in an initiation zone of chromosomal DNA replication, *Proc. Natl. Acad. Sci. USA* **90**, 3695-3699.
91. Marilley, M. and Gassend-Bonnet, G. (1989) Supercoiled loop organization of genomic DNA: a close relationship between loop domains, expression units, and replicon organization in rDNA from Xenopus laevis, *Exp. Cell. Res.* **180**, 475-489.
92. Du, C., Sanzgiri, R.P., Shaiu, W.L., Choi, J.K., Hou, Z., Benbow, R.M., and Dobbs, D.L. (1995) Modular structural elements in the replication origin region of Tetrahymena rDNA, *Nucl. Acids Res.* **23**, 1766-1774
93. Umek, R.M., Linskens, M.H.K., Kowalski, D., and Huberman, J. (1989) New beginnings in the studies of eukaryotic DNA replication origins, *Biochim. Biophys. Acta* **1007**, 1-14.
94. Amati, B. and Gasser, S.M. (1988) Chromosomal ARS and CEN elements bind specifically to the yeast nuclear scaffold, *Cell* **54**, 967-978.
95. Boulikas, T. (1992) Homeotic protein binding sites, origins of replication and nuclear matrix anchorage sites share the ATTA and ATTTA motifs, *J. Cell. Biochem.* **50**, 1-13.
96. Stief, A.C., Winter, D.M., Stratling, W.H., and Sippel, A.E. (1989) A nuclear DNA attachment element mediates elevated and position independent gene activity, *Nature* **341**, 343-345.
97. Phi-Van, L., von Kries, J.P., Ostertag, W., and Stratling, W.H. 1990. The chicken lysozyme 5' matrix attachment region increases transcription from a heterologous promoter in heterologous cells and dampens position effects on the expression of transfected genes, *Mol. Cell. Biol.* **10**, 2302-2307.
98. Yu, J., Bock, J.H., Slightom, J.L., and Villeponteau, B. (1994) A 5' matrix-attachment region and the polyoma enhancer together confer position-independent transcription, *Gene* **139**, 139-145.
99. Kalos, M. and Fournier, R.E. (1995) Position independent transgene expression mediated by boundary elements from the apolipoprotein B chromatin domain, *Mol. Cell. Biol.* **15**,198-207.
100. Poljak, L., Seum, C., Mattioni, T. and Laemmli, U.K. 1994. SARs stimulate but do not confer position independent gene expression, *Nucleic Acids Res.* **22**, 4386-4394.
101. Dorer, D.R. and Henikoff, S. (1994) Expansions of transgene repeats cause heterochromatin formation and gene silencing in Drosophila, *Cell* **77**, 993-1002.

INTERRELATIONSHIPS BETWEEN NUCLEAR STRUCTURE AND TRANSCRIPTIONAL CONTROL OF CELL CYCLE AND TISSUE-SPECIFIC GENES

G.S. STEIN, A.J. VAN WIJNEN, J.L. STEIN, J.B. LIAN, M. MONTECINO
University of Massachusetts Medical Center
Department of Cell Biology & Cancer Center
55 Lake Ave. North
Worcester, MA 01655-0106

Abstract

Three parameters of nuclear structure contribute to transcriptional control. The linear representation of promoter elements provides competency for physiological responsiveness within the contexts of developmental as well as cell cycle and phenotype-dependent regulation. Chromatin structure and nucleosome organization reduce distances between independent regulatory elements providing a basis for integrating components of transcriptional control. The nuclear matrix supports gene expression by imposing physical constraints on chromatin related to three dimensional genomic organization. In addition, the nuclear matrix facilitates gene localization as well as the concentration and targeting of transcription factors. Several lines of evidence are presented which are consistent with involvement of multiple levels of nuclear architecture in cell growth and tissue-specific gene expression during differentiation. Growth factor and steroid hormone responsive modifications in chromatin structure, nucleosome organization and the nuclear matrix are considered which influence transcription of the cell cycle regulated histone gene and the bone tissue-specific osteocalcin gene during progressive expression of the osteoblast phenotype.

1. Introduction: Nuclear Architecture Contributes to Transcriptional Control

There is a growing awareness of functional interrelationships mediating nuclear structure and function. Historically, there was a perceived dichotomy between regulatory mechanisms supporting gene expression and components of nuclear architecture. However, this parochial view is rapidly changing. The emerging concept is that both transcription and DNA synthesis occur in association with structural parameters of the nucleus. Consequently, it has become increasingly evident that the cellular and molecular mechanisms must be defined which contribute to both the regulated and regulatory relationships of nuclear morphology to the expression and replication of genes.

C. Nicolini (ed.), Genome Structure and Function, 57–82.
© 1997 *Kluwer Academic Publishers. Printed in the Netherlands.*

During the past several years, there has been an accrual of insight into the complexities of transcriptional control in eukaryotic cells. Our concept of a promoter has evolved from the initial expectation of a single regulatory sequence which determines transcriptional competency and level of expression. We now appreciate that transcriptional control is mediated by an interdependent series of regulatory sequences which reside 5', 3' and within transcribed regions of genes. Rather than focusing on the minimal sequences required for transcriptional control to support biological activity, efforts are being directed towards defining functional limits. Consequently, contributions of distal flanking sequences to regulation of transcription are being experimentally addressed. This is a necessity for understanding mechanisms by which activities of multiple promoter elements are responsive to a broad spectrum of regulatory signals and the activities of these regulatory sequences are functionally integrated. Crosstalk between a series of regulatory domains must be understood under diverse biological circumstances where expression of genes supports cell and tissue functions. The overlapping binding sites for transcription factors within promoter regulatory elements and protein-protein interactions which influence transcription factor activity provide further components of the requisite diversity to accommodate regulatory options for physiologically responsive gene expression.

As the intricacies of gene organization and regulation are elucidated, the implications of a fundamental biological paradox become strikingly evident. How, with a limited representation of gene-specific regulatory elements and low abundance of cognate transactivation factors, can sequence-specific interactions occur to support a threshold for initiation of transcription within nuclei of intact cells. Viewed from a quantitative perspective, the *in vivo* regulatory challenge is to account for formation of functional transcription initiation complexes with a nuclear concentration of regulatory sequences that is approximately 20 nucleotides per 2.5 yards of DNA and a similarly restricted level of DNA binding proteins.

There is a growing appreciation that nuclear architecture provides a basis for support of stringently regulated modulation of cell growth and tissue specific transcription which is necessary for the onset and progression of differentiation. Here, multiple lines of evidence point to contributions by three levels of nuclear organization to *in vivo* transcriptional control where structural parameters are functionally coupled to regulatory events. The primary level of gene organization establishes a linear ordering of promoter regulatory elements. This representation of regulatory sequences reflects competency for responsiveness to physiological regulatory signals. However, interspersion of sequences between promoter elements that exhibit coordinate and synergistic activities indicates the requirement of a structural basis for integration of activities at independent regulatory domains. Parameters of chromatin structure and nucleosome organization are a second level of genome architecture that reduce the distance between promoter elements thereby supporting interactions between the modular components of transcriptional control. Each nucleosome (approximately 140 nucleotide base pairs wound around a core complex of 2 each of H3, H4, H2 and H2B histone proteins) contracts linear spacing by seven-fold. Higher order chromatin structure further reduces nucleotide distances between regulatory sequences. Folding of nucleosome arrays into solenoid-type structures provides a potential for interactions which support synergism between promoter elements and responsiveness to multiple signaling pathways. A third level of nuclear architecture which contributes to transcriptional control is provided by the nuclear matrix (summarized in Table I).

TABLE I

**CONTRIBUTIONS OF THE NUCLEAR MATRIX TO REGULATION
OF GENE EXPRESSION**

I.   Involvement of the nuclear matrix in DNA replication. [97-104]

II.   Involvement of the nuclear matrix in transcriptional control.

    A. Biologically relevant modifications in representation of nuclear matrix proteins.

        1.   Cell type and tissue-specific nuclear matrix proteins. [94, 105-111]
        2.   Developmental stage-specific nuclear matrix proteins. [88]
        3.   Tumor-specific nuclear matrix proteins. [112-114]
        4.   Steroid hormone responsive nuclear matrix proteins. [115-117]
        5.   Polypeptide hormone responsive nuclear matrix proteins. [118]

    B. Association of actively transcribed genes with the nuclear matrix.

        1.   Matrix associated regions of genes (MARs). [88, 119-143]

        2.   MAR binding proteins. [144-153]

    C. Nuclear matrix localization of transcription factors.

        1.   Steroid hormone receptors. [115, 116, 154, 155]
        2.   Ubiquitous transcription factors. [35, 88, 156]
        3.   Tissue-specific transcription factors. [94, 53]
        4.   Viral regulatory proteins. [121, 157]

    D. Selective partitioning of transcription factors between the nuclear matrix and non-matrix nuclear fractions. [158]

III.   Involvement of the nuclear matrix in post-transcriptional control.

    A. RNA processing. [159-174]

    B. Post-translational modifications of chromosomal proteins. [175]

    C. Phosphorylation [176]

The anastomosing network of fibers and filaments which constitute the nuclear matrix supports the structural properties of the nucleus as a cellular organelle and accommodates structural modifications associated with proliferation, differentiation and changes necessary to sustain phenotypic requirements of specialized cells. Regulatory functions of the nuclear matrix include but are by no means restricted to gene localization, imposition of physical constraints on chromatin structure which support formation of loop domains, RNA processing and transport of gene transcripts, as well as imprinting and modifications of chromatin structure. Taken together these components of nuclear architecture facilitate biological requirements for physiologically responsive modifications in gene expression within the contexts of: 1) homeostatic control involving rapid, short-term and transient responsiveness; 2) developmental control which is progressive and stage-specific and 3) differentiation-related control which is associated with long term phenotypic commitments to gene expression for support of structural and functional properties of cells and tissues.

We are just beginning to comprehend the significance of nuclear domains in the control of gene expression. However it is already apparent that local nuclear environments which are generated by the multiple aspects of nuclear structure are intimately tied to developmental expression of cell growth and tissue-specific genes. From a broader perspective, reflecting the diversity of regulatory requirements as well as the phenotype-specific and physiologically responsive representation of nuclear structural proteins, there is a reciprocally functional relationship between nuclear structure and gene expression. Nuclear structure is a primary determinant of transcriptional control and the expressed genes modulate the regulatory components of nuclear architecture. Thus, the power of addressing gene expression within the three-dimensional context of nuclear structure would be difficult to overestimate. Membrane-mediated initiation of signaling pathways that ultimately influence transcription have been recognized for some time. Here, the mechanisms which sense, amplify, dampen and/or integrate regulatory signals involve structural as well as functional components of cellular membranes. Extending the structure-regulation paradigm to nuclear architecture expands the cellular context in which cell-structure-gene expression interrelationships are operative.

2. Developmental Transcriptional Control During Proliferation and Differentiation: Regulation of the Cell Cycle-Dependent Histone Genes and the Bone-Tissue-Specific Osteocalcin Gene During Progressive Expression of the Osteoblast Phenotype

It has been well documented that differentiation is a multistep process, orchestrated by a complex and interdependent series of stringently controlled regulatory events. Expression of genes supporting proliferation and those which mediate both development and maintenance of cell and tissue phenotypic properties are responsive to intracellular as well as extracellular regulatory cues. Osteoblast differentiation is a striking example where progression of temporally compartmentalized gene expression, together with crosstalk between regulatory events requisite for each developmental period, are necessary for establishment of bone tissue organization.

The sequential expression of cell growth and tissue-specific genes during osteoblast differentiation **(Figure 1)** is supported by developmental transcriptional control [1-3]. Both *in vivo* and *in vitro* cultures of normal diploid osteoblasts, proliferating cells express genes which mediate competency for cell growth as well as extracellular matrix biosynthesis.

Postproliferatively, genes functionally related to the organization and mineralization of the bone extracellular matrix are expressed [1,2]. From the perspective of growth control, there is a requirement to support expression of genes for proliferation in early stage cells. At the same time, expression of genes associated with the postproliferative acquisition of bone cell and tissue phenotypic properties must be repressed. Then, at a key transition point which marks completion of the initial developmental period, proliferation is downregulated and gene expression which supports the establishment and maintenance of bone tissue organization is upregulated.

We will focus on regulatory mechanisms controlling transcription of the cell cycle regulated histone genes in proliferating osteoblasts and those controlling transcription of the bone-specific osteocalcin gene in mature osteoblasts during extracellular matrix mineralization. Transcription of these cell growth and tissue-specific genes will be considered within the context of regulatory contributions from principal components of nuclear architecture.

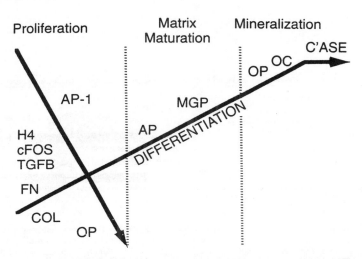

Figure 1: Reciprocal and functionally coupled relationship between cell growth and differentiation-related gene expression in osteoblasts. These relationships are schematically illustrated as arrows representing changes in expression of cell cycle and cell growth regulated genes as well as genes associated with the regulated and regulatory events linked to the onset and progression of differentiation. The three principal periods of the osteoblast developmental sequence are designated within broken vertical lines (proliferation, matrix maturation, and mineralization). These broken lines indicate two experimentally established transition points in the developmental sequence exhibited by normal diploid osteoblasts during sequential acquisition of the bone cell phenotype - the first at the completion of proliferation when genes associated with extracellular matrix development and maturation are upregulated, and the second at the onset of extracellular matrix mineralization. Examples of genes which are selectively expressed during developmental progression of osteoblast differentiation are indicated: histone H4 (H4), cfos, TGFβ, fibronectin (FN), type I collagen (COL), osteopontin (OP), AP1 (fos/jun-related transcription factors), alkaline phosphatase (AP), matrix gla protein (MGP), osteocalcin (OC), and collagenase (C!ase).

2.1. THE HISTONE GENE PROMOTER IS A MODEL FOR THE INTEGRATION OF REGULATORY SIGNALS MEDIATING CELL CYCLE CONTROL AT THE G_1/S PHASE TRANSITION AND PROLIFERATION/DIFFERENTIATION INTERRELATIONSHIPS

The histone gene promoter is a paradigm for cell cycle-mediated transcriptional control [4-6]. Transcription is constitutive throughout the cell cycle, upregulated at the onset of S phase and completely suppressed in quiescent cells or following the onset of differentiation [7-10]. Consequently, activity of the promoter is responsive to regulatory signals which contribute to transcriptional competency for cell cycle progression at the G_1/S phase transition point and to transcriptional downregulation postproliferatively (Figure 2).

62

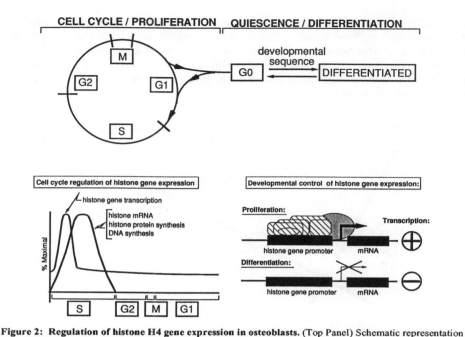

Figure 2: Regulation of histone H4 gene expression in osteoblasts. (Top Panel) Schematic representation of the cell cycle (G_1, S, G_2, Mitosis), indicating the pathway associated with the postproliferative onset of differentiation initiated following completion of mitosis. (Lower Left Panel) Representation of data defining the principal biochemical parameters of histone gene expression, indicating a restriction of histone protein synthesis and the presence of histone mRNA to S phase cells (DNA synthesis). Constitutive transcription of histone genes occurs throughout the cell cycle with an enhanced transcriptional level during the initial 2 hours of S phase. These results establish the combined contribution of transcription and messenger RNA stability to the S phase-specific regulation of histone biosynthesis in proliferating cells, with histone mRNA levels as the rate limiting step. (Lower Right) In contrast, the completion of proliferative activity at the onset of differentiation is mediated by transcriptional downregulation of histone gene expression, supported by a parallel decline in rate of transcription and cellular mRNA levels.

The modularly organized promoter regulatory elements of the histone H4 gene promoter and the cognate transcription factors have been characterized within the context of cell cycle dependent regulatory parameters [11-14] There is a direct indication that the cell cycle regulatory element designated Site II exhibits phosphorylation-dependent modifications in transcription factor interactions which parallel and are functionally related to cell cycle as well as growth control of histone gene expression. The S phase transcription factor complexes assembling at the H4 promoter include cdc2, cyclin A, an RB related protein, CDP1, and IRF-2, [15, 15a, 16]. Association of these factors with the histone H4 cell cycle regulatory domain reflects an integration of phosphorylation-mediated control of histone gene expression as well as growth stimulation, growth suppression and repression of differentiation at the G_1/S phase transition point (Figure 3).

This regulatory mechanism is not restricted to cell cycle dependent transcriptional control of the histone H4 gene. There is an analogous representation and organization of regulatory motifs in the histone H3 and histone H1 gene promoters supporting coordinate control of histone genes which are co-expressed [15, 15a, 17]. In a broader biological context there are similarities of histone gene cell cycle regulatory element sequences with those in the proximal promoter of the thymidine kinase gene which exhibits enhanced transcription during S phase [18]. The possibility may therefore be considered that genes functionally linked to DNA replication may at least in part be coordinately controlled. Support for such a mechanism is provided by analogous promoter domains of the histone and thymidine kinase promoters which both interact with transcription factor complexes that include cyclins, cyclin-dependent kinases and Rb-related proteins.

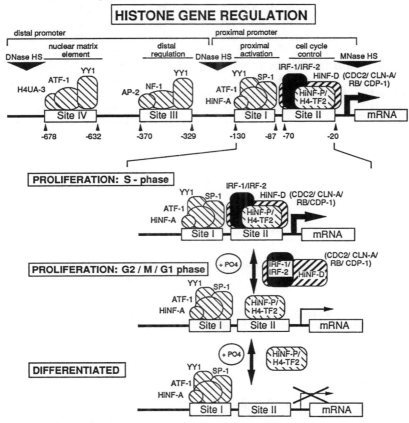

Figure 3: Regulation of histone gene expression during the cell cycle. (Top Two Panels) Organization of the human histone H4 gene promoter regulatory elements (Sites I-IV) is illustrated. The transcription factors which exhibit sequence-specific interactions with these domains are indicated during the S phase of the cell cycle when the gene is maximally transcribed. Site II contributes to cell cycle regulation of transcription. Site IV binds a nuclear matrix protein complex (NMP-1/YY-1) while the protein-DNA interactions at Sites III and I support general transcriptional enhancement. The Site II complex includes cyclin A, cyclin-dependent kinase cdc2, an RB-related protein, CDP-1, and IRF growth regulatory factors, reflecting integration of phosphorylation-mediated control of histone gene expression. (Third Panel) The Site II transcription factor complex is modified by phosphorylation during the G_1/G_2/mitotic periods of the cell cycle resulting in modifications in levels of transcription. Phosphorylation-dependent dissociation of the IRF and HiNF-D (cdc2, cyclin A, CDP-1, and RB-related protein) factors occurs in non S phase cells. (Lower Panel) The complete loss of transcription factor complexes at Sites II, III and IV occurs following exit from the cell cycle with the onset of differentiation. At this time transcription is completely downregulated.

64

2.2. TRANSCRIPTIONAL CONTROL OF THE BONE-SPECIFIC OSTEOCALCIN GENE AT THE ONSET OF EXTRACELLULAR MATRIX MINERALIZATION IN POSTPROLIFERATIVE OSTEOBLASTS

Influences of promoter regulatory elements that are responsive to basal and tissue-restricted transactivation factors, steroid hormones, growth factors and other physiologic mediators has provided the basis for understanding regulatory mechanisms contributing to developmental expression of osteocalcin, tissue specificity and biological activity [1, 2, 19]. These regulatory elements and cognate transcription factors support postproliferative transcriptional activation and steroid hormone (e.g. vitamin D) enhancement at the onset of extracellular matrix mineralization during osteoblast differentiation (Figure 4,5). Thus, the bone-specific osteocalcin gene is to be organized in a manner which supports responsiveness to homeostatic physiologic mediators and developmental expression in relation to bone cell differentiation.

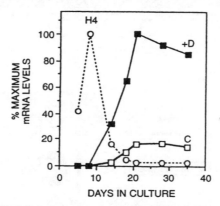

Figure 4: Expression of the osteocalcin gene and vitamin D enhancement during the postproliferative period of the osteoblast developmental sequence. H4 mRNA is an indication of proliferative activity.

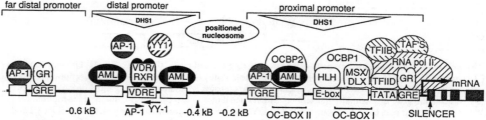

Figure 5: Organization of the osteocalcin gene promoter: indicated are promoter regulatory domains and cognate transcription factors. The OC box is the primary tissue-specific transcriptional element that binds homeodomain proteins (MSX). Fos-jun-related proteins form heterodimers at AP-1 sites or a cryptic tissue-specific complex. HLH proteins bind to the contiguous E box motif. The nuclear matrix protein binding sites for NMP-2 (open ovals; designated A, B and C from left to right) interact with an AML-related transcription factor. Several glucocorticoid response elements (GRE) are indicated as well as the vitamin D response element (VDRE), which is a primary enhancer sequence. Additionally, AP-1 sites are indicated that overlap the TGFβ (TGRE) and the VDRE.

The regulatory sequences illustrated in (Figure 5) have been established in the OC gene promoter and coding region by one or more criteria that includes: (1) demonstration of an influence on transcriptional activity by deletion, substitution or site specific mutagenesis *in vitro* and *in vivo*; (2) identification and characterization of sequence specific regulatory element occupancy by cognate transcription factors *in vitro* and *in vivo*; (3) modifications in protein-DNA interactions as a function of biological activity; and (4)consequential modifications in functional activity following overexpression or suppression of factors which exhibit sequence-specific recognition for regulatory domains.

2.2.1. *Basal/Tissue Specific Factors and Sequences*

A series of elements contributing to basal expression [20] include a TATA sequence (located at -42 to -39) and the osteocalcin box I (OC Box I), a 24 nucleotide element with a CCAAT motif as the central core. Both promoter elements are required for rendering the OC gene transcribable [21, 22]. OC Box 1 defines the threshold for initiation of transcription [23, 24]. OC Box II, which binds an AML related transcription factor spans the -138 to -130 domain and plays a significant role in phenotype-restricted expression of the osteocalcin gene. Because multiple sequences are operative in tissue-specific regulation opportunities are provided for stringent control of osteocalcin gene expression in bone under diverse biological circumstances.

2.2.2. *Multiple Glucocorticoid Responsive Promoter Domains*

Multiple glucocorticoid responsive elements (GRE) with sequences that exhibit both strong and weak affinities for glucocorticoid receptor binding have been identified in the proximal promoter [25-27]. Interactions of other transcription factors with the proximal glucocorticoid responsive elements that include NF-IL6 have been reported [28] further expanding the potential of the OC gene to be transcriptionally regulated by glucocorticoids. It is reasonable to consider that the OC gene GRE'S may be selectively utilized in a developmentally and/or physiologically responsive manner. The possibility of functional interactions with transcription factors other than glucocorticoid receptors under certain conditions should not be dismissed.

2.2.3. *The VDRE*

The vitamin D responsive element (VDRE) functions as an enhancer [29-33]. Maximal activity of the VDRE is mediated by binding of the VDRE/RXR heterodimeric receptor complex. However, under conditions where the osteocalcin gene is inactive or expressed at basal levels in the absence of vitamin D, the VDRE may be associated with transcription factors which include but are not restricted to fos/jun-related proteins [34] and YY1 [35, 36]. Activity of the VDRE transcription factor complex appears to be a target for modifications in vitamin D mediated transcription by other physiologic factors including, TNF-α [37] and retinoic acid [38-42].

2.2.4. *Other Osteocalcin Gene Promoter Regulatory Sequences*

Additional regulatory sequences include an NFkB site also reported to be involved in regulation mediated by TNFα [43]; a series of AP-1 sites [44-48], one of which mediates TGFβ responsiveness [49, 50]; an E box [51] that binds HLH containing transcription factor complexes; and, a sequence in the proximal promoter that binds a multi subunit complex containing CP1/NR-Y/CBF-like CAAT factor complex. Two osteocalcin gene promoter regulatory domains which exhibit recognition for transcription factors that mediate developmental pattern formation are an MSX binding site within the OC box [21, 23, 24, 52] and an AML-1 site (runt homology) sequence [53-56]. These sequences may represent components of regulatory mechanisms that contribute to pattern formation associated with bone tissue organization during initial developmental stages and subsequently during tissue remodelling. Involvement of other homeodomain-related genes that are expressed during skeletal development in control of osteoblast proliferation and differentiation is worthy of consideration. These include but are not restricted to the families of Dlx and PAX [57-65] genes.

2.2.5 Upstream Regulatory Sequences

Although a majority of the responses which have to date been identified reside in the region of the promoter which spans the VDRE domain to the first exxon, upstream sequences that must be further defined, may contribute to both basal and enhancer-mediated control of transcription [49, 33, 66-70]. A GRE residing at -683 to -697 is an example of such an upstream regulatory element [69]. Additionally, intragenic and downstream regulatory sequences are a viable candidate for regulatory involvement in osteocalcin gene expression.

2.2.6. Regulatory Implications of Overlapping and Contiguous Regulatory Domains

The overlapping and contiguous organization of osteocalcin gene regulatory elements (Figure 5), as illustrated by the TATA/GRE, E Box/AP-1/CCAAT/homeodomain and TNFα/VDRE, AP-1/VDRE provide a basis for combined activities that support responsiveness to physiologic mediators [34, 37, 71-80]. Additionally, hormones modulate binding of transcription factors other than the cognate receptor to non-steroid regulatory sequences. For example, vitamin D induced interactions occur at the basal TATA domain [70] and $1,25(OH)_2D_3$ upregulates MSX-2 binding to the OC box homeodomain motif as well as supports increased MSX-2 expression [21, 22]. It is this complexity of OC gene promoter element upregulation that allows for hormone responsiveness in relation to either basal or enhanced levels of expression.

3. Nuclear Structure Supports Cell Cycle Stage-Specific Histone Gene Transcription in Proliferating Osteoblasts

3.1. CHROMATIN STRUCTURE AND NUCLEOSOME ORGANIZATION

A synergistic contribution of activities by Sites I, II, III and IV H4 histone gene promoter elements to the timing and extent of H4-FO108 gene transcription has been established experimentally [81-83, 13]. The integration of intracellular signals that act upon these independent multiple elements may partly reside in the three-dimensional organization of the promoter within the spatial context of nuclear architecture (Figure 6).

The presence of nucleosomes in the H4 promoter when the gene is transcriptionally active [84-87] (Figure 6) may serve to increase the proximity of independent regulatory elements, and supports synergistic and/or antagonistic cooperative interactions between histone gene DNA binding activities. In addition, chromatin structure and nucleosomal organization varies as a function of the cell cycle

[84-87] which may enhance and/or restrict accessibility of transcription factors as well as modulate the extent to which DNA-bound factors are phosphorylated.

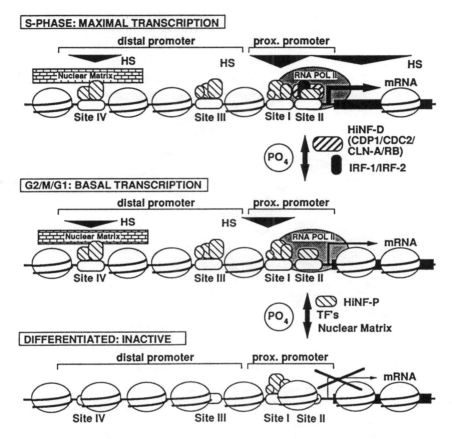

Figure 6A: Schematic illustration of the remodelling of chromatin structure and nucleosome organization which accommodates cell cycle stage-specific and developmental parameters of histone gene promoter architecture to support modifications in level of expression. Placement of nucleosomes and representation as well as magnitude of nuclease hypersensitive sites (solid triangles) are designated. The principal regulatory elements and transcription factors are shown.

Parameters of chromatin structure and nucleosome organization were experimentally established by accessibility of DNA sequences within nuclei of intact cells to a series of nucleases that include micrococcal nuclease for establishing nucleosome placement, DNase I for mapping nuclease hypersensitive sites, S1 nuclease for determination of single stranded DNA sequences and restriction endonucleases to determine protein/DNA interactions within specific sites at single nucleotide resolution. Findings from these approaches illustrate that nucleosome spacing exhibits cell cycle-dependent variations in response to levels of histone gene expression and/or nuclear organization related to mitotic division. Modifications are observed in accessibility of histone gene sequences to micrococcal nuclease, DNase I, S1 nuclease and restriction endonucleases reflecting a remodelling of chromatin architecture which is related to the extent that the histone gene is transcribed to support proliferative activity and cell cycle progression.

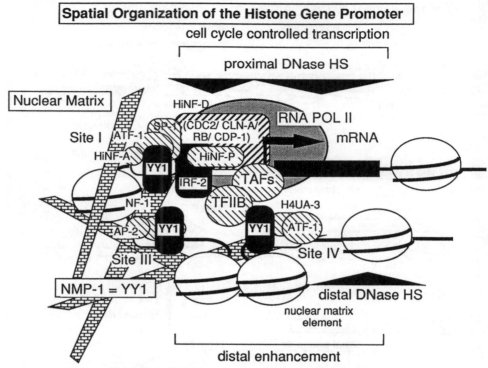

Figure 6B: Three-dimensional spatial organization of the histone gene promoter. A model is schematically presented for the spatial organization of the rat osteocalcin gene promoter based on evidence for nucleosome placement and the interaction of DNA binding sequences with the nuclear matrix. These components of chromatin structure and nuclear architecture restrict mobility of the promoter and impose physical constraints that reduce distances between proximal and distal promoter elements. Such a postulated organization of the osteocalcin gene promoter can facilitate cooperative interactions for crosstalk between elements that mediate transcription factor binding and consequently determine the extent to which the gene is transcribed.

3.2. THE NUCLEAR MATRIX

A nuclear matrix attachment site has been identified in the upstream region (-0.8kB) of the histone H4 gene promoter [88], which may serve two functions: imposing constraints on chromatin structure, and concentrating and localizing transcription factors. Such a role for the nuclear matrix in regulation of histone gene expression is supported by distinct modifications in the composition of nuclear matrix proteins observed when proliferation-specific genes are down-regulated during differentiation [89] and, more directly, by the isolation of ATF-related and YY1 transcription factors from the nuclear matrix [88, 35], which interact with Site IV of the histone H4 gene promoter.

The specific mechanisms by which the 5' histone gene promoter elements and sequence-specific transactivating factors participate in regulating transcription of histone genes during proliferation and differentiation remain to be determined. However, regulation is unquestionably operative within the context of the complex series of spatial interactions, which are responsive to a broad spectrum of biological signals.

4. Nuclear Structure Supports Developmental and Steroid Hormone Responsive Osteocalcin Gene Transcription During Osteoblast Differentiation

4.1. CHROMATIN STRUCTURE AND NUCLEOSOME ORGANIZATION

Modifications in parameters of chromatin structure and nucleosome organization parallel both competency for transcription and the extent to which the osteocalcin gene is transcribed. Changes are observed in response to physiological mediators of basal expression and steroid hormone responsiveness. This remodelling of chromatin provides a conceptual and experimental basis for the involvement of nuclear architecture in developmental, homeostatic and physiologic control of osteocalcin gene expression during establishment and maintenance of bone tissue structure and activity.

In both normal diploid osteoblasts and in osteosarcoma cells basal expression and enhancement of osteocalcin gene transcription are accompanied by two alterations in structural properties of chromatin. DNase I hypersensitivity of sequences flanking the tissue-specific osteocalcin box and the vitamin D responsive element enhancer domain are observed [90-92]. Together with modifications in nucleosome placement [91], a basis for accessibility of transactivation factors to basal and steroid hormone-dependent regulatory sequences can be explained. In early stage proliferating normal diploid osteoblasts when the osteocalcin gene is repressed nucleosomes are placed in the OC box and in VDRE promoter sequences; and nuclease hypersensitive sites are not present in the vicinity of these regulatory elements. In contrast, when osteocalcin gene expression is transcriptionally upregulated post-proliferatively and vitamin D mediated enhancement of transcription occurs, the osteocalcin box and VDRE become nucleosome free and these regulatory domains are flanked by DNase I hypersensitive sites (Figure 7).

Functional relationships between structural modifications in chromatin and osteocalcin gene transcription are observed in response to $1,25(OH)_2D_3$ in ROS 17/2.8 osteosarcoma cells which exhibit vitamin D-responsive transcriptional upregulation. There are marked changes in nucleosome placement at the VDRE and OC box as well as DNase I hypersensitivity of sequences flanking these basal and enhancer osteocalcin gene promoter sequences [90-92]. The complete absence of hypersensitivity and the presence of nucleosomes in the VDRE and osteocalcin box domains of the osteocalcin gene promoter in ROS 24/1 cells which lack the vitamin D receptor and are therefore refractory to the steroid hormone additionally corroborate these findings [90, 92]. These steroid hormone-responsive alterations in chromatin structure have been confirmed by restriction enzyme accessibility of promoter sequences within intact nuclei [93] and by LMPCR [Montecino et al., in preparation] at single nucleotide resolution.

We have recently found that agents which induce histone hyperacetylation (Sodium Butyrate) promote reorganization of the nucleosomal structure in the distal

70

region of the osteocalcin gene promoter (including the VDRE). This transition results in inhibition of the vitamin D-induced upregulation of basal transcription in ROS 17/2.8 cells. Additionally, we have established an absolute requirement for sequences residing in the proximal region of the osteocalcin gene promoter for both formation of the proximal DNase I hypersensitive site and basal transcriptional activity. Our approach was to assay nuclease accessibility (DNase I and restriction endonucleases) in ROS 17/2.8 cell lines stably transfected with promoter deletion constructs driving expression of a CAT reporter gene.

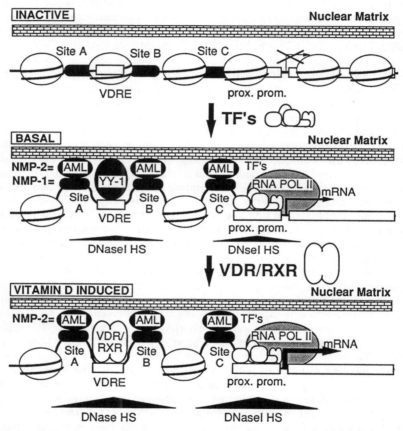

Figure 7: Schematic representation of the osteocalcin gene promoter organization and occupancy of regulatory elements by cognate transcription factors paralleling and supporting functional relationships to either: A) suppression of transcription in proliferating osteoblasts; B) activation of expression in differentiated cells, or C) enhancement of transcription by vitamin D. The placement of nucleosomes is indicated. Remodelling of chromatin structure and nucleosome organization to support suppression, basal and vitamin D induced transcription of the osteocalcin gene is indicated. The representation and magnitude of DNase I hypersensitive sites are designated by solid triangles and gene-nuclear matrix interactions are shown.

4.2. THE NUCLEAR MATRIX

Involvement of the nuclear matrix in control of osteocalcin gene transcription is provided by several lines of evidence. One of the most compelling is association of a bone-specific nuclear matrix protein designated NMP2 with sequences flanking the VDRE of the osteocalcin gene promoter [94]. Initial characterization of the NMP2 factor has revealed that a component is an AML-1 related transactivation protein [53, 54]. These results implicate the nuclear matrix in regulating events that mediate structural properties of the VDRE domain.

It is apparent from available findings that the linear organization of gene regulatory sequences is necessary but insufficient to accommodate the requirements for physiological responsiveness to homeostatic, developmental and tissue-related regulatory signals. It would be presumptive to propose a formal model for the three dimensional organization of the osteocalcin gene promoter. However the working model presented in (Figure 8) represents postulated interactions between OC gene promoter elements that reflect the potential for integration of activities by nuclear architecture to support transcriptional control within a three dimensional context of cell structure and regulatory requirements at the cell and tissue levels.

A role of the nuclear matrix in steroid hormone-mediated transcriptional control of the osteocalcin gene is further supported by overlapping binding domains within the VDRE for the VDR and the NMP-1 nuclear matrix protein which we have recently shown to be a YY1 transcription factor [35]. One can speculate that reciprocal interactions of NMP-1 and VDR complexes may contribute to competency of the VDRE to support transcriptional enhancement. Binding of NMP-2 at the VDRE flanking sequence may establish permissiveness for VDR interactions by gene-nuclear matrix associations which facilitate conformational modifications in the transcription factor recognition sequences.

Taken together, these findings provide a basis for involvement of both the nuclear matrix and chromatin structure in modulating accessibility of promoter sequences to cognate transcription factors and facilitating the integration of activities at multiple regulatory domains. Recent *in vivo* studies support functional contributions of nuclear matrix proteins to steroid hormone-mediated transcription. Overexpression of AML transcription factors which flank the osteocalcin gene VDRE upregulates expression. In contrast, overexpression of YY1 which binds to a site overlapping the osteocalcin gene vitamin D receptor binding sequences abrogates the vitamin D enhancement of transcription and displaces VDR/RXR interactions. Functional data supporting nuclear structure-mediated crosstalk between the osteocalcin gene VDRE and the TATA domain are provided by the recent demonstration that the transcription factor TF2B and the VDR cooperatively coactivate ligand-dependent transcription [95] and are partner proteins by the 2 hybrid system [96]. Functional interrelationships between the VDRE and TATA domains under conditions where YY1 occupancy suppresses enhancer activity are consistent with mutual exclusive binding of YY1 or the VDR to the basic domain of TF2B [36].

72

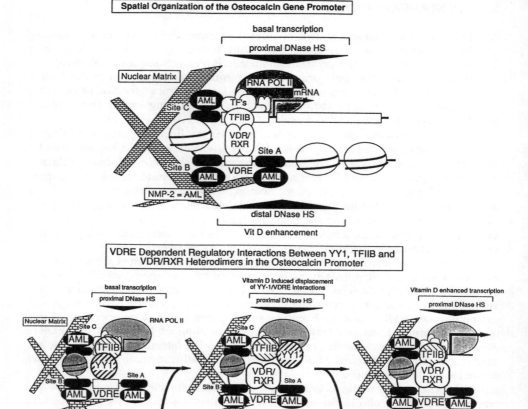

Figure 8: Three-dimensional spatial organization of the rat osteocalcin gene promoter. (Top) A model is schematically presented for the spatial organization of the rat osteocalcin gene promoter based on evidence for nucleosome placement and the interaction of DNA binding sequences with the nuclear matrix. These components of chromatin structure and nuclear architecture restrict mobility of the promoter and impose physical constraints that reduce distances between proximal and distal promoter elements. Such postulated organization of the osteocalcin gene promoter can facilitate cooperative interactions for crosstalk between elements that mediate transcription factor binding and consequently determine the extent to which the gene is transcribed. (Bottom) This model for modifications in osteocalcin gene promoter organization is consistent with protein/protein and protein/DNA interactions at the principal regulatory domains and crosstalk between regulatory elements when the gene is transcribed at basal level (left), maximally expressed following vitamin D treatment (right) and in a transition following vitamin D treatment (middle).

5. Conclusions and Prospects

It is becoming increasingly evident that developmental transcriptional control and modifications in transcription to accommodate homeostatic regulation of cell and tissue function is modulated by the integration of a complex series of physiological regulatory signals. Fidelity of responsiveness necessitates the convergence of activities mediated by multiple regulatory elements of gene promoters. Our current knowledge of promoter organization and the repertoire of transcription factors which mediate activities provides a single dimension map of options for biological control. We are beginning to appreciate the additional structural and functional dimensions provided by chromatin structure, nucleosome organization and subnuclear localization and targeting of both genes and transcription factors. Particularly exciting is increasing evidence for dynamic modifications in nuclear structure which parallel developmental expression of genes. The extent to which nuclear structure regulates and/or is regulated by modifications in gene expression remains to be experimentally established.

There is a necessity to mechanistically define how nuclear matrix-mediated subnuclear distribution of actively transcribed genes is responsive to nuclear matrix association of transcriptional and post-transcriptional regulatory factors. And, we cannot dismiss the possibility that association of regulatory factors with the nuclear matrix is consequential to sequence-specific interactions of transcriptionally active genes with the nuclear matrix. As these issues are resolved, we will gain additional insight into determinants of cause and/or effect relationships which interrelate specific components of nuclear architecture with gene expression at the transcriptional and post-transcriptional levels. However, it is justifiable to anticipate that while nuclear structure-gene expression interrelationships are operative under all biological conditions, situation-specific variations are the rule rather than the exception. As subtleties in the functional component of nuclear architecture are further defined, the significance of nuclear domains to DNA splicing, transcription and processing of RNA transcripts will be additionally understood.

A challenge that we face is to experimentally establish the rate limiting nucleotide sequences and factors which integrate nuclear structure and gene expression. What are the sequence determinants and regulatory proteins which facilitate remodelling of chromatin structure and nucleosome organization to facilitate developmental and homeostatic requirements for transcription? How are structurally and functionally dynamic modifications in chromatin organization related to interactions of genes with the nuclear matrix? There is confidence that resolution to these issues will in part be provided by studies being carried out with lower eukaryotes where the gene content is limited and our knowledge of the genetics is encyclopedic as well as easily applied to mapping structural and regulatory parameters of gene expression. However, caution must be exercised in extrapolating results from such studies to generalizations which apply to all eukaryotic cells and organisms. While the size of the genome and the content of genetic information is lower in yeast and drosophila compared to mammalian cells, the nuclear organization and proteins which package DNA as chromatin are different. These variations may reflect the increased regulatory components in mammalian cells which support both structure and function.

74

6. Acknowledgements

Studies reported in this chapter were supported by grants from the National Institutes of Health (GM32010, AR39588, AR42262) and the March of Dimes Birth Defects Foundation. The authors are appreciative of the editorial assistance from Elizabeth Bronstein with the preparation of this manuscript.

References

1. Stein, G.S., Lian, J.B., and Owen, T.A. (1990) Relationship of cell growth to the regulation of tissue-specific gene expression during osteoblast differentiation. *FASEB J.* **4**, 111-3123.
2. Stein, G.S. and Lian, J.B.(1993) Molecular mechanisms mediating proliferation/differentiation interrelationships during progressive development of the osteoblast phenotype. *Endocrine Reviews,* **14**, 424-442.
3. Stein, G.S. and Lian, J.B. (1995) Molecular mechanisms mediating proliferation-differentiation interrelationships during progressive development of the osteoblast phenotype: update 1995. Endocrine Reviews, 4:290-297.
4. Stein, G.S., Stein, J.L., van Wijnen, A.J., and Lian, J.B. (1992) Regulation of histone gene expression. *Curr. Op. in Cell Biol.* **4**, 166-173.
5. Stein, G.S., Stein, J.L., van Wijnen, A.J., and Lian, J.B. (1994) Histone gene transcription: a model for responsiveness to an integrated series of regulatory signals mediating cell cycle control and proliferation/differentiation interrelationships. *J. Cell. Biochem.* **54**, 393-404.
6. Stein, G.S., Stein, J.L., van Wijnen, A.J., Lian, J.B., Quesenberry, P.J. (1995) Molecular mechanisms mediating cell cycle and cell growth. Experimental Hematology, 23:1053-1061.
7. Baumback, L., Marashi, F., Plumb, M., Stein, G.S., and Stein, J.L. (1984) Inhibition of DNA replication coordinately reduces cellular levels of core and H1 histone mRNAs: requirement for protein synthesis. *Biochem* **23**, 1618-1625.
8. Detke, S., Lichtler, A., Phillips, I., Stein, J., and Stein, G. (1979) Reassessment of histone gene expression during the cell cycle in human cells by using homologous H4 histone cDNA. *Proc. Natl. Acad. Sci. USA* **76**, 4995-4999.
9. Plumb, M.A., Stein, J., and Stein, G. (1983) Coordinate regulation of multiple histone mRNAs during the cell cycle in HeLa cells. *Nucl. Acids Res.* **11**, 2391-2410.
10. Plumb, M., Stein, J., and Stein, G. (1983) Influence of DNA synthesis inhibition on the coordinate expression of core human histone genes. *Nucl. Acids Res.* **11**, 7927-7945.
11. van Wijnen, A.J., van den Ent, F.M.I., Lian, J.B., Stein, J.L., and Stein, G.S. (1992) Overlapping and CpG methylation-sensitive protein/DNA interactions at the histone H4 transcriptional cell cycle domain: distinctions between two human H4 gene promoters. *Mol. Cell. Biol.* **12**, 3273-3287.
12. van Wijnen, A.J., Wright, K.L., Lian, J.B., Stein, J.L., and Stein, G.S. (1989) Human H4 histone gene transcription requires the proliferation-specific nuclear factor HiNF-D. *J. Biol. Chem.* **264**, 15034-15042.
13. Ramsey-Ewing, A., van Wijnen, A., Stein, G.S., and Stein, J.L. (1994) Delineation of a human histone H4 cell cycle element in vivo: the master switch for H4 gene transcription. *Proc. Natl. Acad. Sci. USA* **91**, 4475-4479.
14. Pauli, U., Chrysogelos, S., Stein, J., Stein, G.S., and Nick, H. (1987) Protein-DNA interactions in vivo upstream of a cell cycle regulated human H4 histone gene. *Science* **236**, 1308-1311.
15. van Wijnen, A.J., van Gurp, M.F., de Ridder, M., Tufarelli, C., Last, T.J., Birnbaum, M., Vaughan, P.S., Giordano, A., Krek, W., Neufeld, E.J., Stein, J.L., and Stein, G.S. CDP/*cut* is the DNA binding subunit of transcription factor HiNF-D (cyclin A, cdc2, RB-related complex): a non-E2F mechanism for histone gene regulation at the G1/S phase cell cycle transition point. *Proc. Natl. Acad. Sci. USA,* in press.
15a. van Wijnen, A.J., Aziz, F., Grana, X., De Luca, A., Desai, R.K., Jaarsveld, K., Last, T.J., Soprano, K., Giordano, A., Lian, J.B., Stein, J.L., and Stein, G.S. (1994) Transcription of histone H4, H3 and H1 cell cycle genes: promoter factor HiNF-D contains CDC2, cyclin A and an RB-related protein. *Proc. Natl. Acad. Sci USA* **91**, 12882-12886.
16. Vaughan, P.S., Aziz, F., van Wijnen, A.J., Wu, S., Harada, H., Taniguchi, T., Soprano, K., Stein, G.S., and Stein, J.L. (1995) Activation of a cell cycle regulated histone gene by the oncogenic transcription factor IRF2. *Nature* **377**, 362-365.

17. van den Ent, F.M.I., van Wijnen, A.J., Lian, J.B., Stein, J.L., and Stein, G.S. (1994) Cell cycle controlled histone H1, H3 and H4 genes share unusual arrangements of recognition motifs for HiNF-D supporting a coordinate promoter binding mechanism. *J. Cell. Physiol.* **159**, 515-530.

18. Dou, Q., Markell, P.J., and Pardee, A.B. (1992) Retinoblastoma-like protein and cdc2 kinase are in complexes that regulate a G1/S event. *Proc. Natl. Acad. Sci USA* **89**, 3256-3260.

19. Lian, J.B. and Stein, G.S. (1996) Osteocalcin gene expression: a molecular blueprint for developmental and steroid hormone mediated regulation of osteoblast growth and differentiation. *Endocrine Rev.* in press.

20. Lian, J.B., Stewart, C., Puchacz, E., Mackowiak, S., Shalhoub, V., Collart, D., Zambetti, G., and Stein, G.S. (1989) Structure of the rat osteocalcin gene and regulation of vitamin D-dependent expression. *Proc. Natl. Acad. Sci. USA* **86**, 1143-1147.

21. Towler, D.A., Bennett, C.D., and Rodan, G.A. (1994) Activity of the rat osteocalcin basal promoter in osteoblastic cells is dependent upon homeodomain and CP1 binding motifs. *Mol. Endocrinol.* **8**, 614-624.

22. Kawaguchi, N., DeLuca, H.F., and Noda, M. (1992) Id gene expression and its suppression by 1,25-dihydroxyvitamin D$_3$ in rat osteoblastic osteosarcoma cells. *Proc. Natl. Acad. Sci. USA* **89**, 4569-4572.

23. Heinrichs, A.A.J., Banerjee, C., Bortell, R., Owen, T.A., Stein, J.L., Stein, G.S., and Lian, J.B. (1993) Identification and characterization of two proximal elements in the rat osteocalcin gene promoter that may confer species-specific regulation. *J. Cell. Biochem.* **53**, 240-250.

24. Heinrichs, A.A.J., Bortell, R., Bourke, M., Lian, J.B., Stein, G.S., and Stein, J.L. (1995) Proximal promoter binding protein contributes to developmental, tissue-restricted expression of the rat osteocalcin gene. *J. Cell. Biochem.* **57**, 90-100.

25. Stromstedt, P.E., Poellinger, L., Gustafsson, J.A., and Cralstedt-Duke, J. (1991) The glucocorticoid receptor binds to a sequence overlapping the TATA box of the human osteocalcin promoter: A potential mechanisms for negative regulation. *Mol. and Cell. Biol.* **11**, 3379-3383.

26. Heinrichs, A.A.J., Bortell, R., Rahman, S., Stein, J.L., Alnemri, E.S., Litwack, G., Lian, J.B., and Stein, G.S. (1993) Identification of multiple glucocorticoid receptor binding sites in the rat osteocalcin gene promoter. *Biochem.* **32**, 11436-11444.

27. Aslam, F., Shalhoub, V., van Wijnen, A.J., Banerjee, C., Bortell, R., Shakoori, A.R., Litwack, G., Stein, J.L., Stein, G.S., and Lian, J.B. (1995) Contributions of distal and proximal promoter elements to glucocorticoid regulation of osteocalcin gene transcription. *Mol. Endocrinol.* **9**, 679-690.

28. Towler, D.A. and Rodan, G.A. (1994) Cross-talk between glucocorticoid and PTH signaling in the regulation of the rat osteocalcin promoter. *J. Bone Min. Res.* **9**, S282.

29. Morrison, N.A., Shine, J., Fragonas, J.C., Verkest, V., McMenemy, L., and Eisman, J.A. (1989) 1,25-dihydroxyvitamin D-responsive element and glucocorticoid repression in the osteocalcin gene. *Science* **246**, 1158-1161.

30. Demay, M.B., Gerardi, J.M., DeLuca, H.F., and Kronenberg, H.M. (1990) DNA sequences in the rat osteocalcin gene that bind the 1,25-dihydroxyvitamin D$_3$ receptor and confer responsive to 1,25-dihydroxyvitamin D$_3$. *Proc. Natl. Acad. Sci. USA* **87**, 369-373.

31. Markose, E.R., Stein, J.L., Stein, G.S., and Lian, J.B. (1990) Vitamin D-mediated modifications in protein-DNA interactions at two promoter elements of the osteocalcin gene. *Proc. Natl. Acad. Sci. USA* **87**, 1701-1705.

32. Kerner, S.A., Scott, R.A., and Pike, J.W. (1989) Sequence elements in the human osteocalcin gene confer basal activation and inducible response to hormonal vitamin D$_3$. *Proc. Natl. Acad. Sci. USA* **86**, 4455-4459.

33. Terpening, C.M., Haussler, C.A., Jurutka, P.W., Galligan, M.A., Komm, B.S., and Haussler, M.R. (1991) The vitamin D-responsive element in the rat bone Gla protein gene is an imperfect direct repeat that cooperates with other cis-elements in 1,25-dihydroxyvitamin D$_3$-mediated transcriptional activation. *Mol. Endocrinol.* **5**, 373-385.

34. Owen, T.A., Bortell, R., Yocum, S.A., Smock, S.L., Zhang, M., Abate, C., Shalhoub, V., Aronin, N., Wright, K.L., van Wijnen, A.J., Stein, J.L., Curran, T., Lian, J.B., and Stein, G.S. (1990) Coordinate occupancy of AP-1 sites in the vitamin D responsive and CCAAT box elements by fos-jun in the osteocalcin gene: a model for phenotype suppression of transcription. *Proc. Natl. Acad. Sci. USA* **87**, 9990-9994.

35. Guo, B., Odgren, P.R., van Wijnen, A.J., Last, T.J., Fey, E.G., Penman, S., Stein, J.L., Lian, J.B., and Stein, G.S. (1995) The nuclear matrix protein NMP-1 is the transcription factor YY-1. *Proc. Natl. Acad. Sci. USA* **92**, 10526-10530.

76

36. Guo, B., Aslam, F., van Wijnen, A., Roberts, S.G.E., Frenkel, B., Green, M., DeLuca, H., Lian, J.B., Stein, G.S., Stein, J.L. YY1 regulates VDR/RXR mediated transactivation of the vitamin D responsive osteocalcin gene. Manuscript submitted.

37. Kuno, H., Kurian, S.M., Hendy, G.N., White, J., DeLuca, H.F., Evans, C-O., and Nanes, M.S. (1994) Inhibition of 1,25-dihydroxyvitamin D_3 stimulated osteocalcin gene transcription by tumor necrosis factor-α: Structural determinants within the vitamin D response element. *Endocrinol.* **134**, 2524-2531.

38. Scule, R., Umesono, K., Mangelsdorf, D.J., Bolado, J., Pike, J.W., and Evans, R.M. (1990) Jun-Fos and receptors for vitamins A and D recognize a common response element in the human osteocalcin gene. *Cell* **61**, 497-504.

39. Bortell, R., Owen, T.A., Shalhoub, V., Heinrichs, A., Aronow, M.A.B., and Stein, G.S. (1993) Constitutive transcription of the osteocalcin gene in osteosarcoma cells is reflected by altered protein-DNA interactions at promoter regulatory elements. *Proc. Natl. Acad. Sci. USA* **90**, 2300-2304.

40. Schrader, M., Bendik, I., Becker-Andre, M., and Carlberg, C. (1993) Interaction between retinoic acid and vitamin D signaling pathways. *J. Bio. Chem.* **268**, 17830-17836.

41. MacDonald, P.N., Dowd, D.R., Nakajima, S., Galligan, M.A., Reeder, M.C., Haussler, C.A., Ozato, K., and Haussler, M.R. (1993) Retinoid X Receptors stimulate and 9-*cis* retinoic acid inhibits 1,25-dihydroxyvitamin D_3-activated expression of the rat osteocalcin gene. *Mol. Cell. Biol.* **13**, 5907-5917.

42. Kliewer, S.A., Umesono, K., Mangelsdorf, D.J., and Evans, R.M. (1992) Retinoic X receptor interacts with nuclear receptors in retinoic acid, thyroid hormone an vitamin D signaling. *Nature* **355**, 441-446.

43. Li, Y. and Stashenko, P. (1993) Characterization of a tumor necrosis factor-responsive element which down-regulates the human osteocalcin gene. *Molec. & Cell. Biol.* **13**, 3714-3721.

44. Ozono, K., Sone, T., and Pike, J.W. (1991) The genomic mechanism of action of 1,25-dihydroxyvitamin D_3. *J. Bone Min. Res.* **6**, 1021-1027.

45. Lian, J.B., Stein, G.S., Bortell, R., and Owen, T.A. (1991) Phenotype suppression: a postulated molecular mechanism for mediating the relationship of proliferation and differentiation by fos/jun interactions at AP-1 sites in steroid responsive promoter elements of tissue-specific genes. *J. Cell. Biochem.* **45**, 9-14.

46. Demay, M.B., Kiernan, M.S., DeLuca, H.F., and Kronenberg, H.M. (1992) Characterization of 1,25-dihydroxyvitamin D_3 receptor interactions with target sequences in the rat osteocalcin gene. *Molec. Endocrinol.* **6**, 557-562.

47. McCabe, L.R., Kockx, M., Lian, J.B., Stein, J.L., and Stein, G.S. (1995) Selective expression of FOS & JUN related genes during osteoblast proliferation and differentiation. *Exp. Cell Res.* **218**, 255-262.

48. McCabe, L.R., Banerjee, C., Kundu, R., Harrison, R.J., Dobner, P.R., Stein, J.L., Lian, J.B., and Stein, G.S. (1996) Developmental expression and activities of specific fos and jun proteins are functionally related to osteoblast maturation: role of fra-2 and jun-D during differentiation. *Endocrinology,* in press.

49. Lian, J.B. and Stein, G.S. (1993) Proto-oncogene mediated control of gene expression during osteoblast differentiation. *Ital. J. Min. Elect. Metab.* **7**, 175-83.

50. Banerjee, C., Stein, J.L., van Wijnen, A.J., Frenkel, B., Lian, J.B., and Stein, G.S. (1996) TGF-β1 responsiveness of the rat osteocalcin gene is mediated by an AP-1 binding site. *Endocrinology,* **137**, 1991-2000.

51. Tamura, M. and Noda, M. (1994) Identification of a DNA sequence involved in osteoblast-specific gene expression via interaction with helix-loop-helix (HLH)-type transcription factors. *J. Cell. Biol.* **126**, 773-782.

52. Hoffmann, H.M., Catron, K.M., van Wijnen, A.J., McCabe, L.R., Lian, J.B., Stein, G.S., and Stein, J.L. (1994) Transcriptional control of the tissue-specific developmentally regulated osteocalcin gene requires a binding motif for the MSX-family of homeodomain proteins. *Proc. Natl. Acad. Sci. USA* **91**, 12887-12891.

53. Merriman, H.L., van Wijnen, A.J., Hiebert, S., Bidwell, J.P., Fey, E., Lian, J.B., Stein, J.L., and Stein, G.S. (1995) The tissue-specific nuclear matrix protein, NMP-2, is a member of the AML/PEBP2/*runt domain* transcription factor family: interactions with the osteocalcin gene promoter. *Biochemistry* **34**, 13125-13132.

54. van Wijnen, A.J., Merriman, H., Guo, B., Bidwell, J.P., Stein, J.L., Lian, J.B., and Stein, G.S. (1994) Nuclear matrix interactions with the osteocalcin gene promoter: multiple binding sites for the runt-

homology related protein NMP-2 and for NMP-1, a heteromeric CREB-2 containing transcription factor. *J. Bone Min. Res.* **9**, S148.

55. Banerjee, C., Hiebert, S.W., Stein, J.L., Lian, J.B., and Stein, G.S. (1996) An AML-1 consensus sequence binds an osteoblast-specific complex and transcriptionally activates the osteocalcin gene. *Proc. Natl. Acad. Sci. USA*, **93**, 4968-4973.

56. Geoffroy, V., Ducy, P., and Karsenty, G. (1995) A PEBP2α/AML-1-related factor increases osteocalcin promoter activity through its binding to an osteoblast-specific cis-acting element. *J. Biol. Chem.*, **270**, 30973-30979.

57. Lufkin, T., Mark, M., Hart, C.P., LeMeur, M., and Chambon, P. (1992) Homeotic transformation of the occipital bones of the skull by ectopic expression of a homeobox gene. *Nature* **359**, 835-841.

58. Ryoo, H.-M., Stein, J.L., Lian, J.B., Stein, G.S. Detection of a proliferation specific gene during development of the osteoblast phenotype by mRNA differential display. J. Cell. Biochem., in press.

59. Krumlauf, R. (1993) *Hox* genes and pattern formation in the branchial region of the vertebrate head. *Trends in Genetics* **9**, 106-112.

60. Gruss, P. and Walther, C. (1992) Pax in development. *Cell* **69**, 719-722.

61. Tabin, C.J. (1991) Retinoids, homeoboxes, and growth factors: towards molecular models for limb development. *Cell* **66**, 199-217.

62. Niehrs, C. and DeRobertis, E.M. (1992) Vertebrate axis formation. *Current Opinion in Genetics and Dev.* **2**, 550-555.

63. McGinnis, W. and Krumlauf, R. (1992) Homeobox genes and axial patterning. *Cell* **68**, 283-302.

64. Simeone, A., Acampora, D., Pannese, M., D'Esposito, M., Stornaiuolo, A., Gulisano, M., Mallamaci, A., Kastury, K., Druck, T., Huebner, K., and Boncinelli, E. (1994) Cloning and characterization of two new members of the vertebrate Dlx family. *Proc. Natl. Acad. Sci. USA* **91**, 2250-2254.

65. Cohen, S.M., Broner, G., Kuttner, F., Jurgens, G., and Jackle, H. (1989) *Distal-less* encodes a homeodomain protein required for limb development in *Drosophila. Nature* **338**, 432-434.

66. Yoon, K., Rutledge, S.J.C., Buenaga, R.F., and Rodan, G.A. (1988) Characterization of the rat osteocalcin gene: stimulation of promoter activity by 1,25-dihydroxyvitamin D_3. *Biochem.* **17**, 8521-8526.

67. Bortell, R., Owen, T.A., Bidwell, J.P., Gavazzo, P., Breen, E., van Wijnen, A.J., DeLuca, H.F., Stein, J.L., Lian, J.B., and Stein, G.S. (1992) Vitamin D-responsive protein-DNA interactions at multiple promoter regulatory elements that contribute to the level of rat osteocalcin gene expression. *Proc. Natl. Acad. Sci. USA* **89**, 6119-6123.

68. Morrison, N. and Eisman, J. (1993) Role of the negative glucocorticoid regulatory element in glucocorticoid repression of the human osteocalcin promoter. *J. Bone Miner. Res.* **8**, 969-975.

69. Aslam, F., Lian, J.B., Stein, G.S., Stein, J,L,, Litwack, G., van Wijnen, A.J., and Shalhoub, V. (1994) Glucocorticoid responsiveness of the osteocalcin gene by multiple distal and proximal elements. *J. Bone Min. Res.* **9**, S125.

70. Owen, T.A., Bortell, R., Shalhoub, V., Heinrichs, A., Stein, J.L., Stein, G.S., and Lian, J.B. (1993) Postproliferative transcription of the rat osteocalcin gene is reflected by vitamin D-responsive developmental modifications in protein-DNA interactions at basal and enhancer promoter elements. *Proc. Natl. Acad. Sci. USA* **90**, 1503-1507.

71. Nanes, M.S., Rubin, J., Titus, L., Hendy, G.N., and Catherwood, B.D. (1991) Tumor necrosis factor alpha inhibits 1,25-dihydroxyvitamin D_3-stimulated bone gla protein synthesis in rat osteosarcoma cells (ROS 17/2.8) by a pretranslational mechanism. *Endocrinol.* **128**, 2577-2582.

72. Taichman, R.S. and Hauschka, P.V. (1992) Effects of interleukin-1 beta and tumor necrosis factor-alpha on osteoblastic expression of osteocalcin and mineralized extracellular matrix in vitro. *J. Inflam.* **16(6)**, 587-601.

73. Li, Y. and Stashenko, P. (1992) Proinflammatory cytokines tumor necrosis factor and IL-6, but not IL-1, down-regulate the osteocalcin gene promoter¹. *J. of Immun.* **148**, 788-794.

74. Evans, D.B., Thavarajah, M., and Kanis, J.A. (1990) Involvement of prostaglandin E_2 in the inhibition of osteocalcin synthesis by human osteoblast-like cells in response to cytokines and systemic hormones. *Biochem. Biophys. Res. Comm.* **167**, 194-202.

75. Nanes, M.S., Rubin, J., Titus, L., Hendy, G.N., and Catherwood, B.D. (1990) Interferon-γ inhibits 1,25-dihydroxyvitamin D_3-stimulated synthesis of bone Gla protein in rat osteosarcoma cells by a pretranslational mechanism. *Endocrinol.* **127**, 588-594.

76. Guidon, P.T., Salvatori, R., and Bockman, R.S. (1993) Gallium nitrate regulates rat osteoblast expression of osteocalcin protein and mRNA levels. *J. Bone Min. Res.* **8**, 103-110.

77. Jenis, L.G., Waud, C.E., Stein, G.S., Lian, J.B., and Baran, D.T. (1993) Effect of gallium nitrate in vitro and in normal rats. *J. Cell. Biochem.* **52**, 330-336.

78

78. Vaishnav, R., Beresford, J.N., Gallagher, J.A., and Russell, R.G.G. (1988) Effects of the anabolic steroid stanozolol on cells derived from human bone. *Clin. Sci.* **74**, 455-460.
79. Fanti, P., Kindy, M.S., Mohapatra, S., Klein, J., Colombo, G., and Malluche, H.H. (1992) Dose-dependent effects of aluminum on osteocalcin synthesis in osteoblast-like ROS 17/2 cells in culture. *Amer. J. Physio.* **263**, E1113-8.
80. Schedlich, L.J., Flanagan, J.L., Crofts, L.A., Gillies, S.A., Goldberg, D., Morrison, N.A., and Eisman, J.A. (1994) Transcriptional activation of the human osteocalcin gene by basic fibroblast growth factor. *J. Bone Min. Res.* **9**, 143-152.
81. Kroeger, P., Stewart, C., Schaap, T., van Wijnen, A., Hirshman, J., Helms, S., Stein, G.S., and Stein, J.L. (1987) Proximal and distal regulatory elements that influence in vivo expression of a cell cycle dependent H4 histone gene. *Proc. Natl. Acad. Sci. USA* **84**, 3982-3986.
82.. Wright, K.L., Dell'Orco, R.T., van Wijnen, A.J., Stein, J.L., and Stein, G.S. (1992) Multiple mechanisms regulate the proliferation specific histone gene transcription factor, HiNF-D, in normal human diploid fibroblasts. *Biochem.* **31**, 2812-2818.
83. Wright, K.L., Birnbaum, M.J., van Wijnen, A., Stein, G.S. and Stein, J.L. (1995) Bipartite structure of the proximal promoter of a human H4 histone gene. *J. Cell. Biochem.* **58**, 372-379.
84. Moreno, M.L., Chrysogelos, S.A., Stein, G.S., and Stein, J.L. (1986) Reversible changes in the nucleosomal organization of a human H4 histone gene during the cell cycle. *Biochemistry* **25**, 5364-5370.
85. Moreno, M.L., Stein, G.S., and Stein, J.L. (1987) Nucleosomal organization of a BPV minichromosome containing a human H4 histone gene. *Mol. Cell. Biochem.* **74**, 173-177.
86. Chrysogelos, S., Riley, D.E., Stein, G.S., and Stein, J.L. (1985) A human histone H4 gene exhibits cell cycle-dependent changes in chromatin structure that correlate with its expression. *Proc. Natl. Acad. Sci. USA* **82**, 7535-7539.
87. Chrysogelos, S., Pauli, U., Stein, G., and Stein, J. (1989) Fine mapping of the chromatin structure of a cell cycle-regulated human H4 histone gene. *J. Biol. Chem.* **264**, 1232-1237.
88. Dworetzky, S.I., Wright, K.L., Fey, E.G., Penman, S., Lian, J.B., Stein, J.L., and Stein, G.S. (1992) Sequence-specific DNA binding proteins are components of a nuclear matrix attachment site. *Proc. Natl. Acad. Sci. USA* **89**, 4178-4182.
89. Dworetzky, S.I., Fey, E.G., Penman, S., Lian, J.B., Stein, J.L., and Stein, G.S. (1990) Progressive changes in the protein composition of the nuclear matrix during rat osteoblast differentiation. *Proc. Natl. Acad. Sci. USA* **87**, 4605-4609.
90. Montecino, M., Pockwinse, S., Lian, J.B., Stein, G.S., and Stein, J.L. (1994) DNase I hypersensitive sites in promoter elements associated with basal and vitamin D dependent transcription of the bone-specific osteocalcin gene. *Biochemistry* **33**, 348-353.
91. Montecino, M., Lian, J.B., Stein, G.S., and Stein, J.L. (1994) Specific nucleosomal organization supports developmentally regulated expression of the osteocalcin gene. *J. Bone Min. Res.* **9**, S352.
92. Breen, E.C., van Wijnen, A.J., Lian, J.B., Stein, G.S., and Stein, J.L. (1994) In Vivo Occupancy of the Vitamin D Responsive Element in the Osteocalcin Gene Supports Vitamin D Dependent Transcriptional Upregulation in Intact Cells. *Proc. Natl. Acad. Sci. USA* **91**, 12902-12906.
93. Montecino, M., Lian, J.B., Stein, G.S., and Stein, J.L. Changes in chromatin structure support constitutive and developmentally regulated transcription of the bone specific osteocalcin gene in osteoblastic cells. Manuscript submitted.
94. Bidwell, J.P., van Wijnen, A.J., Fey, E.G., Dworetzky, S., Penman, S., Stein, J.L., Lian, J.B., and Stein, G.S. (1993) Osteocalcin gene promoter-binding factors are tissue-specific nuclear matrix components. *Proc. Natl. Acad. Sci. USA* **90**, 3162-3166.
95. Blanco, J.C.G., Wang, I.-M., Tsai, S.Y., Tsai, M.-J., O'Malley, B.W., Jurutka, P.W., Haussler, M.R., and Ozato, K. (1995) Transcription factor TFIIB and the vitamin D receptor cooperatively activate ligand-dependent transcription. *Proc. Natl. Acad. Sci. USA* **92**, 1535-1539.
96. MacDonald, P.N., Sherman, D.R., Dowd, D.R., Jefcoat, S.C.Jr., and DeLisle, R.K. (1995) The vitamin D receptor interacts with general transcription factor IIB. *J. Biol. Chem.* **270**, 4748-4752.
97. Pardoll, D.M., Vogelstein, B., and Coffey, D.S. (1980) A fixed site of DNA replication in eukaryotic cells. *Cell* **19**, 527-536.
98. Belgrader, P., Siegel, A.J., and Berezney, R. (1991) A comprehensive study on the isolation and characterization of the HeLa S3 nuclear matrix. *J. Cell Sci.* **98**, 281-291.
99. Berezney, R. and Coffey, D.S. (1974) Identification of a nuclear protein matrix. *Biochem. Biophys. Res. Commun.* **60**, 1410-1417.
100. Berezney, R. and Coffey, D.S. (1975) Nuclear protein matrix: association with newly synthesized DNA. *Science* **189**, 291-292.

101. Berezney, R. and Coffey, D.S. (1977) Nuclear matrix: isolation and characterization of a framework structure from rat liver nuclei. *J. Cell Biol.* **73**, 616-637.
102. Vaughn, J.P., Dijkwel, P.A., Mullenders, L.H.F., and Hamlin, J.L. (1990) Replication forks are associated with the nuclear matrix. *Nucl. Acids Res.* **18**, 1965-1969.
103. Jackson, D.A. and Cook, P.R. (1986) Replication occurs at a nucleoskeleton. *EMBO J.* **5**, 1403-1410.
104. Nakayasu, H. and Berezney, R. (1989) Mapping replicational sites in the eukaryotic nucleus. *J. Cell Biol.* **108**, 1-11.
105. Fey, E.G., Krochmalnic, G., and Penman, S. (1986) The nonchromatin substructures of the nucleus: the ribonucleoprotein (RNP)-containing and RNP-depleted matrices analyzed by sequential fractionation and resinless section electron microscopy. *J. Cell Biol.* **102**, 1654-1665.
106. Fey, E.G. and Penman, S. (1988) Nuclear matrix proteins reflect cell type of origin in cultured human cells. *Proc. Natl. Acad. Sci. USA* **85**, 121-125.
107. Fey, E., Bangs, P., Sparks, C., and Odgren, P. (1991) The nuclear matrix: defining structural and functional roles. *Crit. Rev. Eukaryotic Gene Exp.* **1**, 127-143.
108. Pienta, K.J., Getzenberg, R.H., and Coffey, D.S. (1991) Cell structure and DNA organization. *Crit. Rev. Eukaryotic Gene Exp.* **1**, 355-385.
109. He, D., Nickerson, J.A., and Penman, S. (1990) Core filaments of the nuclear matrix. *J. Cell Biol.* **110**, 569.
110. Capco, D.G., Wan, K.M., and Penman, S. (1982) The nuclear matrix: three-dimensional architecture and protein composition. *Cell* **29**, 847-858.
111. Nickerson, J.A., Krockmalnic, G., He, D., and Penman, S. (1990) Immunolocalization in three dimensions: immunogold staining of cytoskeletal and nuclear matrix proteins in resinless electron microscopy sections. *Proc. Natl. Acad. Sci. USA* **87**, 2259-2263.
112. Bidwell, J.P., Fey, E.G., van Wijnen, A.J., Penman, S., Stein, J.L., Lian, J.B., and Stein, G.S. (1994) Nuclear matrix proteins distinguish normal diploid osteoblasts from osteosarcoma cells. *Cancer Research* **54**, 28-32.
113. Getzenberg, R.H. and Coffey, D.S. (1991) Identification of nuclear matrix proteins in the cancerous and normal rat prostate. *Cancer Res.* **51**, 6514-6520.
114. Pienta, K.J. and Coffey, D.S. (1991) Correlation of nuclear morphometry with progression of breast cancer. *Cancer* **68**, 2012-2016.
115. Barrack, E.R. and Coffey, D.S. (1983) Hormone receptors and the nuclear matrix, in A.K. Roy and J.H. Clark (eds.), *Gene Regulation by Steroid Hormones II,* Springer-Verlag, New York, NY, pp. 239-266.
116. Kumara-Siri, M.H., Shapiro, L.E., and Surks, M.I. (1986) Association of the 3,5,3'-triiodo-L-thyronine nuclear receptor with the nuclear matrix of cultured growth hormone-producing rate pituitary tumor cells (GC cells). *J. Biol. Chem.* **261**, 2844-2852.
117. Getzenberg, R.H. and Coffey, D.S. (1990) Tissue specificity of the hormonal response in sex accessory tissues is associated with nuclear matrix protein patterns. *Mol. Endocrinol.* **4**, 1336-1342.
118. Bidwell, J.P., van Wijnen, A.J., Banerjee, C., Fey, E.G., Merriman, H., Penman, S., Stein, J.L., Lian, J.B., and Stein, G.S. (1994) PTH-responsive modifications in the nuclear matrix of ROS 17/2.8 rat osteosarcoma cells. *Endocrinology* **134**, 1738-1744.
119. Nelkin, B.D., Pardoll, D.M., and Vogelstein, B. (1980) Localization of SV40 genes with supercoiled loop domains. *Nucl. Acids Res.* **8**, 5623-5633.
120. Robinson, S.I., Nelkin, B.D., and Vogelstein, B. (1982) The ovalbumin gene is associated with the nuclear matrix of chicken oviduct cells. *Cell* **28**, 99-106.
121. Schaack, J., Ho, W.Y.-W., Friemuth, P., and Shenk, T. (1990) Adenovirus terminal protein mediates both nuclear-matrix association and efficient transcription of adenovirus DNA. *Genes Dev.* **4**, 1197-1208.
122. Stief, A., Winter, D.M., Stratling, W.H., and Sippel, A.E. (1989) A nuclear attachment element mediates elevated and position-independent gene activity. *Nature* **341**, 343-345.
123. Bode, J. and Maass, K. (1988) Chromatin domain surrounding the human interferon-β gene as defined by scaffold-attached regions. *Biochemistry* **27**, 4706-4711.
124. Ciejek, E.M., Tsai, M.-J., and O'Malley, B.W. (1983) Actively transcribed genes are associated with the nuclear matrix. *Nature* **306**, 607-609.
125. Cockerill, P.N. and Garrard, W.T. (1986) Chromosomal loop anchorage of the kappa immunoglobulin gene occurs next to the enhancer in a region containing topoisomerase II sites. *Cell* **44**, 273-282.

80

126. Gasser, S.M. and Laemmli, U.K. (1986) Cohabitation of scaffold binding regions with upstream/enhancer elements of three developmentally regulated genes of D. melanogaster. *Cell* **46**, 521-530.
127. Jackson, D.A. and Cook, P.R. (1985) Transcription occurs at a nucleoskeleton. *EMBO J.* **4**, 919-925.
128. Jarman, A.P. and Higgs, D.R. (1988) Nuclear scaffold attachment sites in the human globin gene complexes. *EMBO J.* **7**, 3337-3344.
129. Kas, E. and Chasin, L.A. (1987) Anchorage of the chinese hamster dihydrofolate reductase gene to the nuclear scaffold occurs in an intragenic region. *J. Mol. Biol.* **198**, 677-692.
130. Keppel, F. (1986) Transcribed human ribosomal RNA genes are attached to the nuclear matrix. *J. Mol. Biol.* **187**, 15-21.
131. Mirkovitch, J., Mirault, M-.E. and Laemmli, U.K. (1984) Organization of the higher-order chromatin loop: specific DNA attachment sites on nuclear scaffold. *Cell* **39**, 223-232.
132. Phi-Van, L., Von Kries, J.P., Ostertag, W., and Stratling, W.H. (1990) The chicken lysozyme 5' matrix attachment region increases transcription from a heterologous promoter in heterologous cells and dampens positional effects on the expression of transfected genes. *Mol. Cell Biol.* **10**, 2302-2307.
133. Phi-Van, L., and Stratling, W.H. (1988) The matrix attachment regions of the chicken lysozyme gene co-map with the boundaries of the chromatin domain. *EMBO J.* **7**, 655-664.
134. Thorburn, A., Moore, R., and Knowland, J. (1988) Attachment of transcriptionally active sequences to the nucleoskeleton under isotonic conditions. *Nucl. Acids Res.* **16**, 7183.
135. Farache, G., Razin, S.V., Rzeszowska-Wolny, J., Moreau, J., Targa, F.R., and Scherrer, K. (1990) Mapping of structural and transcription-related matrix attachment sites in the a-globin gene domain of avian erythroblasts and erythrocytes. *Mol. Cell. Biol.* **10**, 5349-5358.
136. De Jong, L., van Driel, R., Stuurman, N., Meijne, A.M.L., and van Renswoude, J. (1990) Principles of nuclear organization. *Cell Biol. Int. Rep.* **14**, 1051-1074.
137. Jackson, D.A. (1991) Structure-function relationships in eukaryotic nuclei. *BioEssays* **13**, 1-10.
138. Mirkovitch, J., Gasser, S.M., and Laemmli, U.K. (1988) Scaffold attachment of DNA loops in metaphase chromosomes. *J. Mol. Biol.* **200**, 101-109.
139. Nelson, W.G., Pienta, K.J., Barrack, E.R., and Coffey, D.S. (1986) The role of the nuclear matrix in the organization and function of DNA. *Nucl. Acids Res.* **14**, 6433-6451.
140. Targa, F.R., Razin, S.V., de Moura Gallo, C.V., and Scherrer, K. (1994) Excision close to matrix attachment regions of the entire chicken alpha-globin gene domain by nuclease S1 and characterization of the framing structures. *Proc. Natl. Acad. Sci. USA* **91**, 4422-4426.
141. Kay, V. and Bode, J. (1994) Binding specificity of a nuclear scaffold: supercoiled, single-stranded, and scaffold-attached-region DNA. *Biochem.* **33**, 367-374.
142. Dietz, A., Kay, V., Schlake T., Landsmann, J., and Bode, J. (1994) A plant scaffold attached region detected close to a T-DNA integration site is active in mammalian cells. *Nucl. Acids Res.* **22**, 2744-2751.
143. Klehr, D., Maass, K., and Bode, J. (1991) Scaffold-attached regions from the human interferon beta domain can be used to enhance the stable expression of genes under the control of various promoters. *Biochem.* **30**, 1264-1270.
144. von Kries, J.P., Buhrmester, H., and Stratling, W.H. (1991) A matrix/scaffold attachment region binding mprotein: identification, purification and mode of binding. *Cell* **64**, 123-135.
145. von Kries, J.P., Buck, F., and Stratling, W.H. (1994) Chicken MAR binding protein p120 is identical to human heterogeneous nuclear ribonucleoprotein (hnRNP]U). *Nucl. Acids Res.* **22**, 1215-1220.
146. Buhrmester, H., von Kries, J.P., and Strätling (1995) Nuclear matrix protein ARBP recognizes a novel DNA sequence motif with high affinity. *Biochem.* **34**, 4108-4117.
147. Nakagomi, K., Kohwi, Y., Dickinson, L.A., and Kohwi-Shigematsu, T. (1994) A novel DNA-binding motif in the nuclear matrix attachment DNA-binding protein SATB1. *Mol. Cell Biol.* **14**, 1852-1860.
148. Fishel, B.R., Sperry, A.O., and Garrard, W.T. (1993) Yeast calmodulin and a conserved nuclear protein participate in the in vivo binding of a matrix association region. *Proc. Natl. Acad. Sci. USA* **90**, 5623-5627.
149. Tsutsui, K., Tsutsui, K., Okada, S., Watarai, S., Seki, S., Yasuda, T., and Shohmori, T. (1993) Identification and characterization of a nuclear scaffold protein that binds the matrix attachment region DNA. *J. Biol. Chem.* **268**, 12886-12894.
150. Dickinson, L.A., Joh, T., Kohwi, Y., and Kohwi-Shigematsu, T. (1992) A tissue-specific MAR/SAR DNA-binding protein with unusual binding site recognition. *Cell* **70**, 631-645.

151. Dickinson, L.A. and Kohwi-Shigematsu, T. (1995) Nucleolin is a matrix attachment region DNA-binding protein that specifically recognizes a region with high base-unpairing potential. *Mol. Cell Biol.* **15**, 456-465.
152. Hakes, D.J. and Berezney, R. (1991) Molecular cloning of matrin F/G: a DNA binding protein of the nuclear matrix that contains putative zinc finger motifs. *Proc. Natl. Acad. Sci. USA* **88**, 6186-6190.
153. Cunningham, J.M., Purucker, M.E., Jane, S.M., Safer, B., Vanin, E.F., Ney, P.A., Lowrey, C.H., Nienhuis, A.W. (1994) The regulatory element 3' to the A gamma-globin gene binds to the nuclear matrix and interacts with special A-T-rich binding protein 1 (SATB1), an SAR/MAR-associating region DNA binding protein. *Blood* **84**, 1298-1308.
154. Landers, J.P. and Spelsberg, T.C. (1992) New concepts in steroid hormone action: transcription factors, proto-oncogenes, and the cascade model for steroid regulation of gene expression. *Crit. Rev. Eukaryotic Gene Expression* **2**, 19-63.
155. van Steensel, B., Jenster, G., Damm, K., Brinkmann, A.O., and van Driel, R. (1995) Domains of the human androgen receptor and glucocorticoid receptor involved in binding to the nuclear matrix. *J. Cell. Biochem.* **57**, 465-478.
156. Zenk, D.W., Ginder, G.D., and Brotherton, T.W. (1990) A nuclear-matrix protein binds very tightly to DNA in the avian β-globin gene enhancer. *Biochemistry* **29**, 5221-5226.
157. Abulafia, R., Ben-Ze'Ev, A., Hay, N., and Aloni, Y. (1984) Control of late SV40 transcription by the attenuation mechanism and transcriptionally active ternary complexes are associated with the nuclear matrix. *J. Mol. Biol.* **172**, 467-487.
158. van Wijnen, A.J., Bidwell, J.P., Fey, Edward G., Penman, S., Lian, J.B., Stein, J.L., and Stein, G.S. (1993) Nuclear matrix association of multiple sequence specific DNA binding activities related to SP-1, ATF, CCAAT, C/EBP, OCT-1 and AP-1. *Biochemistry* **32**, 8397-8402.
159. van Eeklen, C.A.G. and van Venrooij, W.J. (1981) hnRNA and its attachment to a nuclear-protein matrix. *J. Cell Biol.* **88**, 554-563.
160. Zeitlin, S., Parent, A., Silverstein, S., and Efstratiadis, A. (1987) Pre-mRNA splicing and the nuclear matrix. *Mol. Cell. Biol.* **7**, 111-120.
161. Nickerson, J.A. and Penman, S. (1992) The nuclear matrix: structure and involvement in gene expression, in G.S. Stein and J.B. Lian (eds.), *Molecular and Cellular Approaches to the Control of Proliferation and Differentiation*, Academic Press, San Diego, pp. 434-380.
162. Xing, Y., Johnson, C.V., Dobner, P.R., and Lawrence, J.B. (1993) Higher level organization of individual gene transcription and RNA splicing. *Science* **259**, 1326-1330.
163. Ben-Ze'Ev, A. and Aloni, Y. (1983) Processing of SV40 RNA is associated with the nuclear matrix and is not followed by the accumulation of low-molecular weight RNA products. *Virology* **125**, 475-479.
164. Mariman, E.C.M., van Eekelen, C.A.G., Reinders, J., Berns, A.J.M., and van Venrooij, W.J. (1982) Adenoviral heterogenous nuclear RNA is associated with the host nuclear matrix during splicing. *J. Mol. Biol.* **154**, 103-119.
165. Ross, D.A., Yen, R.W., and Chae, C.B. (1982) Association of globin ribonucleic acid and its precursors with the chicken erythroblast nuclear matrix. *Biochemistry* **21**, 764-771.
166. Schroder, H.C., Trolltsch, D., Friese, U., Bachmann, M., and Muller, W.E.G. (1987) Mature mRNA is selectively released from the nuclear matrix by an ATP/dATP-dependent mechanism sensitive to topoisomerase inhibitors. *J. Biol. Chem.* **262**, 8917-8925.
167. Schroder, H.C., Trolltsch, D., Wenger, R., Bachmann, M., Diehl-Seifert, B., and Muller, W.E.G. (1987) Cytochalasin B selectively releases ovalbumin mRNA precursors but not the mature ovalbumin mRNA from hen oviduct nuclear matrix. *Eur. J. Biochem.* **167**, 239-245.
168. Jackson, D.A., McCready, S.J., and Cook, P.R. (1981) RNA is synthesized at the nuclear cage. *Nature* **292**, 552-555.
169. Nickerson, J.A., He, D., Fey, E.G., and Coffey, D.S. (1990) The nuclear matrix, in P.R. Strauss and S.H. Wilson (eds.), *The Eukaryotic Nucleus, Molecular Biochemistry and Macromolecular Assemblies,* Telford Press, Caldwell, New Jersey, pp. 763.
170. Herman, R., Weymouth, L., and Penman, S. (1978) Heterogeneous nuclear RNA-protein fibers in chromatin-depleted nuclei. *J. Cell Biol.* **78**, 663-674.
171. Durfee, T., Mancini, M.A., Jones, D., Elledge, S.J., and Lee, W.H. (1994) The amino-terminal region of the retinoblastoma gene product binds a novel nuclear matrix protein that co-localizes to centers for RNA processing. *J. Cell Biol.* **127**, 609-622.
172. Blencowe, B.J., Nickerson, J.A., Issner, R., Penman, S., and Sharp, P.A. (1994) Association of nuclear matrix antigens with exon-containing splicing complexes. *J. Cell Biol.* **127**, 593-607.

82

173. Huang, S. and Spector, D.L. (1992) U1 and U2 small nuclear RNAs are present in nuclear speckles (published erratum appears in Proc Natl Acad Sci USA 1992 May 1;89(9):4218-9). *Proc. Natl. Acad. Sci. USA* **89**, 305-308.
174. O'Keefe, R.T., Mayeda, A., Sadowski, C.L., Krainer, A.R., and Spector, D.L. (1994) Disruption of pre-mRNA splicing in vivo results in reorganization of splicing factors. *J. Cell Biol.* **124**, 249-260.
175. Hendzel, M.J., Sun, J.-M., Chen, H.Y., Rattner, J.B., and Davie, J.R. (1994) Histone acetyltransferase is associated with the nuclear matrix. *J. Biol. Chem.* **269**, 22894-22901.
176. Tawfic, S. and Ahmed, K. (1994) Growth stimulus-mediated differential translocation of casein kinase 2 to the nuclear matrix. *J. Biol. Chem.* **269**, 24615-24620.

TRANSCRIPTIONAL REGULATION IN A CHROMATIN ENVIRONMENT

A.P. Wolffe
Laboratory of Molecular Embryology
National Institute of Child Health and Human Development, NIH
Bethesda, MD 20892

DNA is severely compacted within chromatin, yet complex events such as replication, transcription, recombination and repair all occur with remarkable efficiency. Evolution has been successful in shaping chromatin such that it does not prevent trans-acting factors from gaining access to specific DNA sequences, or hinder polymerases from progressing along the chromatin fibre. Eukaryotic trans-acting factors have evolved to operate in a chromatin environment and histones have evolved to let them function.

1. Problems for nuclear processes in chromatin

A consideration of the complexity of nuclear processes such as transcription suggests that the formation of the large nucleoprotein complexes required to control these events might be incompatible with the packaging of DNA into chromatin and the chromosome. For example, how can several 100-500 bp transcription factor-binding regions exist, each requiring the association of multiple proteins to ensure faithful transcription, when this DNA may also be wrapped around the core histones and folded into the chromatin fibre? The same question can be asked for replication, recombination and repair. Moreover how can DNA polymerase and RNA polymerase progress through arrays of nucleosomes and the chromatin fibre once access to the DNA duplex is achieved? Methodologies to approach these questions have only recently become available. Potential molecular mechanisms to explain the access of trans-acting factors to DNA and the progression of polymerases through the chromatin fibre are beginning to be elucidated [1].

Since the packaging of DNA into nucleosomes and the chromatin fibre initially had been thought to make DNA refractory for any process of interest in the nucleus, experimental analysis through much of the 1980s focused either on the regulation of naked DNA templates in vitro or on transiently transfected templates that were uncharacterized with respect to chromatin structure. Results obtained through these analyses are increasingly recognized to be gross over-simplifications of the subtlety and complexity with which genes are regulated in their natural environment - the chromosome. Progress in several experimental systems has clearly shown that promoter elements are specifically organized within and between the nucleosomes of a nucleosomal array, and that the regulation of a gene depends upon the organization of DNA in a chromatin template [2]. It has also been conclusively demonstrated that

C. Nicolini (ed.), Genome Structure and Function, 83–109.
© 1997 *Kluwer Academic Publishers. Printed in the Netherlands.*

nucleosomes, including histone H1, are present on the majority of transcribed genes [3]. Understanding how transcription events occur in the context of chromatin will be seen to have regulatory significance important for all nuclear processes.

2. Interaction of trans-acting factors with chromatin

The difficulties inherent in having non-histone proteins gain access to a chromatin packaged template were recognized even before the nucleosome model was developed. Experimental approaches to this problem have continually been refined as the first non-histone proteins were purified and their binding sites on DNA defined. More recently, methodologies for determining specific chromatin structures have been developed. There has been a gradual trend from away from studying non-representative naked DNA-protein interactions towards recognition of the role of specific chromatin structures in mediating the interactions and functions of trans-acting factors.

2.1 ACCESSIBILITY OF DNA IN CHROMATIN: AN HISTORICAL PERSPECTIVE

Our knowledge of the accessibility of non-histone proteins to DNA in chromatin has progressed slowly. It has long been known that RNA synthesis using a bacterial or bacteriophage RNA polymerase is more efficient from naked DNA than from chromatin, and that the histones are responsible for this reduced efficiency [4]. These observations led to the idea that it was the uniform coating of DNA with histones that prevented RNA polymerase from reaching the template. The first experiments to suggest that DNA was not uniformly covered with histones were those of Felsenfeld and colleagues [5]. Polylysine precipitation of DNA, either naked or as chromatin, revealed that as much as 50% of the DNA in chromatin was accessible to the polycation and therefore presumably naked. This number was very similar to the amount of DNA that could be made acid soluble by nucleases. Clearly some DNA sequences in chromatin were more accessible than others.

The development of the nucleosome concept led investigators to explore the relative accessibility to various DNA binding proteins of linker DNA compared to that DNA tightly associated with the core histones (core DNA). It follows from the initial definition of the nucleosome via the action of endogenous nucleases that linker DNA is more readily cleaved by and hence more accessible to these enzymes. Bacterial and bacteriophage RNA polymerases do not initiate transcription of eukaryotic genes with any specificity, however they will initiate transcription non-specifically at AT-rich sequences resembling natural prokaryotic promoter elements [6, 7]. These enzymes have been very useful in assessing the relative accessibility of core versus linker DNA in chromatin. Early studies suggested that DNA in the nucleosome is not accessible to E. coli RNA polymerase [8-10]. A detailed analysis by Gould and colleagues revealed that linker DNA was more accessible than core DNA to E. coli RNA polymerase [11]. Titration of linker DNA availability through the addition of histone H1 revealed a rapid decline in accessibility, probably reflecting not only occlusion of linker DNA but also folding of the chromatin fibre. This extensive occlusion of DNA through relatively small changes in linker histone concentration may have significant consequences for the access of other trans-acting factors.

These studies were extended to the problem of how a DNA binding protein (E. coli RNA polymerase) might search for its binding sites in a nucleosomal array [12]. Surprisingly this search occurred with equivalent efficiency in both naked DNA and chromatin that had been depleted of histone H1. Two mechanisms have been envisaged for such a search, either 'sliding' of the DNA binding protein from site to site or 'hopping' between sites [13]. As E. coli RNA polymerase was known not to be able to efficiently slide or progress through nucleosomes, it was concluded that the enzyme was able to hop between sites efficiently in a chromatin template. These sites are the regions of relatively accessible linker DNA. Removal of histones H2A/H2B from chromatin increases the accessibility of DNA to RNA polymerase even more [14, 15]. In the chromosomal context, the search by RNA polymerase for binding sites is probably an accurate reflection of the search of trans-acting factors for recognition sequences.

Several interesting biological examples exist of changes in RNA polymerase accessibility to chromatin during development. Brown and colleagues were able to document that the normal somatic form of histone H1 was responsible for maintaining the repression of certain types of class III genes in Xenopus somatic cells [16]. This repressed state is established gradually during development as the amount of somatic histone H1 found in chromatin increases [17, 18]. The accumulation of histone H1 causes a general decline in the accessibility of RNA polymerase III to DNA [19, 20]. This decline in access to trans-acting factors is due to changes in chromatin structure. This change also correlates with the cessation of rapid cell division events and the imposition of a normal cell cycle.

Experiments with prokaryotic RNA polymerases were responsible for the first demonstration that the chromatin structure of the active form of a gene differed from that present when it was repressed. This conclusion followed from the relatively easy access of these polymerases to the DNA of transcriptionally active chromatin. Similar results were later obtained using nucleases. Unfortunately, the specificity of transcription was not improved upon using purified eukaryotic RNA polymerases (I, II and III). No eukaryotic RNA polymerase faithfully transcribes specific genes in the form of purified DNA templates. Roeder and colleagues were responsible for the fundamental demonstration that either a natural chromatin template isolated from a cell nucleus or a template reconstituted with transcription factors was necessary for recognition of a genes by RNA polymerase [21]. It was at this point that the focus of research on gene regulation shifted from the properties of the chromatin template to the properties of the promoter-specific transcription factors. As will be described below, the focus is now beginning to return to chromatin.

2.2 INTERACTION OF SPECIFIC TRANS-ACTING FACTORS WITH NON-SPECIFIC CHROMATIN

The availability of in vitro transcription systems that employ both specific cis-acting elements and trans-acting factors for the initiation of transcription by RNA polymerase has led to an increasing number of experiments in which the influence of chromatin structure on transcription has been investigated. A popular experiment with a long history has been to first mix a DNA template with histones or a nucleosome assembly system and then ask whether transcription could still occur. This type of experiment is responsible for the general belief that histone-DNA interactions repress transcription, since the usual result is that the addition of histones inhibits the given process. Although some investigators have undertaken numerous experimental controls to eliminate artifacts, there are often several possible explanations for the observed inhibitory effects that must be excluded.

DNA can precipitate or aggregate following non-specific association with histones. For example, linker histones (histone H1 or H5) are notorious for forming aggregates on DNA, which sometimes cause precipitation [22]. Most investigators attempt to exclude this possibility by examining the supercoiling or micrococcal nuclease cleavage patterns of their DNA template after nucleosome assembly. Each nucleosome should introduce one negative superhelical turn in the presence of topoisomerase into a closed circular DNA molecule, and protect approximately 146 bp of DNA from micrococcal nuclease. If these events occur some fraction of the template must be in solution and contain nucleosomes. However, subnucleosomal particles and proteolysed nucleosomes will also supercoil DNA and protect it from nucleases, therefore these assays give no guarantee of nucleosome integrity. Unfortunately it is all too easy to detect a few superhelical turns or to detect a single nucleosome length fragment of DNA after micrococcal nuclease, but more difficult to prove that the DNA molecule is efficiently (> 50%) assembled with nucleosomes.

Another difficulty with these reconstitution experiments is the inherent possibility that the template is also associated with non-specific DNA binding proteins. These proteins are often present in crude nucleosome assembly extracts and might also lead to occlusion of cis-acting sequences [23]. Even if efficient nucleosome assembly does occur, the various systems do not always position nucleosomes as found in the chromosome, nor do they always correctly space nucleosomes as would be found in vivo. These discrepancies might be explained by the facts that in vivo, chromatin assembly is coupled to the replication of DNA, special chromatin assembly factors are employed, the histones are post-translationally modified, and nucleosome assembly is staged. Thus it is not surprising that the prior association of unmodified histones with DNA under artificial conditions often leads to repressive effects. As a consequence the physiological significance of the repression observed in these experiments may be questionable.

An additional problem is that the unusual composition of the chromatin assembled by Xenopus and Drosophila oocyte, egg or embryo extracts might lead to significant differences in the transcriptional properties of a promoter relative to those found with the chromatin of normal somatic nuclei. The chromatin assembled in the extracts contains unusual core histone and linker histone variants and much larger amounts of proteins like HMG2 than are normally found. Proteins such as HMG2 might repress transcription independent of chromatin assembly [24]. With these reservations in mind it is better possible to critically evaluate the large body of data on interaction of specific transcription factors with non-specific chromatin.

2.2.1 *Experiments with class III genes*

Several experiments showed that mixing histones with class III genes would prevent their transcription [25, 26]. however, the organization of the chromatin template was not characterized. In contrast, it was found that in vivo, packaging of the genes into correctly spaced nucleosomes actually correlated with efficient 5S RNA gene transcription [27, 28]. An important conclusion from these and related studies was that the transcription factors had to gain access to DNA before the histones if transcription was to occur. Subsequent studies have shown that a complete transcription complex (TFIIIA, B and C) assembled onto a 5S RNA gene is more resistant to chromatin-mediated repression than the TFIIIA-5S RNA gene complex alone [29, 30]. The experiments discussed to this point used chromatin assembly systems that were deficient in histone H1. Reconstitution of histone H1 into chromatin such that the interaction alters nucleosome spacing (albeit with questionable physiological relevance) allows inhibition of 5S RNA gene transcription at a lower density of nucleosomes per length of DNA sequence than is otherwise required. In the absence of histone H1, very high nucleosome densities (one every 160-180 bp) or removal of free DNA with restruction endonuclease are required to repress transcription [31-34]. Chromatin appears to be repressive, yet if transcription factors gain access to DNA first, RNA polymerase III has no problems finding the transcription complex in chromatin.

2.2.2 *Experiments with class II genes*

A similar series of experiments has examined the effect of chromatin structure on the transcription of class II genes [35]. Luse and colleagues showed histones restricted transcription from the adenovirus type 2 major late promoter. A major problem of this type of experiment is that only a small percentage (< 5%) of templates are actually transcribed. As long as the promoter elements were free it was possible to demonstrate that the small number of active class II genes were in fact assembled into chromatin suggesting that nucleosomes did not completely inhibit RNA polymerase from transcribing once it had initiated the process [36, 37]. As with class III genes, transcription complexes formed prior to chromatin assembly resisted repression. Roeder and colleagues extended this analysis to suggest that TFIID binding alone was sufficient to relieve the inhibition of transcription due to nucleosome assembly. These experiments however have used crude assembly extracts supplemented with mixtures of histones including histone H1; and analysis of the resulting chromatin assembly has not been extensive [38, 39]. An important point established from these experiments is that promoter-specific factors that stimulate TFIID binding to promoters facilitate transcription of the promoter in the face of whatever is inhibiting transcription in the extract, including chromatin assembly [40, 41]. In contrast to this simple competition for binding to the promoter, Wu and colleagues have shown that TFIID binding is insufficient for transcriptional activation of the Drosophila hsp 70 promoter [42]. However a `potentiated' chromatin template is assembled that can respond to the presence of an `activated' heat shock transcription factor (HSTF). It has been suggested that HSTF requires TFIID in order to bind to nucleosomal templates.

Kingston and colleagues examined the influence of particular activation domains, especially regions rich in acidic, amino acids in regulating the transcription process of DNA packaged in chromatin. It was suggested that the presence of a transcription factor containing these regions could perturb nucleosome structure in the local region (< 100 bp) around the factor binding site [43].

Under appropriate conditions a variety of transcription factors have the potential to associate with nucleosomal DNA even in the absence of activation domains. These experiments generally use short DNA fragments reconstituted with histone octamers, where the DNA is not rotationally positioned with respect to the histones. Consequently, a wide range of DNA conformations within the nucleosome is exposed to the transcription factors. The factors used include Gal4, SP1, USF, TBP and Myc/Max [44-50]. In general, the binding of trans-acting factors to nucleosomal DNA occurs at an excess of protein that is approximately 100-1000 fold greater than that necessary to saturate the same binding site when present as naked DNA. Several parameters facilitate trans-acting factor association within the nucleosome. These include multiple binding sites for the factor being present within the nucleosomal DNA, positioning of the recognition site at the edge of the nucleosome, proteolysis of the core histones [51], the presence of histone binding proteins like nucleoplasmin to facilitate histone exchange from DNA [45] and the SWI/SNF general activator complex [48-50]. These experiments provide a useful illustration of the difficulty of binding trans-acting factors to DNA in which substantially random histone-DNA interactions occur. These nucleosomal templates also provide insight into how histone-DNA interactions might prevent inappropriate trans-acting factor access within regulatory positioned nucleosomes.

Histone H1 is a contaminant of many crude in vitro transcription extracts. Kadonaga and colleagues have shown that the action of several promoter-specific DNA binding proteins is to relieve this non-specific inhibitory process. Thus, antirepression of transcription in in vitro transcription reactions is as important as the actual activation process [23]. These experiments were extended to the reconstitution of close-packed arrays of nucleosomes with histone H1 [52] and most recently with a Drosophila in vitro chromatin assembly extract supplemented with exogenous H1 [53]. There is an increase in repressive character as the type of chromatin assembled more closely achieves a physiologically correct H1 association. High affinity histone H1 binding to chromatin requires both core histones and linker DNA [54]. Histone H1 associated with naked DNA has no resemblance to a native chromatin template.

It is important to note that under the in vitro transcription conditions used by many investigators, particularly those that include high concentrations of divalent cations, even H1-depleted nucleosomal arrays are highly folded and can form insoluble aggregates [55]. Folding and aggregation effects can inhibit both transcription initiation and elongation within chromatin [56, 57] (see Figure 1).

Any results involving non-specific chromatin structures and specific trans-acting factors should be treated with caution, especially if the properties and organization of the chromatin template are not documented. A general conclusion from a large number of experiments is that prior assembly of a template with nucleosomes inhibits trans-acting factor access to DNA whereas if the factors bind first, subsequent chromatin assembly will not be inhibitory. RNA polymerase can both find and associate with a transcription complex in a chromatin background without any problem.

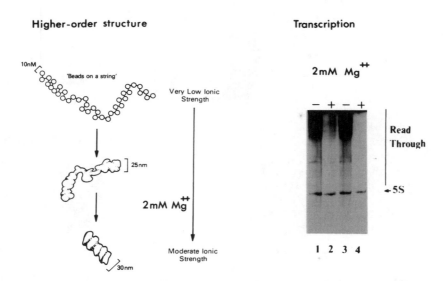

Figure 1. Folding of chromatin as visualized in the electron microscope. As ionic strength increases chromatin changes from a beads-on-a-string configuration to a flat fiber, 30 mn in diameter. It is possible to correlate the Mg^{2+}-dependent folding of chromatin as determined in the analytical ultracentrifuge with the transcriptional properties of chromatin (Reproduced with permission from Hansen, J.C., Wolffe, A.P. Proc. Natl. Acad. Sci. USA 1994, 91, 2339. Copyright 1994 by the National Academy of Sciences).

3. Specific trans-acting factors and specific chromatin

Although the functional importance of association of trans-acting factors with specific DNA sequences was readily accepted, the significance of the `sequence specific' organization of DNA into nucleosomes has taken longer to be acknowledged. Most investigators accept that there is no logical necessity to organize the vast majority of DNA into chromatin structures that have specific DNA sequences organized in a precise way. However it is also clear that nucleosomes are often specifically positioned around certain DNA sequences and that they have important functional roles [58-60]. Formation of such specific chromatin structures is true for the vast majority of genes for which the appropriate assays have been carried out to assess nucleosome positioning. Incorporation of cis acting elements into a positioned nucleosome has significant consequences for its accessibility of DNA to trans-acting factors. DNA in the nucleosome is highly bent and the helical periodicity of the double helix changes from an average of 10.5 bp/turn to 10.2 bp/turn. Thus, not only is one face of the DNA helix occluded by the histone core, but DNA has an entirely different structure from that in solution.

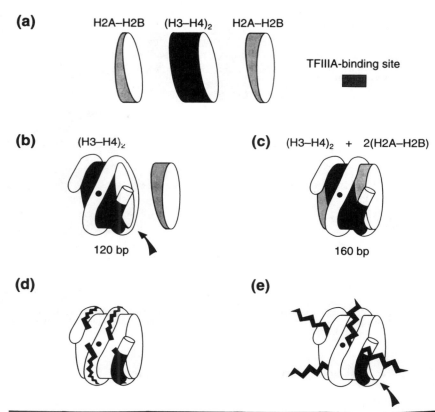

(a) H2A–H2B   (H3–H4)₂   H2A–H2B

TFIIIA-binding site

(b) (H3–H4)₂   **(c)** (H3–H4)₂ + 2(H2A–H2B)

120 bp   160 bp

(d)   **(e)**

Figure 2. Modifications of a positioned nucleosome containing the 5S rRNA gene that facilitate TFIIIA binding. **(a)** The histone octamer has a tripartite organization. It consists of a central kernel of a tetramer [(H3-H4)₂] surrounded by two dimers (H2A-H2B). **(b)** Each (H3-H4)₂ tetramer organizes 120 bp of DNA and recognizes the DNA sequences that direct nucleosome positioning within the 5S rRNA gene. The tetramer-5S DNA complex leaves the key binding site for TFIIIA exposed (arrow). **(c)** Each H2A-H2B dimer organizes an additional 20 bp to either side of the tetramer-5S DNA complex. Association of H2A-H2B with the key binding site for TFIIIA inhibits TFIIIA binding. **(d)** A nucleosome core consists of the DNA wrapped around the globular domains of the core histones (cylinder). The histone tails (black zig zags) interact with the phosphodiester backbone of DNA on the outside of the nucleosome. **(e)** Acetylation of the tails weakens their interaction with DNA and facilitates TFIIIA binding (arrow). (Reproduced with permission from Wolffe, A.P. Trends Biochem. Sci. 1994, 19, 240. Copyright 1994 by Elsevier Science Ltd).

3.1 TRANS-ACTING FACTOR ACCESS TO DNA IN POSITIONED NUCLEOSOMES IN VITRO

Martinson and colleagues were the first to examine the issue of trans-acting factor access to a specific DNA sequence incorporated into a nucleosome. Their experiments made use of a prokaryotic repressor (the lac repressor) and a 144 base pair DNA fragment containing binding sites for the repressor. This short DNA fragment is able to assemble a single nucleosome structure following reconstitution with equimolar amounts of the four core histones. The lac repressor is a helix-turn-helix protein that binds to B-form DNA on one side of the double helix in the major groove. Based on sedimentation studies the authors concluded that both lac repressor and the histone octamer could simultaneously occupy the same DNA fragment. This implied that a triple complex of the DNA, histone proteins and trans-acting factor formed [61, 62]. While these early studies lacked the resolution necessary to be absolutely sure of either nucleosome positioning or specific association of the DNA-binding proteins, they clearly demonstrated the correct approach to this problem.

Transcription factor (TF) IIIA was the first sequence-specific eukaryotic DNA binding protein to be purified. This protein (38 KDa) consists of a chain of zinc fingers which bind in clusters over the 50 bp internal control region of the 5S RNA gene. Simpson had demonstrated that 5S RNA genes contain strong nucleosome positioning sequences [63, 64]. The interaction of TFIIIA with a Xenopus 5S RNA gene that had been incorporated into a nucleosome was investigated. TFIIIA could form a triple complex with the 5S RNA gene and histones dependent on stoichiometry and post-transcriptional modification of the histone proteins [65, 66]. If the nucleosome is deficient in histones H2A/H2B, more DNA is free at the edge of the nucleosome for interaction with TFIIIA. Unlike the lac repressor, which can interact with DNA that is actually bound to the histone octamer, TFIIIA binds primarily to free DNA at the edge of the octamer, not to DNA actually contacting the histones. Interestingly, acetylation of the core histones facilitates the binding of TFIIIA to the 5S RNA gene within the nucleosome [66]. Since acetylation of the core histones does not apparently alter DNA conformation in the nucleosome directly, or the position of the nucleosome relative to the 5S RNA gene, it is likely that a small change in the conformation of the histone octamer or in the stability of the core histone-DNA contacts allows TFIIIA to bind [67], (summarized in Figure 2).

3.2 TRANSCRIPTION FACTOR ACCESS TO DNA IN SPECIFIC CHROMATIN STRUCTURES IN VIVO

In vitro experiments attempt to reconstruct phenomena believed to occur in vivo, thereby offering mechanistic insights into a process. However many of our clearest insights into gene regulation events in chromosomes come directly from the documentation of in vivo events. The general observation from such studies is that promoters that are organized into chromatin in the living cell are often accessible to trans-acting factors, even when DNA regulatory elements are adjacent to nucleosomes or actually incorporated into them. This conclusion is in contrast with the vast majority of in vitro experiments.

One of the best examples of in vivo control of gene expression in a chromosomal context concerns the regulation of the PHO5 gene of Saccharomyces cerevisiae [68]. The yeast PHO5 gene encodes an acid phosphatase that is induced by a reduction in inorganic phosphate concentration. Two nucleosomes are positioned to either side of an essential promoter element recognized by the trans-acting factor PHO4. A second binding site for PHO4 and a site for another trans-acting factor PHO2 are incorporated into one of the positioned nucleosomes. On induction, all four positioned nucleosomes are disrupted. Nucleosome disruption is independent of replication or transcription events [69, 70]. Analysis of mutants reveals that PHO4 is essential for both the transcriptional activation of the PHO5 gene and the rearrangement of chromatin structure. PHO4 activity is regulated by phosphorylation. Under conditions of transcriptional repression PHO4 is phosphorylated [71]. The DNA binding affinity of PHO4 is substantially increased by phosphate starvation. PHO4 binds to the site between the two nucleosomes and facilitates the disruption of the nucleosome containing the other PHO4 binding site [72]. The transactivation domain of PHO4 is necessary to mediate this nucleosome disruption process [73]. Changes in DNA sequence in the adjacent nucleosomes can influence transcriptional activation and chromatin rearrangement, suggesting that the chromosomal organization of the whole promoter region is essential for correct regulation [74-78]. The precise placement of regulatory elements between or within nucleosomes is clearly important for the regulation of this promoter.

An important point is that it is difficult in all of the above experiments to distinguish between nucleosome displacement, i.e., removal of a complete histone octamer from DNA, and nucleosome disruption, or a conformational change or displacement of a H2A/H2B dimer, on induction of PHO5. The only certain way to do this is to use protein-DNA cross-linking reagents and to examine the association of specific histones with particular DNA fragments containing the promoter element of interest. Using this technique Varshavsky, Mirzebekov and colleagues have presented evidence for continued histone-DNA contacts in actively transcribed genes [79, 80]. Related to this issue are the experiments of Grunstein and colleagues, who have shown through genetic manipulation of histone stoichiometry that disruption of the chromatin structure of the PHO5 promoter even in the absence of factor-dependent induction can significantly activate transcription [81, 82]. This further establishes that the repression of PHO5 transcription is related to the chromosomal structure of the gene, and that correct PHO5 gene regulation requires a chromatin template, as is true for many other yeast genes.

Although experiments with yeast are useful for establishing the major players in the transcriptional regulation of chromatin templates, significant differences exist between the chromosomal architecture of yeast and vertebrates. The Saccharomyces cerevisiae core histones are relatively divergent from those of vertebrates and a protein resembling the histone H1 of vertebrate somatic cells does not appear to exist (although HMG1 homologs are present). Yeast histones are more heavily acetylated than those found in normal vertebrate cells and nucleosomes are relatively tightly packed together. A yeast cell has to regulate most of its genes rapidly and continually through the cell cycle in a very different way compared to a differentiated cell from a metazoan. The only truly stable state of gene expression in a yeast cell that is analogous to those found in larger eukaryotes is that which controls mating-type. Here multiple chromatin-dependent control mechanisms regulate gene activity. It is therefore important to examine the general applicability of any models established in yeast. Preeminent among the systems exploited to this end is the regulation by glucocorticoids of transcription of the mouse mammary tumor virus (MMTV) long terminal repeat (LTR) [83, 84] (see Chapter by Beato in this volume).

The MMTV LTR is incorporated into six positioned nucleosomes in both episomes and within a mouse chromosome. The positioned nucleosomes serve to prevent the basal transcriptional machinery from associating with the promoter under non-inducing circumstances, i.e., in the absence of glucocorticoids. Induction of transcription by glucocorticoids requires binding of the liganded glucocorticoid receptor (GR) to the LTR, followed by disruption of the local chromatin structure upon GR binding to recognition sequences within nucleosomes, and finally the assembly of a transcription complex over the TATA box [85, 86]. Thus, comparable events occur on both the PHO5 and MMTV LTR promoters: an inducible transcription factor binds, chromatin structure is rearranged, a transcription complex is assembled and transcription is activated.

Vigorous attempts have been made to reconstruct the transcriptional regulation and concomitant chromatin structural changes of the MMTV LTR in vitro. The GR, which is a zinc-finger protein, appears to bind its nucleosome bound cognate DNA sequence with only a slight reduction in affinity relative to the naked DNA sequence. This interaction is dependent on the precise position of the nucleosome and hence the translational position of the GR binding site within the nucleosome. GR binding is reported to occur when the nucleosome is at -188 to -45 [87] and at -219 to -76 [88] or -221 to -78 [89] relative to the start site of transcription (+1). In these instances the rotational orientation of the individual GR binding sites on the surface of the histone octamer will be similar due to the separation of nucleosome boundaries by almost exactly three helical turns of DNA. The DNA sequence containing the GR binding sites has regions of intrinsic flexibility and curvature that direct the histones to occupy these specific positions [90]. Four GR recognition elements (GREs) are within this DNA segment, located at -175, -119, -98 and -83; they have different rotational and translational positions within the nucleosome. The two elements at -119 and -98 face toward the core histones and remain unbound in the presence of GR, whereas the elements at -175 and -83 face toward solution and are bound by the GR. These sites are separated by 92bp which places them together on one side of the nucleosome. This proximity might facilitate both the binding and subsequent nucleosome-destibilizing activity of GR [91].

Experiments that compare the affinity of the GR for a recognition element present as free DNA compared to one facing towards solution but present in the nucleosome at different translational positions, show that binding affinity is reduced 3-11 fold in the nucleosome [92]. This is a remarkably small reduction in binding affinity compared to the complete absence of binding observed when the recognition element faces towards the histones. This suggests that the key variable determining the accessibility of the GR to a nucleosome-bound DNA binding site is not the translational position of the binding site, but the rotational position of the DNA sequence with respect to the surface of the histones.

The GR is well suited to interact specifically with nucleosomal DNA. GR binds to DNA using a domain containing two zinc fingers. An a-helix in one of the two fingers interacts with a short 6bp region in the major groove of the double helix, while the other finger is involved in protein-protein interaction [93]. The GR associates with DNA as a dimer; the second molecule of the receptor has interactions on the same side of the double helix as the first receptor, but is one helical turn away. Thus, the GR binds to DNA on the side exposed towards solution in the nucleosome, thereby circumventing steric interference by the histone core. Moreover, the GR dimer can bind specifically to DNA containing only one GRE half site, presumably by making both specific and non specific contacts to DNA in each half of the dimer. Thus, highly bent nucleosomal DNA might still provide enough precisely aligned contacts for at least one specific half-site interaction, which could then be supplemented by non-specific contacts. This type of non-specific interaction could account in part for the reduction in affinity of the GR for nucleosomal recognition elements. Surprisingly, association of the GR with the nucleosome containing its binding site appears to have no effect on the integrity of the structure in vitro, unlike the apparent consequence in vivo. Binding of the other promoter-specific transcription factor (NF1), which is facilitated by the GR in vivo, does not occur in vitro on nucleosomal templates. Certain nuclear components that presumably facilitate chromatin structural changes together with the GR in vivo have been so far lacking in the in vitro system.

What general molecular processes cause chromatin structure to be disrupted in vivo? DNA replication is the one event certain to disrupt chromatin and provide access of transcription factors to their cognate sequences. However, DNA replication is not required for chromatin disruption and transcriptional activation of the MMTV LTR [86].

If replication is not involved in chromatin rearrangement, nucleosomes must be disrupted in some alternative way. Trans-acting factors might displace histones from nucleosomes directly or might wait for histones to passively exchange out of chromatin before binding to their recognition sequences. In most assays, nucleases are used to examine the incorporation of specific promoter elements into nucleosomes. It is possible that nucleosomes that have altered their position or composition would lose the sharp boundaries attributed to nuclease protection, even though DNA could remain associated with histones. Protein-DNA cross-linking reagents and antibodies against histones are being used to explore this possibility. Histone H1 and histones H2A/H2B are known to readily exchange in and out of chromatin under physiological conditions. A certain amount of histone-dependent DNA packaging would occur even in the absence of these proteins. The capacity to reassemble a complete nucleosome would also remain if the histone H3/H4 tetramer stays bound to DNA. Histones H1, H2A and H2B are not able to correctly assemble into chromatin unless the histone H3/H4 tetramer is first bound to the DNA. It is also possible that complete histone octamers might exchange from DNA in vivo. Only very basic arginine-rich proteins, that are unlike known transcription factors, can remove histones in a natural context. This integrity of the nucleosome is almost entirely due to the stability of the interaction of the arginine-rich histones H3/H4 with DNA. A partial disruption of the nucleosome seems by far the most likely mechanism by which trans-acting factors might gain access to DNA in chromatin. Removal of histone H1 from the nucleosome might facilitate some mobility of the contacts made by the histone octamer with DNA [94]. This local nucleosome sliding could facilitate transcription factor access to recognition sites otherwise constrained within the nucleosome [95, 96].

The proteins that GR recruits to the MMTV chromatin are the mammalian homologs of the SWI (Switch) 1, 2, 3 proteins of S. cerevisiae [97, 98]. The exact mechanism whereby they disrupt chromatin is unknown, but they initiate a chain of events that cause the removal of histone H1 from the linker DNA in MMTV chromatin [99] and a substantial increase in the nuclease accessibility of the DNA that is within the positioned nucleosomes [84]. The precise positioning of the nucleosomes may help facilitate the displacement of histone H1, since the linker region between the nucleosomes contains the binding sites for histone H1, NF1 and the octamer factor. It is possible that histone H1 and the transcription factors might compete for binding to this linker region. In any event, the binding sites fo the transcription factors NF1, the octamer factor and TFIID lie in either the linker DNA or at the periphery of the positioned nucleosomes, and these factors are recruited to their binding sites in this disrupted chromatin and subsequently assemble an active transcription complex. Transcriptional activation by GR is only transient. After a few hours the basal transcription complex, NF1, octamer factor and GR are displaced from the MMTV LTR, their binding sites are reincorporated into the positioned nucleosomes, and the promoter becomes repressed [100]. The molecular mechanisms responsible for displacement of the transcription factors have not been determined, however these results indicate that complexes of both transcription factors and histones with promoters are likely to be dynamic.

As our information concerning the molecular basis of sequence-directed nucleosome positioning grows, it becomes possible to manipulate specific chromatin structure around the recognition site for a sequence-specific DNA binding protein. Wolffe and Drew made use of the selective affinity of histone octamers for synthetic DNA curves to manipulate nucleosome positioning relative to a promoter for bacteriophage T7 RNA polymerase [101]. They found that small changes of less than 10 bp in the association of the promoter (~ 20 bp in length) with the nucleosome could affect the repression of transcription by 10-20 fold. In this case even interactions of DNA binding proteins with the periphery of a nucleosome severely inhibit function in vitro.

Simpson has carried out analagous experiments in vivo. Normally, in the TRP1ARS1 minichromosome, nucleosomes do not incorporate the ARS1 sequence element that is required for DNA replication. However, using a sequence-directed nucleosome positioning element to move nucleosomes, it was possible to show that incorporation of the 11 bp ARS1 core sequence domain into the centre of nucleosomal DNA, but not into a peripheral region, inhibited replication of TRP1ARS1 DNA [102]. Subsequent work from the Simpson lab has systematically defined a hierarchy of DNA sequence and protein directed contributions to the assembly of specific chromatin structures involving to ARSI sequence [103]. Although certain proteins, such as those associated with class III gene transcription, can function effectively in a variety of chromatin structures [104, 105], others cannot [106-109]. Using similar approaches, the heat shock transcription factor, which appears to bind to nucleosomes weakly and non-specifically in vitro, associates efficiently with rotationally positioned DNA within a nucleosome in vivo [110]. Alteration of the context of the regulatory elements within a heat-shock gene promoter can result in a complete restriction in heat shock factor access to DNA in a nucleosome [111]. It is clear that the context in which a DNA recognition element is organized within chromatin can have major consequences for the regulation of gene activity.

Thus, several studies have shown that certain specific trans-acting factors can recognize their binding sites, even when these sites are wrapped around the histone octamer. The histone octamer is positioned such that the binding sites are accessible to trans-acting factors. This suggests that specific chromatin structures assemble in order to fulfill the contrasting needs of DNA compaction and accessibility. Of course, in agreement with experiments using non-specific chromatin systems, certain positioned nucleosomes will inhibit trans-acting factor access to DNA. The hypothesis that specifically configured chromatin structures are compatible with nuclear processes gains strength through observations in vivo. In yeast and mammalian cells, excellent examples exist of promoter elements being regulated through trans-acting factors that function in a chromatin environment and modify chromatin structure.

4. Relationships between transacting factors, DNaseI sensitivity, DNaseI hypersensitive sites and chromosomal architecture

After the early experiments that demonstrated the selective association of prokaryotic RNA polymerases with transcriptionally active chromatin, Weintraub, Felsenfeld and colleagues were able to show a comparable general accessibility to nucleases [112, 113, 114]. This general sensitivity to nucleases includes the coding region of a gene and may extend several kilobases to either side of it. DNaseI normally introduces double strand breaks over 100 times more frequently in transcriptionally active chromatin than in inactive chromatin. However the exact structural basis of this generalized sensitivity is unknown. Later we will discuss structural changes in the transcribed regions of genes, however the dispersed nature of the generalized sensitivity implies additional contributory factors. Careful analysis reveals the non-transcribed regions are just as sensitive to DNaseI digestion as are the transcribed regions, provided the last-cut approach to the measurement of DNaseI sensitivity is used. This is defined as digestion by DNaseI to fragment sizes so small that DNA no longer hybridizes to complimentary strands after denaturation. A certain length of DNA is necessary to allow specific recognition (hybridization) of two separated single-stranded regions. An important question that is not yet completely resolved is whether transcription is required to generate generalized nuclease sensitivity in certain instances or whether sensitivity always precedes transcription.

Experiments on the action of mitogens on quiescent cells reveal that a subset of genes, called the immediate-early genes, are rapidly induced [115]. The most studied examples of such genes are c-myc and c-fos. These two proto-oncogenes are transcriptionally activated within minutes. Coincident with transcription, the chromatin structure of the proto-oncogenes becomes more accessible to nucleases. Once proto-oncogene transcription ceases, preferential nuclease accessibility is lost [116-118]. Possible conformational changes in nucleosome structure might account for such effects. It has been proposed that histone H3 sulphydryl residues might become accessible in nuclease sensitive chromatin, however transcriptionally active S. cerevisiae chromatin, which has no cysteine and hence no sulphydryls in the core histones, is also retained on the organomercurial agarose columns used to assess nucleosome conformational changes. This suggests that RNA polymerase or HMGs might provide the sulphydryls that bind to the organomercurial agarose columns that retains active chromatin [119]. However, the rapidity of the changes in nuclease sensitivity (< 90s) and their propagation in both directions 5' and 3' to the promoter means that transcription, and hence RNA polymerase or HMGs, cannot account for all of the observed changes. This is consistent with some column retention due to histone H3 in higher eukaryotes. In fact the speed of the response suggests that changes in nuclease sensitivity precede transcription, and may play a role in regulating c-fos expression.

It is interesting that one of the earliest mitogen-induced nuclear signalling events coincident with proto-oncogene induction is the rapid phosphorylation of histone H3 on serine residues within its highly charged, basic N-terminal domain. Whether these changes are localized to chromatin regions containing either c-fos, c-myc or the other immediate early genes has not yet been determined [120]. An additional component that contributes to the prior sensitization of the proto-oncogenes to nucleases may come from the existence of trans-acting factors that are already associated with the promoter [121]. Such interactions are responsible for the second landmark in chromatin: DNase I hypersensitive sites. These sites are the first place DNase I introduces a double strand break in chromatin. They usually involve small segments of DNA sequences (100-200 bp) and are two or more orders of magnitude more accessible to cleavage than when in inactive chromatin [122-125]. DNase I hypersensitive sites are often flanked by positioned nucleosomes; both DNase I hypersensitive sites and nucleosome arrays are detected by DNase I or micrococcal nuclease digestion and indirect end-labelling methodologies. As the most accessible regions of chromatin to non-histone DNA binding proteins, DNaseI hypersensitive sites generally denote DNA sequences with important functions in the nucleus.

DNaseI hypersensitive sites were first detected in the SV40 minichromosome (at the ORI region), and in Drosophila chromatin [126]. In general these sites appear to be accessible to all enzymes or reagents that cut duplex DNA. Higher resolution studies have shown these sites often represent clusters of recognition sites for promoter-specific DNA binding proteins [127]. These sites have been mapped to a large number of functional segments of DNA, including promoters, enhancers, locus control elements, transcriptional silencers, origins of replication, recombination elements and structural sites within or around telomeres [128].

Around regulated genes DNase I hypersensitive sites often fall into a hierarchy of patterns. In the chicken ß-globin locus, which contains 4 globin genes (5'ρ-ß^H-ß^A-ε-3') covering over 65Kb, twelve DNaseI hypersensitive sites are found. One site is present in all cells independent of whether the genes are transcriptionally active or not. Three sites, upstream of the ρ globin gene were present only in erythroid cells destined to express the globin genes; these sites have no known functional significance. However a similar site was found between the ß^A and ε genes that corresponded to an enhancer element. Four sites were found over the promoters of each gene depending on whether the gene was transcriptionally active, and three sites were found downstream of the genes corresponding to transcription termination elements (the ß^A gene excluded). It is important to note that the formation of DNaseI hypersensitive sites at the promoters of the globin genes is a relatively late step in the commitment of these genes to become transcriptionally active. However it is clear that the formation of such sites precedes the actual initiation of transcription by RNA polymerase; indeed the generation of these sites may account for a component of the general nuclease sensitivity of a gene [129, 130].

One of the most thorough dissections of a DNaseI hypersensitive site has been carried out by Elgin and colleagues [126]. Transcription of the Drosophila heat shock protein (hsp) 26 gene is very rapidly activated by raising the temperature of a fly to a stressful level (a heat shock of 34°C). Two DNaseI hypersensitive sites exist at the promoter of the hsp 26 gene, which include recognition sequences for the promoter-specific heat shock transcription factor (HSTF, a leucine zipper protein). Following heat shock, HSTF binds to these sites. In contrast, TFIID is bound to the TATA box both before and after heat shock. TFIID alone is insufficient to cause the hsp 26 gene to be transcribed, the specific association of the HSTF protein is also required. High resolution analysis has revealed that a nucleosome is positioned between the proximal and distal binding sites for HSTF, i.e. between the two DNaseI hypersensitive sites. In this case the exact position occupied by this nucleosome depends not only on the DNA sequence to which the histones bind (from -300 to -140), but also on adjacent DNA sequences. These are repeats of the type (CT)n.(GA)n which bind a specific trans-acting factor, the GAGA protein [131-134]. The (CT)n.(GA)n repeat regions are located to either side of the positioned nucleosome at -347 to -341 and at -135 to -85. The GAGA factor bound to these repeats may function as a `bookend' to determine exactly where the nucleosome will be positioned [134]. Recent evidence suggests that the GAGA factor might function through ATP- dependent mechanisms to actively direct the assembly of a particular chromatin architecture [135]. Transcription of the gene is regulated through the association of the heat shock transcription factor with recognition elements at -51, -170, -269 and -340. The sites at -170 and -269 will be wrapped around the core histones in rotational frames that prevent heat shock transcription factor association. However the wrapping of DNA around the nucleosome will bring the HSTF molecules bound to the sites at -340 and -51 into juxtaposition, and the histones may potentially facilitate transcription through causing a clustering of HSTF activation domains [136]. Direct evidence for transcriptional activation mediated by this nucleosome is yet to be established; however it is clear that the positioning of a nucleosome in this particular way allows key transcription factors to obtain access to essential regulatory elements in spite of the assembly of the gene into chromatin. Lis and colleagues have shown that RNA polymerase II is also bound to the promoter, at a site next to the TFIID protein. Heat shock and the binding of HSTF allows RNA polymerase II to begin transcriptional elongation by an unknown process [137, 138].

There are several important conclusions that can be derived from this analysis. The proximal hypersensitive site exists because there is a region of naked DNA between the specific protein complex at the TATA box and the positioned nucleosome. This region is too small for an additional nucleosome, but contains the binding sites for a promoter specific-factor (HSTF). Thus HSTF can bind to DNA without hindrance from chromatin structure; in fact the presence of the nucleosome may create a constrained loop which facilitates interactions between HSTF proteins and TFIID. This is another excellent in vivo example of how specific proteins can regulate gene expression in combination with a specific chromatin structure.

Similar situations as that described above for the hsp26 promoter have been determined to exist for the Drosophila alcohol dehydrogenase gene and the Xenopus vitellogenin genes [139, 140]. For the vitellogenin genes, nucleosome positioning directed by DNA sequence occurs between -300 and -140. The binding sites for the stimulatory transcription factors, the estrogen receptor and nuclear factor 1 lie outside the region of DNA that is wrapped around the histones at -300 to -330 and at -120 to -110 respectively. When these sites are brought together by the wrapping of DNA around a positioned nucleosome (or artificially by deleting the intervening DNA), transcription is enhanced about 5 to 10 fold. This moderate stimulatory effect is much more significant than it might appear since the assembly of a non-specific chromatin structure would normally lead to a > 20 fold repression of transcription as the binding sites for transcription factors are occluded by the histones. Thus, in the vitellogenin gene nucleosome positioning has two roles: 1) to provide a scaffold that allows transcription factors to communicate more effectively; 2) to prevent the formation of repressive histone-DNA interactions that may prevent any transcription factor from gaining access to a chromatin template (see Figure 3).

DNaseI hypersensitive sites have been useful in defining important DNA sequences that have no apparent function (at least when first discovered). Among these were four strongly nuclease-sensitive sites located 10-20 Kbp upstream of the cluster of human ß-globin genes [141-143]. These sites, which have come to known as locus control regions (LCRs), represent cis-acting elements that allow genes that are integrated into a chromosome to be expressed in a way that is independent of chromosomal position, i.e. position effects are abolished. A consequence of this is that LCRs allow each copy of a gene integrated in multiple copies to be expressed equivalently, so that gene expression is copy-number dependent.

When all four DNaseI hypersensitive sites comprising the LCR are placed adjacent to reporter genes, they appear to function like enhancers. However, three of the four sites do not function as enhancers in transient expression sites, but will do so only after incorporation into the chromosome. This suggests that the LCRs may play a special role in the stabilization of an accessible chromatin structure distinct from the function of a normal enhancer element [144-147]. Recent evidence suggests that certain enhancers/LCRs have to interact with promoter sequences in order to confer general DNase I sensitivity on chromatin [148], whereas others can facilitate local factor access within chromatin without the influence of a normal eukaryotic promoter [149].

Each LCR hypersensitive site contains multiple binding sites for promoter-specific proteins. In the case of the human ß-globin genes, one of these sites contains four recognition elements for an erythroid tissue specific DNA-binding protein known as GATA-1. In spite of this information, how these sites function is unknown. Several models have been proposed, including: unravelling of the chromatin fibre, normal enhancer functions, and stabilization of specific nucleoprotein complexes during replication. Evidence consistent with the propagation of an altered chromatin structure comes from transgenic experiments in which LCRs confer hypersensitive sites and general DNaseI sensitivity whereas promoter elements do not. This suggests that LCRs may be able to overcome the inhibitory influences of localized regions of heterochromatin.

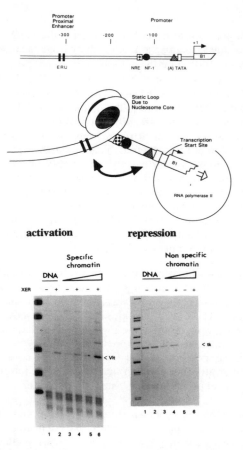

Figure 3. Role of a static loop created by a nucleosome including part of the Xenopus vitellogenin B1 promoter in potentiating transcription. Transcription from a specific chromatin template that positions nucleosomes increases with increasing numbers of nucleosomes. Transcription from a non-specific chromatin template in which nucleosomes are not positioned decreases with increasing numbers of nucleosomes (see ref. 31 for details). (Reproduced, with permission, from Schild, C., et al., EMBO J. 1993, 12, 423. Copyright 1993 by IRL Press Ltd).

Schedl and colleagues have used transposable elements (P-elements) that can introduce stable transformants into the germ line of Drosophila melanogaster. Using the hsp 70 gene, these investigators defined DNA sequence elements that contained DNaseI hypersensitive sites and that were present to either side of the hsp 70 genes. These specialized chromatin structure (SCS) elements conferred position-independent, copy number-dependent transcription from the white promoter, with the exception of one insertion into heterochromatin within the Drosophila X chromosome. Importantly, the SCS elements do not behave as scaffold attachment regions, establishing a functional separation between the two DNA elements [150]. The Drosophila suppressor of hairy-wing protein and the complex it assembles with the gypsy transposable element has been recognized as having comparable `insulating' properties [151].

An important observation concerning the functions of SCS and gypsy elements is that they block enhancer function if the element is between the enhancer and the promoter [151]. This capacity to block activation or silencing effects has led to the current definition of these elements as insulators. Enhancers normally act over very long distances (many kilobases). It is hard to imagine how a short DNA sequence (100-200 bp) could inhibit DNA looping, suggesting that enhancers in this case are functioning by an alternate mechanism, perhaps related to chromatin folding or nuclear compartmentalization. Additional insight comes from experiments that suggest that gene regulation can occur by progressively altering the structure of domains of chromatin that are continuous in the chromosome. In the bithorax complex, three genes (Ubx, abd-A and Abd-B) are aligned 5' to 3', both in the order in which they are expressed during development and remarkably in the structures whose origin they control in an anterior to posterior direction in the fly. It has been suggested that these three large (> 20 Kb) segments of DNA are each activated in succession [152]. Several mutations in the bithorax complex are consistent with this hypothesis, since by removing a boundary between two adjacent chromatin domains, the anterior-posterior boundary between different structures is also removed [152]. In this model, boundary elements prevent the propagation of a particular modification of chromatin by functioning as a barrier. The propagation of repressive changes in chromatin or chromosome structure on the bithorax complex is mediated via the Polycomb group of proteins [153]. Once again, studies have provided strong evidence for the existence of chromosomal domain boundaries.

5. Trans-acting factors and the local organization of chromatin structure

The presence of promoter-specific factors or basal transcription complexes prevents nucleosomes from forming on these regions. These protein-DNA interactions contribute to generating nuclease hypersensitive sites in the chromatin fibre. Often such hypersensitive sites are found in the midst of otherwise ordered arrays of nucleosomes. In these genes the nucleosome arrays exist even within the coding region when transcription is repressed; specific regions of the arrays are altered or lost when transcription is activated. Similar arrays can be found in centromeres and telomeres [154-156]; these arrays contain not only hypersensitive sites but regions that are refractory to nuclease cleavage.

Three components clearly contribute to the placement of nucleosomes relative to nuclease hypersensitive or refractory sites. One is the sequence directed positioning of nucleosomes. The second is statistical positioning of nucleosomes [157], which relies on the generation of boundaries to detric the borders of the nucleosome arrays. The third is the directed positioning of nucleosomes that occurs through the interaction of trans-acting factors and nucleosomes [106]. Statistical positioning depends on nucleosomes packing adjacent to a boundary and then being phased to either side of that boundary due to spacing constraints. The influence of the boundary should decay with distance from the boundaries. Kornberg and colleagues examined nucleosome positioning with respect to the GAL1-GAL10 intergenic region that had been inserted into Saccharomyces cerevisiae minichromosomes. Mutation of the DNA sequences that flanked the DNaseI hypersensitive sites left the nucleosomal array unchanged. This indicated that nucleosome positioning was not a consequence of sequence-directed histone-DNA interactions, but instead depended on proximity to the boundary hypersensitive site. A particular promoter-specific DNA binding protein was found to function as a boundary. These experiments were interpreted as representing a passive role of sequence-specific DNA binding proteins in organizing chromatin structure. However more recent experiments suggest that highly selective interactions between histones and sequence-specific DNA binding proteins can also occur.

In Saccharomyces cerevisiae, activation of cell-type specific genes depends on expression of the MAT locus (mating type) [158, 159]. Two of the proteins regulating gene expression, a2 and a1, have helix-turn-helix domains. a2 represses specific gene expression in a cells through cooperative interactions with another promoter-specific protein, MCM1. The a2-MCM1 complex associates with a 32 bp sequence located approximately 100 bp upstream of the TATA sequence of five different a cell-type specific genes. Repression can also be effected from a comparable distance downstream of the TATA box. In the presence of the a2-MCM1 complex, nucleosomes are positioned directly over the TATA box of at least two of these six genes [160]. The placement of nucleosomes in similar positions relative to the TATA box despite very different DNA sequences suggested that the a2-MCM1 repressor complex might direct this nucleaosome positioning. The inclusion of the TATA box into a nucleosome does not independently repress transcription, but the assembly of an extended nucleosomal array within a repressive chromosome domain might account for transcriptional repression [161].

Interactions between nucleosomal histones and the a2/MCM1 complex have been suggested by studies with mutants. Although mutations of the amino terminal tails of histones H2A and H2B have little effect on growth or mating ability, deletions of the amino terminal tail of histone H4 leads to a lengthening of the cell cycle and sterility due to aberrant expression of the silent mating loci [162]. Activation of these mating loci is specific as a consequence of deletion of the histone H4 tail, since repressed genes such as PHO5 are not transcribed under these conditions. Individual point mutations of amino acids 16-19 in the amino terminal tail of histone H4 have effects similar to the deletions [163-165]. One of these point mutations alters a lysine residue that could potentially be acetylated. The effect of the point mutations can be suppressed by mutations in a known regulator of the silent mating loci, SIR3, suggesting direct interaction of the H4 histone tail with trans-acting factors [166]. Aside from the influence of the histone H4 tail on repression of the silent mating type loci, an intact amino terminus is required for the efficient activation of a number of inducible promoters [167].

Simpson and colleagues have extended these primarily genetic-based observations to ask direct questions about how a trans-acting factor, the a2/MCM1 repressor, influences chromatin organization. Reexamining the positioning of nucleosomes adjacent to the a2-MCM1 complex [160], it was found that deletions of the histone H4 amino terminal tail prevented formation of a stably positioned nucleosome next to the a2-MCM1 complex. Individual point mutations in the tail region had a comparable effect. An a-cell specific promoter was derepressed in the mutant strains, suggesting the positioning of nucleosomes directed by a2 is essential for stringent gene regulation. The general transcriptional repressors of yeast TupI and SSN6 are also essential for directing the positioning of nucleosomes next to the a2/MCM1 complex. Mutations in these proteins also both disrupts positioning and activates transcription, further cementing the correlation between specific chromatin structure and gene silencing.

Other examples of sequence specific DNA binding proteins that appear to actively contribute to the assembly of specific chromatin structures include the role of the GAGA factor at the Drosophila hsp70 and hsp26 promoters [135]. On the hsp70 promoter the GAGA factor directs the repositioning of nucleosomes in a process that is dependent on ATP hydrolysis. On the hsp26 promoter, interaction of the GAGA factor with its binding sites in chromatin that had been reconstituted using a Drosophila embryo chromatin assembly extract resulted in a structure closely resembling that seen in vivo. Regulatory elements are kept accessible to factors and a positioned nucleosome is assembled between these sites. The active process indicated by the requirement for ATP hydrolysis has not yet been defined. These remodeling experiments might reflect a role for the Drosophila homologs of the SWI/SNF general activator complex or they might reflect residual association of chromatin assembly proteins eg. CAF1. These molecular chaperones might retain the capacity to cause a dynamic exchange of histone from assembled nucleosomes.

The HNF-3 protein has also been shown to direct the formation of positioned nucleosomes on the serum albumin enhancer [168]. This observation is especially interesting because the structural identity between linker histones and transcription factors such as HNF-3 provides a potential mechanism for how nucleosome positioning might be achieved. Clearly HNF-3 might be able to substitute for the linker histones in the nucleosome and at the same time still recognize DNA specifically.

6. Conclusion

Chromatin is revealed to be highly differentiated with respect to DNA sequence with many contributory factors determining both structure and function. In the past biophysicists and biochemists have often focussed on chromatin structure simply as an engineering problem. The highly conserved histones are proposed to wrap DNA around them to assemble monotonous arrays of nucleosomes, which, in turn, self-associate to form semi-crystalline higher-order structures. This image has proven very convenient for those scientists who require structurally uniform samples in order to interpret their experiments on the physical properties of biological material. However, this homogenous packaging of DNA within the chromosome is not consistent with the experience of those molecular geneticists and cell biologists who investigate chromatin function. These investigators find the organization of DNA within chromatin be highly variant, reflecting the functional requirements at particular promoters. Individual core histones are recognized by a variety of regulatory proteins. These interactions can be targeted by particular DNA sequences and sequence specific DNA binding proteins thereby providing an explanation for the highly selective activation or repression of particular genes following mutation of individual histones. The inclusion of histones as architectural components within regulatory nucleoprotein complexes further strengthens

104

the evidence for their essential role in eukaryotic transcription. Therefore, the reasons for the conservation of the primary sequence of the core histones go beyond merely conservation of the primary sequence of the core histones that merely conserves the internal architecture of the nucleosome and include the functional requirement of retaining interactions with the regulatory proteins that modulate chromatin function. Reconstructing, and thus understanding, the correct regulation of genes clearly requires the assembly of a natural chromosomal context.

7. References

1. Wolffe, A.P. (1995) *Chromatin: structure and function*, 2nd ed.
2. Simpson, R.T. (1991) Nucleosome positioning: occurrence, mechanisms and functional consequences, *Progress in Nucleic Acids Res. and Mol. Biol.* **40**, 143-184.
3. Morse, R.H. (1992) Transcribed chromatin, *Trends in Biochem. Sci.* **17**, 23-26.
4. Georgiev, G.P. (1969) Histones and the control of gene action, *Ann. Rev. Genet.* **3**, 155-180.
5. Clark, R.J. and Felsenfeld, G. (1971) Structure of chromatin, *Nature New Biology* **229**, 101-106.
6. Maryanka, D., Cowling, G.J., Allan, J., Fey, S.J., Huvos, P. and Gould, H. (1979) Transcription of globin genes in reticulocyte chromatin, *FEBS Letts* **105**, 131-136.
7. Pays, E., Donaldson, D. and Gilmour, R.S. (1979) Specificity of chromatin transcription in vitro anomalies due to RNA-dependent RNA synthesis, *Biochim. Biophys. Acta* **562**, 112-130.
8. Cedar, H. and Felsenfeld, G. (1973) Transcription of chromatin in vitro. *J. Mol. Biol.* **77**, 237-254.
9. Williamson, P. and Felsenfeld, G. (1978) Transcription of histone-covered T7 DNA by Escherichia coli RNA polymerase, *Biochemistry* **17**, 5695-5705.
10. Wasylyk, B. and Chambon, P. (1979) Transcription by eukaryotic RNA polymerases A and B of chromatin assembled in vitro. *Eur. J. Biochem.* **98**, 317-327.
11. Hannon, R., Bateman, E., Allan, J., Harborne, N. and Gould, H. (1984) Control of RNA polymerase binding to chromatin by variations in linker histone composition. *J. Mol. Biol.* **180**, 131-149.
12. Hannon, R., Richards, E.G. and Gould, H. (1986) Facilitated diffusion of a DNA binding protein on chromatin. *EMBO J.* **5**, 3313-3319.
13. Berg, O.G., Winter, R.B. and von Hippel, P.H. (1981) Diffusion-driven mechanisms of protein translocation on nucleic acids. *Biochemistry* **20**, 6929-6977.
14. Baer, O.G., B.W. and Rhodes, D. (1983) Eukaryotic RNA polymerase II binds to nucleosome cores from transcribed genes, *Nature (London)* **301**, 482-488.
15. Gonzalez, P.J. and Palacian, E. (1989) Interaction of RNA polymerase II with structurally altered nucleosomal particles, *J. Biol. Chem.* **264**, 18457-18462.
16. Schlissel, M.S. and Brown, D.D. (1984) The transcriptional regulation of Xenopus 5S RNA genes in chromatin : the roles of active stable transcription complexes and histone H1, *Cell* **37**, 903-911.
17. Wolffe, A.P. (1989) Dominant and specific repression of Xenopus oocyte 5S RNA genes and satellite I DNA by histone H1, *EMBO J.* **8**, 527-437.
18. Bouvet, P., Dimitrov, S. and Wolffe, A.P. (1994) Specific regulation of chromosomal 5S rRNA gene transcription in vivo by histone H1. *Genes and Develops.* **8**, 1147-1159.
19. Andrews, M.T., Loo, S. and Wilson, L.R. (1991) Coordinate inactivation of class III genes during the gastrula-neurula transition in Xenopus, *Develop. Biol.* **146**, 250-254.
20. Wolffe, A.P. (1991) Developmental regulation of chromatin structure and function, *Trends in Cell Biol.* **1**, 61-66.
21. Parker, C.S. and Roeder, R.G. (1977) Selective and accurate transcription of the Xenopus laevis 5S RNA genes in isolated chromatin by purified RNA polymerase III, *Proc. Natl. Acad. Sci. USA* **74**, 44-48.
22. Jerzmanowski, A. and Cole, R.D. (1990) Flanking sequences of Xenopus 5S RNA genes determine differential inhibition of transcription by H1 histone in vitro, *J. Biol. Chem.* **265**, 10726-10732.
23. Croston, G.E., Kerrigan, L.A., Liva, L.M., Marshak, D.R. and Kadonaga, J.T. (1991), Sequence-specific antirepression of histone H1-mediated inhibition of basal RNA polymerase II transcription. *Science* **251**, 643-649.
24. Ge, H. and Roeder, R.G. (1994) The high mobility protein HMG1 can reversibly inhibit class II gene transcription by interaction with the TATA-binding protein, *J. Biol. Chem.* **269**, 17136.
25. Bogenhagen, D.F., Wormington, W.M., and Brown, D.D. (1982) Stable transcription complexes of Xenopus 5S RNA genes: a means to maintain the differentiated state, *Cell* **28**, 413-421.
26. Gargiulo, G., Razvi, F., Ruberti, I., Mohr, I. and Worcel, A. (1984) Chromatin-specific hypersensitive sites are assembled on a Xenopus histone gene injected into Xenopus oocytes, *J. Mol. Biol.* **181**, 333-349.
27. Weisbrod, S., Wickens, M.P., Whytock, S. and Gurdon, J.B. (1982) Active chromatin of oocytes injected with somatic cell nuclei or cloned DNA, *Develop. Biol.* **94**, 216-229.

28. Gargiulo, G. and Worcel, A. (1983) Analysis of chromatin assembled in germinal vesicles of Xenopus oocytes, *J. Mol. Biol.* **170**, 699-722.
29. Felts, S.J., Weil, P.A. and Chalkley, R. (1990) Transcription factor requirements for in vitro formation of transcriptionally competent 5S rRNA gene chromatin. *Mol. Cell. Biol.* **10**, 2390-2401.
30. Tremethick, D., Zucker, D. and Worcel, A. (1990) The transcription complex of the 5S RNA gene, but not the transcriptional factor TFIIIA alone, prevents nucleosomal repression of transcription. *J. Biol. Chem.* **265**, 5014-5023.
31. Shimamura, A., Tremethick, D. and Worcel, A. (1988) Characterization of the repressed 5S DNA minichromosomes assembled in vitro with a high-speed supernatant of Xenopus laevis oocytes, *Mol. Cell Biol.* **8**, 4257-4269.
32. Shimamura, A., Sapp, M., Rodriquez-Campos, A. and Worcel, A. (1989) Histone H1 represses transcription from minichromosomes assembled in vitro, *Mol. Cell Biol.* **9**, 5573-5584.
33. Morse, R.H. (1989) Nucleosomes inhibit both transcriptional initiation and elongation by RNA polymerase III in vitro. *EMBO J.* **8**, 2343-2351.
34. Clark, D.J. and Wolffe, A.P. (1991) Superhelical stress and nucleosome mediated repression of 5S RNA gene transcription in vitro, *EMBO J.* **10**, 3419-3428.
35. Paranjape, S.M., Kamakaka, R.T. and Kadonaga, J.T. (1994) Role of chromatin structure in the regulation of transcription by RNA polymerase II, *Annu. Rev. Biochem.* **63**, 265-297.
36. Knezetic, J.A. and Luse, D.S. (1986) The presence of nucleosomes on a DNA template prevents initiation by RNA polymerase II in vitro, *Cell* **45**, 95-104.
37. Knezetic J.A., Jacob, G.A. and Luse, D.S. (1988) Assembly of RNA polymerase II preinitiation complexes before assembly of nucleosomes allows efficient initiation of transcription on nucleosomal templates, *Mol. Cell Biol.* **8**, 3114-3121.
38. Workman, J.L. and Roeder, R.G. (1987) Binding of transcription factor TFIID to the major late promoter during in vitro nucleosome assembly potentiates subsequent initiation by RNA polymerase II, *Cell* **51**, 613-622.
39. Meisterernst, M., Horikoshi, M. and Roeder, R.G. (1990) Recombinant yeast TFIID, a general transcription factor, mediates activation by the gene specific factor USF in a chromatin assembly assay. *Proc. Natl. Acad. Sci. USA* **87**, 9153-9157
40. Workman, J.L., Abmayr, S.M., Cromlish, W.A. and Roeder, R.G. (1988) Transcriptional regulation of the immediate early protein of pseudorabies virus during in vitro nucleosome assembly, *Cell* **55**, 211-219.
41. Workman, J.L., Roeder, R.G. and Kingston, R.E. (1990) An upstream transcription factor, USF (MLTF), facilitates the formation of preinitiation complexes, *EMBO J.* **9**, 1299-1308.
42. Becker, P.B., Rabindran, S.K. and Wu, C. (1991) Heat shock-regulated transcription in vitro from a reconstituted chromatin template, *Proc. Natl. Acad. Sci. USA* **88**, 4109-4113.
43. Workman, J.L., Taylor, I.C.A. and Kingston, R.E. (1991) Activation domains of stably bound GAL4 derivaties alleviate repression of promoters by nucleosomes, *Cell* **64**, 533-544.
44. Workman, J.L. and Kingston, R.E. (1992) Nucleosome core displacement in vitro via a metastable transcription factor/nucleosome complex, *Science* **258**, 1780-1784.
45. Chen, H., Li, B. and Workman, J.L. (1994) A histone-binding protein nucleoplasmin stimulates transcription factor binding to nucleosomes and factor-induced nucleosome disassembly, *EMBO J.* **13**, 380-390.
46. Li, B., Adams, C.C. and Workman, J.L. (1994) Nucleosome binding by the constitutive factor Sp1, *J. Biol. Chem.* **269**, 7756-7763.
47. Wechsler, D.S., Papoulas, O., Dang, C.V. and Kingston, R.E. (1994) Differential binding of c-Myc and Max to nucleosomal DNA, *Mol. Cell Biol.* **14**, 4097-4107.
48. Coté, J., Quinn, J., Workman, J.L. and Peterson, C.L. (1994) Stimulation of GAL4 derivative binding to nucleosomal DNA by the yeast SWI/SNF complex, *Science* **265**, 53-60.
49. Kwon, H., Imbalzano, A.N., Khavarl, P.A., Kingston, R.E. and Green, M.R. (1994) Nucleosome disruption and enhancement of activator binding by a human SWI/SNF complex, *Nature* **370**, 477-481.
50. Imbalzano, A.M., Kwon, H., Green, M.R. and Kingston, R.E. (1994) Facilitated binding of TATA-binding protein to nucleosomal DNA, *Nature* **370**, 481-485.
51. Vettesse-Dadey, M., Walter, P., Chen, H., Juan, L-J. and Workman, J.L. (1994) Role of the histone amino termini in facilitated binding of a transcription factor, GAL4-AH to nucleosome cores, *Mol. Cell. Biol.* **14**, 970-981.
52. Laybourn, P. and Kadonaga, J.T. (1991) Role of nucleosomal cores and histone H1 in the regulation of transcription by RNA polymerase II, *Science* **254**, 238-245.
53. Kamakaka, R.T., Bulger, M. and Kadonaga, J.T. (1993) Potentiation of RNA polymerase II transcription by Gal4-VP16 during but not after DNA replication and chromatin assembly, *Genes and Develop.* **7**, 1779-1795.
54. Hayes, J.J. and A.P. Wolffe. (1993) Preferential and asymmetric interaction of linker histones with 5S DNA in the nucleosome, *Proc. Natl. Acad. Sci. USA* **90**, 6415-6419.
55. Clark, D.J. and Kimura, T. (1990) Electrostatic mechanism of chromatin folding, *J. Mol. Biol.* **211**, 883-896.
56. Hansen, J.C. and Wolffe, A.P. (1992) The influence of chromatin folding on transcription initiation and elongation by RNA polymerase III, *Biochemistry* **31**, 7977-7988.

57. Hansen, J.C. and Wolffe, A.P. (1994) A role for histones H2A/H2B in chromatin folding and transcriptional repression, *Proc. Natl. Acad. Sci. USA* **91**, 2339-2343.
58. Becker, P.B. (1994) The establishment of active promoters in chromatin, *BioEssays* **16**, 541-547.
59. Wallrath, L.L., Lu, Q., Granok, H. and Elgin, S.C.R. (1994) Architectural variations of inducible eukaryotic promoters: present and remodeling chromatin structures, *BioEssays* **16**, 165-170.
60. Wolffe, A.P. (1994) The role of transcription factors, chromatin structure and DNA replication in 5S RNA gene regulation, *J. Cell. Sci.* **107**, 2055-2063.
61. Chao, M.V., Gralla, J.D. and Martinson, H.G. (1980) Lac operator nucleosomes I. Repressor binds specifically to operators within the nucleosome core, *Biochemistry* **19**, 3254-3260.
62. Chao, M.V., Martinson, H.G. and Gralla, J.D. (1980) Lac operator nucleosomes can change conformation to strengthen binding by Lac repressor, *Biochemistry* **19**, 3260-3269.
63. Simpson, R.T. and Stafford, D.W. (1983) Structural features of a phased nucleosome core particle, *Proc. Natl. Acad. Sci. USA* **80**, 51-55.
64. FitzGerald, P.C. and Simpson, R.T. (1985) Effects of sequence alterations in a DNA segment containing the 5S rRNA gene from Lythechinus Yariegatus on positioning of a nucleosome core particle in vitro. *J. Biol. Chem.* **260**, 15318-15324.
65. Rhodes, D. (1985) Structural analysis of a triple complex between the histone octamer, a Xenopus gene for 5S RNA and transcription factor IIIA, *EMBO J.* **4**, 3473-3482.
66. Lee, D.Y., Hayes, J.J., Pruss, D. and Wolffe, A.P. (1993) A positive role for histone acetylation in transcription factor binding to nucleosomal DNA, *Cell* **72**, 73-84.
67. Bauer, W.R., Hayes, J.J., White, J.H. and Wolffe, A.P. (1994) Nucleosome structural changes due to acetylation, *J. Mol. Biol.* **236**, 685-690.
68. Svaren, J. and Horz, W. (1993) Histones, nucleosomes and transcription, *Curr. Opin. Genetics and Develop.* **3**, 219-225.
69. Schmid, A., Fascher, K.D, and Horz, W. (1992) Nucleosome disruption at the PHO5 promoter upon PHO5 induction occurs in the absence of DNA replication, *Cell* **71**, 853-864.
70. Fascher, K.D., Schmitz, J., and Horz, W. (1993) Structural and functional requirements for the chromatin transition at the PHO5 promoter in Saccharomyces cerevisiae upon PHO5 activation, *J. Mol. Biol.* **231**, 658-667.
71. Kaffman, A., Herskowitz, I., Tjian, R. and O'Shea, E.K. (1994) Phosphorylation of the transcription factor PHO4 by a cyclin - CDK complex, PHO80-PHO85. *Science* **263**, 1153-1156.
72. Venter, U., Svaren, J., Schmitz, J., Schmid, A. and Horz, W. (1994) A nucleosome precludes binding of the transcription factor Pho4 in vivo to a critical target site in the PHO5 activation, *J. Mol. Biol.* **231**, 658-667.
73. Svaren, J., Schmitz, J. and Horz, W. (1994) The transactivation domain of Pho4 is required for nucleosome disruption at the PHO5 promoter, *EMBO J.* **13**, 4856-4862.
74. Almer, A. and Horz, W. (1986) Nuclease hypersensitive regions with adjacent positioned nucleosomes mark the gene boundaries of the PHO5/PHO3 locus in yeast, *EMBO J.* **5**, 2681-2687.
75. Almer, A., Rudolph, H., Hinnen, A., Horz, W. (1986) Removal of positioned nucleosomes from the yeast PHO5 promoter upon PHO5 induction releases additional activating DNA elements, *EMBO J.* **5**, 2689-2696.
76. Fascher, K.D., Schmitz, J. and Horz, W. (1990) Role of trans-activating proteins in the generation of active chromatin at the PHO 5 promoter in S. cerevisiae, *EMBO J.* **9**, 2523-2528.
77. Straka, C. and Horz, W. (1991) A functional role for nucleosomes in the repression of a yeast promoter, *EMBO J.* **10**, 361-368.
78. Pavlovic, B. and Horz, W. (1988) The chromatin structure at the promoter of a glyceraldehyde phosphate dehydrogenasce gene from Saccharomyces cerevisiae reflects its functional state, *Mol. Cell. Biol.* **8**, 5513-5520.
79. Solomon, M.J., Larsen, P.L. and Varsharsky, A. (1988) Mapping protein-DNA interactions in vivo with formaldehyde: evidence that histone H4 is retained on a highly transcribed gene, *Cell* **53**, 937-947.
80. Nacheva, G.A., Guschin, D.Y., Preobrazhenskaya, O.V., Karpov, V.L., Elbradise, K.K. and Mirzabekov, A.D. (1989) Change in the pattern of histone binding to DNA upon transcriptional activation, *Cell* **58**, 27-36.
81. Han, M., Kim, U.-J., Kayne, P. and Grunstein, M. (1988) Depletion of histone H4 and nucleosomes activates the PHO5 gene in Saccharomyces cerevisiae, *EMBO J.* **7**, 2221-2228.
82. Han, M. and Grunstein, M. (1988) Nucleosome loss activates yeast downstream promoters in vivo, *Cell* **55**, 1137-1145.
83. Cordingley, M.G., Riegel, A.T. and Hager, G.L. (1987) Steroid dependent interaction of transcription factors with the inducible promoter of mouse mammary tumor virus in vivo, *Cell* **48**, 261-270.
84. Zaret, K.S. and Yamamoto, K.R. (1984) Reversible and persistent changes in chromatin structure accompany activation of a glucocorticoid dependent enhancer element, *Cell* **38**, 29-38.
85. Archer, T.K., Lefebvre, P., Wolford, R.G. and Hager, G.L. (1992) Transcription factor loading on the MMTV promoter: a bimodal mechanism for promoter activation, *Science* **255**, 1573-1576.

86. Archer, T.K., Cordingley, M.G., Marsaud, V., Richard-Foy, H. and Hager, G.L. (1989) Steroid transactivation at a promoter organized in a specifically positioned array of nucleosomes. *In Proceedings: Second International CBT Symposium on the Steroid/Thyroid Receptor Family and Gene Regulation* pp. 221-238.

87. Pina, B., Bruggemeier, U. and Beato, M. (1990) Nucleosome positioning modulates accessibility of regulatory proteins to the mouse mammary tumor virus promoter, *Cell* **60**, 719-731.

88. Perlmann, T. and Wrange, O. (1988) Specific glucocorticoid receptor binding to DNA reconstituted in a nucleosome, *EMBO J.* **7**, 3073-3083.

89. Archer, T.K., Cordingley, M.G., Wolford, R.G. and Hager, G.L. (1991) Transcription factor access is mediated by accurately positioned nucleosomes on the mouse mammary tumor virus promoter, *Mol. Cell. Biol.* **11**, 688-698.

90. Pina, B., Barettino, D., Truss, M. and Beato, M. (1990) Structural features of a regulatory nucleosome, *J. Mol. Biol.* **216**, 975-990.

91. Perlmann, T. and Wrange, O. (1991) Inhibition of chromatin assembly in Xenopus oocytes correlates with derepression of the mouse mammary tumor virus promoter, *Mol. Cell Biol.* **11**, 5259-5265.

92. Li, Q. and Wrange, O. (1993) Translational positioning of a nucleosomal glucocorticoid response element modulates glucocorticoid receptor affinity, *Genes and Dev.* **7**, 2471-2482.

93. Luisi, B.F., Xu, W.X., Otwinowski, Z., Freedman, L.P., Yamamoto, K.R. and Sigler, P. (1991) Crystallographic analysis of the interaction of the glucocorticoid receptor with DNA, *Nature* **352**, 497-502.

94. Pennings, S., Meersseman, G. and Bradbury, E.M. (1991) Mobility of positioned nucleosomes on 5S rDNA, *J. Mol. Biol.* **220**, 101-110.

95. Chipev, C.C. and Wolffe, A.P. (1992) The chromosomal organization of Xenopus laevis 5S ribosomal RNA genes in vivo, *Mol. Cell. Biol.* **12**, 45-55.

96. Ura, K., Wolffe, A.P. and Hayes, J.J. (1994) Core histone acetylation does not block linker histone binding to a nucleosome including a Xenopus borealis 5S rRNA gene, *J. Biol. Chem.* **269**, 27171-27174.

97. Yoshinaga, S.K., Peterson, S.L., Herskowitz, I. and Yamamoto, K.R. (1992) Roles of SWI1, SWI2 and SWI3 proteins for transcriptional enhancement by steroid receptors, *Science* **258**, 1598-1604.

98. Muchardt, C. and Yaniv, M. (1993) A human homolog of Saccharomyces cerevisiae SNF2/SWI2 and Drosophila brm genes potentiates transcriptional activation by the glucocorticoid receptor, *EMBO J.* **12**, 4279-4290.

99. Bresnick, E.H., Bustin, M., Marsaud, V., Richard-Foy, H. and Hager, G.L. (1992) The transcriptionally-active MMTV promoter is depleted of histone H1, *Nucl. Acids Res.* **20**, 273-278.

100. Lee, H.H. and Archer, T.K. (1994) Nucleosome-mediated disruption of transcription factor-chromatin initiation complexes at the mouse mammary tumor virus long terminal repeat in vivo, *Mol. Cell. Biol.* **14**, 32-41.

101. Wolffe, A.P. and Drew, H.R. (1989) Initiation of transcription on nucleosomal templates, *Proc. Natl. Acad. Sci. USA* **86**, 9817-9821.

102. Simpson, R.T. (1990) Nucleosome positioning can affect the function of a cis-acting DNA element in vivo, *Nature (London)* **343**, 387-389.

103. Simpson, R.T., Roth, S.Y., Morse, R.H., Patteron, H.G., Cooper, J.P., Murphy, M., Kladde, M.P. and Shimizu, M. (1993) Nucleosome positioning and transcription, *Cold Spring Harbor Symp. Quant. Biol.* **58**, 237-245.

104. Morse, R.H., Roth, S.Y. and Simpson, R.T. (1992) A transcriptionally active tRNA gene interferes with nuclwosome positioning in vivo, *Mol. Cell. Biol.* **12**, 4015-1025.

105. Morse, R.H. (1993) Nucleosome disruption by transcription factor binding in yeast, *Science* **262**, 1563-1566.

106. Roth, S.Y., Dean, A. and Simpson, R.T. (1990) Yeast a2 repressor positions nucleosomes in TRP1/ARS1 chromatin, *Mol. Cell. Biol.* **10**, 2247-2260.

107. Roth, S.Y., Shimizu, M., Johnson, L., Grunstein, M. and Simpson, R.T. (1992) Stable nucleosome positioning and complete repression by the yeast a2 repressor are disrupted by amino-terminal mutations in histones H4, *Genes and Development* **6**, 411-425.

108. Cooper, J.P., Roth, S.Y. and Simpson, R.T. (1994) The global transcriptional regulators SSN6 and Tup1, play distinct roles in the establishment of a repressive chromatin structure, *Genes Dev.* **8**, 1400-1410.

109. Patterton, H.G. and Simpson, R.T. (1994) Nucleosomal location of the STE6 TATA box and Mata2p-mediated repression, *Mol. Cell Biol.* **14**, 4002-4010.

110. Pederson, D.S. and Fidrych, T. (1994) Heat shock factor can activate transcription while bound to nucleosomal DNA in Saccharomyces cerevisiae, *Mol. Cell Biol.* **14**, 189-199.

111. Gross, D.S., Adams, C.C., Lee, S. and Stentz, B. (1993) A critical role for heat shock transcription factor in establishing a nucleosome free region over the TATA-initiation site of the yeast HSP82 heat shock gene, *EMBO J.* **12**, 3931-3945.

112. Weintraub, H. and Groudine, M. (1976) Chromosomal subunits in active genes have an altered conformation, *Science* **193**, 848-856.

113. Wood, W.I. and Felsenfeld, G. (1982) Chromatin structure of the chicken ß-globin gene region: sensitivity to DNaseI, micrococcal nuclease, and DNaseII, *J. Biol. Chem.* **257**, 7730-7736.

108

114. Felsenfeld, G. (1978) Chromatin, *Nature* **271**, 115-122.
115. Lau, L.F. and Nathans, D. (1987) Expression of a set of growth-related immediate early genes in BALBc/3T3 cells: coordinate regulation with c-fos and c-myc, *Proc. Natl. Acad. Sci. USA* **84**, 1182-1189.
116. Chen, T.A. and Allfrey, V.G. (1987) Rapid and reversible changes in nucleosome structure accompany the activation, repression and superinduction of the murine proto-oncogenes c-fos and c-myc, *Proc. Natl. Acad. Sci. USA* **84**, 5252-5256.
117. Chen, T.A., Sterner, R., Cozzolino, A. and Allfrey, V.G. (1990) Reversible and irreversible changes in nucleosome structure along the c-fos and c-myc oncogenes following inhibition of transcription, *J. Mol. Biol.* **212**, 481-493.
118. Feng, J. and Villeponteau, B. (1990) Serum stimulation of the c-fos enhancer induces reversible changes in c-fos chromatin structure, *Mol. Cell. Biol.* **10**, 1126-1132.
119. Walker, J., Chen, T.A., Sterner, R., Berger, M., Winston, F. and Allfrey, V.G. (1990) Affinity chromatography of mammalian and yeast nucleosomes: two modes of binding of transcriptionally active mammation nucleosomes to organomercucial columns and contrasting behaviour of the active nucleosomes of yeast, *J. Biol. Chem.* **265**, 5736-5746.
120. Mahadevan, L.C., Willis, A.C. and Barrah, M.J. (1991) Rapid histone H3 phosphorylation in response to growth factors, phorbol esters, okadaic acid and protein synthesis inhibitors, *Cell* **65**, 775-783.
121. Herrera, R.E., Shaw, P.E. and Nordheim, A. (1989) Occupation of the c-fos serum response element <u>in</u> <u>vivo</u> by a multi-protein complex is unaltered by growth factor induction, *Nature (London)* **340**, 68-70.
122. Wu, C., Binham, P.M., Livak, K.J., Holmgren, R. and Elgin, S.C.R. (1979) The chromatin structure of specific genes: evidence for higher order domains of defined DNA sequence, *Cell* **16**, 797-806.
123. Wu, C. and Gilbert, W. (1981) Tissue-specific exposure of chromatin structure at the 5' terminus of the rat prepro insulin II gene, *Proc. Natl. Acad. Sci. USA* **78**, 1577-1580.
124. McGhee, J.D., Wood, W.I., Dolan, M., Engel, J.D. and Felsenfeld, G. (1981) A 200 base pair region at the 5' end of the chicken adult ß-globin gene is accessible to nuclease digestion, *Cell* **27**, 45-55.
125. Burch, J.B.E. and Weintraub, H. (1983) Temporal order of chromatin structural changes associated with activation of the major chicken vitellogenin gene, *Cell* **33**, 65-76.
126. Elgin, S.C.R. (1988) The formation and function of DNase I hypersensitive sites in the process of gene activation, *J. Biol. Chem.* **263**, 19259-19262.
127. Emerson, B.M., Lewis, C.D. and Felsenfeld, G. (1985) Interaction of specific nuclear factors with the nuclease-hypersensitive region of the chicken adult ß-globin gene: nature of the binding domain, *Cell* **41**, 21-30.
128. Gross, D.S. and Garrard, W.T. (1988) Nuclease hypersensitive sites in chromatin, *Ann. Rev. Biochem.* **57**, 159-197.
129. Weintraub, H., Beug, H., Groudine, M. and Graf, T (1982) Temperature sensitive changes in the structure of globin chromatin in lines of red cell precursors transformed by ts-AEV, *Cell* **28**, 931-940.
130. Reitman, M. and Felsenfeld, G. (1990) Developmental regulation of topoisomerase II sites and DNase I-hypersensitive sites in the chicken ß-globin locus, *Mol. Cell Biol.* **10**, 2774-2786.
131. Lu, Q., Wallrath, L.L., Allan, B.D., Glaser, R.L., Lis, J.T. and Elgin, S.C.R. (1992) Promoter Sequence Containing (CT)n (GA)n Repeats is Critical for the Formation of the DNaseI Hypersensitive Sites in the <u>Drosophila hsp26</u> Gene, *J. Mol. Biol.* **225**, 985-998.
132. Lu, Q., Wallrath, L.L., Granok, H. and Elgin, S.C.R. (1993) (CT)n (GA)n Repeats and Heat Shock Elements Have Distinct Roles in Chromatin Structure and Transcriptional Activation of the <u>Drosophila</u> <u>hsp26</u> Gene, *Mol. Cell. Biol.* **13**, 2802-2814.
133. Kerrigan, L.A., Croston, G.E., Lira, L.M. and Kadonaga, J.T. (1991) Sequence-specific antirepression of the <u>Drosophila</u> Kruppel gene by the GAGA factor, *J. Biol. Chem.* **266**, 574-582.
134. Fedor, M.J., Lue, N.F. and Kornberg, R.D. (1988) Statistical positioning of nucleosomes by specific protein-binding to an upstream activating sequence in yeast, *J. Mol. Biol.* **204**, 109-127.
135. Tsukiyama, T., Becker, P.B. and Wu, C. (1994) ATP-dependent nucleosome disruption at a heat-shock promoter mediated by binding of GAGA transcription factor, *Nature* **367**, 525-532.
136. Thomas, G.H. and Elgin, S.C.R. (1988) Protein/DNA architecture of the DNase I hypersensitive region of the <u>Drosophila</u> hsp26 promoter, *EMBO J.* **7**, 2191-2201.
137. Gilmour, D.S. and Lis, J.T. (1986) RNA polymerase II interacts with the promoter region of the non-induced hsp 70 gene in <u>Drosophila melanogaster</u> cells, *Mol. Cell. Biol.* **6**, 3984-3989.
138. Rougvie, A.E. and Lis, J.T. (1988) The RNA polymerase II molecule at the 5' end of the uninduced hsp 70 gene of D. melanogaster is transcriptionally engaged, *Cell* **54**, 795-804.
139. Schild, C., Claret, F-X., Wahli, W. and Wolffe, A.P. (1993) A nucleosome-dependent static loop potentiates estrogen-regulated transcription from the <u>Xenopus</u> vitellogenin B1 promoter <u>in vitro</u>, *EMBO J.* **12**, 423-433.
140. Jackson, J.R. and Benyajati, C. (1993) Histone interactions are sufficient to position a single nucleosome juxtaposing <u>Drosophila Adh</u> adult enhancer and distal promoter, *Nucl. Acids Res.* **21**, 957-967.
141. Grosveld, F., van Assendelft, G.B., Greaves, D.R. and Kollias, G. (1987) Position independent, high level expression of the human ß-globin gene in transgenic mice, *Cell* **51**, 975-985.

142. Forrester, W., Takagawa, S., Papayannopoulou, T., Stamatoyannopoulos, G., Groudine, M. (1987) Evidence for a locus activation region: The formation of developmentally stable hypersensitive sites in globin-expressing hybrids, *Nucl. Acids. Res.* **15**, 10159-10177.
143. Tuan, D., Solomon, W., Li, Q. and London, I.M. (1985) The "ß-like-globin" gene domain in human erythroid cells, *Proc. Natl. Acad. Sci. USA* **82**, 6384-6388.
144. Ryan, T.M., Rehringer, R.R., Martin, N.C., Townes, T.M. , Palmiter, R.D. and Brinster, R.L. (1989) A single erythroid-specific DNase I super-hypersensitive site activates high levels of human ß-globin gene expression in transgenic mice, *Genes Develop.* **3**, 314-323.
145. Talbot, D., Philipsen, S., Fraser, P. and Grosveld, F. (1990) Detailed analysis of the site 3 region of the human ß-globin dominant control region, *EMBO J.* **9**, 2169-2178.
146. Philipsen, S., Talbot, O., Fraser, P. and Grosveld, F. (1990) The ß-globin dominant control region: hypersensitive site, *EMBO J.* **9**, 2159-2168.
147. Talbot, D. and Grosveld, F. (1991) The 5' HS 2 of the globin locus control region enhances transcription through the interaction of a multimeric complex binding at two functionally distinct NF-E2 binding sites, *EMBO J.* **10**, 1391-1398.
148. Reitman, M., Lee, E., Westphal, H. and Felsenfeld, G. (1993) An enhancer/locus control region is not sufficient to open chromatin, *Mol. Cell. Biol.* **13**, 3990-3998.
149. Jenuwein, T., Forrester, W.C., Qui, R.-G. and Grosschedl, R. (1993) The immunoglobulin μ enhancer core establishes local factor access in nuclear chromatin independent of transcriptional stimulation, *Genes Dev.* **7**, 2016-2032.
150. Kellum, R. and Schedl, P. (1991) A position effect assay for boundaries of higher order chromosomal domains, *Cell* **64**, 941-950.
151. Geyer, P.K. and Corces, V.G. (1992) DNA Position-Effect Repression of Transcription by a <u>Drosophila</u> Zinc-Finger Protein, *Genes and Dev.* **6**, 1865-1873.
152. Moehrle, A. and Paro, R. (1994) Spreading the silence: epigenetic transcriptional regulation during <u>Drosophila</u> development. *Dev. Gent.* **15**, 478-484.
153. Orlando, V. and Paro, R. (1993) Mapping Polycome-repressed domains in the bithorax complex using in vivo formaldehyde cross-linked chromatin, *Cell* **75**, 1187-1198.
154. Bloom, K.S. and Carbon, J. (1982) Yeast centromere DNA is a unique and highly ordered structure in chromosomes and small circular minichromosomes, *Cell* **29**, 305-317.
155. Gottschling, D.E. and Cech, T.R. (1984) Chromatin structure of the molecular ends of Oxytricha macronuclear DNA: phased nucleosomes and a telomeric complex, *Cell* **38**, 501-510.
156. Budarf, M.L. and Blackburn, E.H. (1986) Chromatin structure of the telomeric region and 3'-nontranscribed spacer of <u>Tetrahymena</u> ribosomal RNA genes, *J. Biol. Chem.* **261**, 363-369.
157. Kornberg, R.D. (1981). The location of nucleosomes in chromatin: specific or statistical? *Nature (London)* **292**, 579-580.
158. Dranginis, A.M. (1986) Regulation of cell type in yeast by the mating type locus, *Trends in Biochem. Sci.* **11**, 328-331.
159. Herskowitz, I. (1989) A regulatory hierarchy for cell specialization in yeast, *Nature (London)* **342**, 749-757.
160. Shimizu, M., Roth, S.Y., Szent-Gyorgi, C. and Simpson, R.T. (1991) Nucleosome are positioned with base pair precision adjacent to the a2 operator in <u>Saccharomyces cerevisiae</u>, *EMBO J.* **10**, 3033-3041.
161. Simpson, R.T., Roth, S.Y., Morse, R.H., Patterton, H.G., Cooper, J.P., Murphy, M., Kladde, M.P.and Shimizu, M. (1993) Nucleosome positioning and transcription, *Cold Spring Harbor Symp. Quant. Biol.* **58**, 237-245.
162. Kayne, P.S., Kim, U.-J., Han, M., Mullen, J.R., Yoshizaki, F. and Grunstein, M. (1988) Extremely conserved histone H4 N terminus is dispensable for growth but essential for repressing the silent mating loci in yeast, *Cell* **55**, 27-39.
163. Johnson, L.M., Kayne, P.S., Kahn, E.S. and Grunstein, M. (1990) Genetic evidence for an interaction between SIR3 and histone H4 in the repression of silent mating loci in <u>Saccharomyces cerevisiae</u>, *Proc. Natl. Acad. Sci. USA* **87**, 6286-6290.
164. Megee, P.C., Morgan, B.A., Mittman, B.A. and Smith, M.M. (1990) Genetic analysis of histone H4: essential role of lysines subject to acetylation, *Science* **247**, 4932-4934.
165. Park, E.C. and Szostak, J.W. (1990) Point mutations in the yeast histone H4 gene prevent silencing of the silent mating type locus HML, *Mol. Cell. Biol.* **10**, 4932-4934.
166. Hecht, A., Laroche, T., strahl-Baslinger, S., Gasser, S.M. and Grunstein, M. (1995) Histone H3 and H4 N-termini interact with SIR 3 and SIR 4 proteins: a molecular model for the formation of heterochromatin in yeast. *Cell* **80**, 583-592.
167. Durrin, L.K., Mann, R.K., Kayne, P.S. and Grunstein, M. (1991) Yeast histone H4 N-terminal sequence is required for promoter activation <u>in vivo</u>, *Cell* **65**, 1023-1031.
168. McPherson, C.E., Shim, E.Y., Friedman, D.S. and Zaret, K.S. (1993) An active tissue-specific enhancer and bound transcription factors existing in a precisely positioned nucleosomal array, *Cell* **75**, 387-398.

NUCLEOSOME AND CHROMATIN STRUCTURES AND FUNCTIONS

Sari PENNINGS
Department of Biochemistry
Hugh Robson Building
George Square
University of Edinburgh
Edinburgh EH8 9XD
Scotland, U.K.

E. Morton BRADBURY
Los Alamos National Laboratory
MS M888
Los Alamos, New Mexico 87545
Department of Biological Chemistry
School of Medicine
UC-Davis
Davis, California 95616

The diploid human genome contains 6×10^9 bp of DNA of total length 2.04 m packaged into cell nuclei 6-8 μm in diameter. Despite decades of intensive research we are still far from understanding the rules that govern the packaging of these enormously long eukaryotic DNA molecules into chromosomes and cell nuclei. Some answers will come from the sequence data generated by the Human Genome Project, particularly the identification of sequence motifs involved in both the long range organization of chromosomes and in nuclear architecture. Such sequence motifs probably bind to proteins in the chromosomal scaffold, the nuclear matrix and nuclear membrane. How chromosome organization and nuclear architecture are involved in chromosome function is not well understood. In an attractive working model for long-range order in metaphase chromosomes the DNA is constrained by scaffold proteins into loops of average size 50 kbps [1]. These loops are packaged by the histones H1, H2A, H2B, H3 and H4 into nucleosomes and higher order chromatin structures.

1. Histones

Histones H3 and H4 are among the most rigidly conserved proteins in nature which implies that every residue in these proteins is essential for their functions. Histones H2A and H2B are more variable and each comprises a family of proteins. The syntheses of some members of the H2A and H2B families are cell cycle dependent e.g., H2A1, H2A2 and others are not, e.g., H2AX, H2AZ. The very lysine rich histones are the most variable of the histones and also comprise a protein family, some members of

C. Nicolini (ed.), Genome Structure and Function, 111–126.
© 1997 *Kluwer Academic Publishers. Printed in the Netherlands.*

which are cell specific. These different histone subtypes provide considerable potential for variability in nucleosome structures and functions.

Histones are the major structural proteins found in chromosomes. The highly conserved histones H3 and H4 are involved in essential interactions in generating the structural framework of the nucleosome which is then completed by the binding of the more variable histones H2A and H2B and the most variable very lysine rich H1 histones [see 2,3]. Histones are multidomain proteins [3]: i) each of histones H3 and H4 has a flexible basic N-terminal domain and a structural apolar central and C-terminal domain which is involved in interactions between H3 and H4 as shown by nuclear magnetic resonance (NMR) spectroscopy [4] and x-ray crystallography [5]; ii) histones H2A and H2B have flexible basic N-terminal domains and C-terminal tails [6]. Their conserved apolar central domains are structured and involved in interactions between H2A and H2B [5,6]; and iii) all H1 subtypes and H1° and H5 have three well-defined domains; a flexible basic N-terminal domain, a central globular domain and a flexible basic C-terminal half of the molecule [7-9]. Histones have been shown to form specific complexes [see 2]. These are the (H2A, H2B) dimer, the $(H3_2, H4_2)$ tetramer and the histone octamer $[(H2A, H2B)_2 (H3, H4_2)]$ that forms the protein core of the nucleosome. The nucleosome is completed by the binding of the fifth histone H1.

1.1 HISTONE MODIFICATIONS

Histones are subjected to reversible chemical modifications that change the chemical nature of the modified residues; acetylations of lysines in the N-terminal domains of H2A, H2B, H3 and H4 and phosphorylations of serines and threonines in the basic N and C-terminal domains of H1, H3 and H2A [see 3, 10-12]. In addition H2A and H2B are modified by the reversible covalent attachment of ubiquitin to lysines in the C-terminal tails of H2A and H2B [see 3,13]. Acetylations and ubiquitinations modify only about 5% of the core histones and thus can affect only small subcomponents of chromatin. In contrast, all of the H1 and H3 protein molecules are phosphorylated at metaphase and these phosphorylations appear to be required for general chromosome functions at mitosis.

The acetylations of the core histones have been associated with chromatin replication [10-12], transcriptionally active and potentially active genes [see 3,10,11] and the replacement of histones by protamines during spermiogenesis [14], i.e., all aspects of DNA processing. Ubiquitinations of histones have been associated with potentially active chromatin [13,15] and with nucleosomes containing heat shock genes in the non-induced state [16]. Ubiquitinated H2A is absent in metaphase chromosomes [17] and we have shown that uH2A and uH2B are deubiquitinated shortly before metaphase and are reubiquitinated in anaphase leading to the suggestion that ubiquitin labels an important subset of genes, and has to be removed prior to metaphase to allow the correct packaging of nucleosomes into metaphase chromosomes [18]. The phosphorylations of the very lysine rich histones have been strongly implicated in the initiation and control of chromosome condensation [3, 19-21], a process which also requires, as a later event, the phosphorylation of histone H3 [21]. In relating chromatin structure and function it is significant that all of the reversible chemical modifications are located in the basic, flexible N and C-terminal domains of the histones [see 3]. Reversible histone modifications most probably provide the mechanisms for modulating chromatin structure in response to cell functions. However, the effects of reversible chemical modifications on chromatin structure, stability and accessibility are largely unknown. Also unknown are the locations of the basic flexible N- and C-terminal regions of histones in nucleosomes and higher order chromatin structures.

2. Nucleosome Structure

Since its discovery in 1973, the nucleosome has been the focus of studies directed towards an understanding of chromatin structure and function [see 2]. For most somatic cells, the nucleosome contains 195 ± 10 bp DNA, the histone octamer and histone H1. Nucleosomes from some specialized cells contain different DNA repeats; chicken erythrocytes (212 bp), sea urchin sperm 241 bp, rabbit neuronal cells 165 bp [see 22-24]. The ends of the DNA can be further trimmed by micrococcal nuclease digestion to give well-defined sub-nucleosome particles. These are the chromatosome with 168 bp DNA and the full histone complement [25] and the core particle with 146 bp DNA and the histone octamer [see 3]. The well-defined core particle has been subjected to intensive structural studies. Neutron scatter studies of these particles in aqueous solution proved that DNA was coiled around the histone octamer core [26-31]. From the neutron scatter curves and pair distance distribution functions, it could be deduced that in solution the core particle was a disc 11.0 nm in diameter by 5.5 to 6.0 nm thick with 1.7 ± 0.2 turns of DNA of mean radius 4.5 nm coiled with a pitch of 3.0 nm on the outside of the particle [27-31]. Low resolution x-ray (2.0 nm) and neutron (1.6 nm) diffraction of core particle crystals gave a model of a wedge shaped disc 11.0 x 5.7 nm with 1.8 turns of DNA of mean radius 4.4 nm and pitch of 2.8 nm [32-35] showing that at low resolution the solution and crystal structures are very similar. The resolution of the core particle crystal structure was extended by x-ray diffraction to 0.7 nm [30]. In the 0.7 nm structure the DNA is not uniformly bent around the octamer but follows a more irregular path with bends. The calculated radius of gyration, Rg, for the observed histone octamer electron density in the core particle crystal structure of 2.97 nm is substantially lower than that determined by neutron scatter contrast matching of 3.3 nm [26,28-31]. However, not all of the electron density is accounted for in the core particle crystal structure indicating the presence of disordered regions in the histones. This difference between the neutron scatter octamer Rg of 3.3 nm and the calculated Rg of 2.97 nm for the observed histone electron density in the crystal structure, has been attributed to disorder in the N- and C-terminal flexible, basic domains of the histones in the core particle [31]. Further support for this proposal comes from our findings [37] that controlled proteolytic removal of the N- and C-terminal domains from the histone octamer reduces its Rg from 3.35 nm to 2.98 nm. More recently the crystal structure of the histone octamer has been solved to 0.33 nm resolution [5]. This shows the modes of interactions between the structured, apolar, central regions of the core histones in the $(H2A,H2B)$ dimer, $(H3_2,H4_2)$ tetramer and the histone octamer. However, no electron density was observed for the flexible N- and C-terminal domains of the core histones [5] most probably because of static or dynamic disorder of these regions. These disordered regions correspond exactly with the disordered, mobile regions identified by NMR spectroscopic studies of the $(H2A,H2B)$ dimer and the $(H3_2,H4_2)$ tetramer [4,6]. An NMR study of the histone octamer and trypsin-trimmed histone octamer showed clearly that the N- and C-terminal tails are mobile [38]. These regions correspond exactly with the disordered regions in the 0.33 nm solution crystal structure of the histone octamer [5]. An understanding of chromatin structure/function relationships will require details of the molecular interaction of these flexible basic N-terminal domains and C-terminal tails in chromatin and the effects of the reversible chemical modifications on these interactions. Previously, we have shown that histone hyperacetylation has little effect on the shape of the 146 bp core particle [39]. Based on these observations it was predicted that histone hyperacetylation would exert its effects on the DNA entering and leaving the nucleosome i.e., on the DNA regions outside of the core particle 146 bp DNA. Recent x-ray scatter studies of fully defined 195 bp acetylated nucleosome particles support this prediction [manuscript in preparation]. Understandings of the modes of interactions and the functions of the flexible basic N-terminal domains of the core histones and the C-terminal tails of histones H2A and H2B are central to understanding nucleosome structures and function. Recently, S.I. Usachenko in our group has mapped the histone DNA contacts in nucleosome core particles from which the C- and N-terminal regions of histone H2A were selectively trimmed by trypsin or clostripain [40]. It was found that the flexible trypsin sensitive C-terminal "tail" of H2A contacted the DNA at the dyad axis, whereas its globular domain contacts the end of the 146 bp DNA in the core particle. The appearance of the H2A contact at the dyad axis was found only in the absence of linker DNA and did not depend on the absence of linker histones. In the absence of linker histones no contact of H2A with the DNA at the dyad axis was observed. It was presumed under these conditions in native H1 depleted chromatin that the C-terminal "tail" bound to the linker DNAs entering and leaving the nucleosome. These results demonstrate the ability of the histone H2A C-terminal tails to rearrange. This rearrangement may play a role in nucleosome disassembly and reassembly and the retention of the H2A/H2B dimer or octamer during the passage of polymerases through the nucleosome.

The current models for the chromatosome and the nucleosome are based on the crystal and solution structures of the 146 bp core particle. The model for the chromatosome contains two full turns of DNA that are coiled around the histone octamer and complexed with the fifth histone H1 [24,41] most probably through the binding of the H1 globular domain [42,43]. The nucleosome model contains in addition the linker DNAs which join adjacent chromatosomes. The paths of these linker DNAs in the different order of chromatin structures are not known. Major outstanding questions are: i) the locations and modes of binding of the N-terminal domains which through reversible chemical modification are involved in chromatin functions; ii) the mode(s) of binding of the very lysine rich histones, and iii) the DNA paths entering and leaving the nucleosome.

3. Chromatin Structure

Chromatin is made up of repeating subunits, the nucleosome, joined by the continuity of the DNA molecule. At low ionic strength, chromatin is in an extended form first described as the 10 nm fibril. Neutron scatter studies of extended chromatin gave a mass per unit length equivalent to one nucleosome/10 ± 2 nm i.e., a DNA packing ratio of about 6 to 7 [44]. The measured cross-section radii of gyration of the DNA and histone components required an arrangement of flat discs with their faces roughly parallel to the axis of the fibril, i.e., edge-to-edge. An edge-to-edge zig-zag arrangement of nucleosome discs has also been proposed from E.M. studies [41].

On increasing the ionic strength, the extended form of chromatin undergoes a transition to the 30 nm fibril. Neutron scatter studies of this transition suggest a family of supercoils undergoing increasing compaction with increasing ionic strength. In its most compact form, the hydrated supercoil has a mass per unit length equivalent to 6-7 nucleosomes per turn of a coil of pitch 11.0 nm and outer diameter of 34 nm [44]. It is described by the previously proposed supercoil [45] or solenoid [46] of nucleosomes. In the fiber diffraction pattern of chromatin, the small number of diffuse diffraction features and their orientation are explained at low resolution by this supercoil of nucleosomes [45,47]. This supercoil or solenoid is a one start helix. An alternative proposal is based on the E.M. observations of an intermediate unfolded state of chromatin in which adjacent nucleosomes form a close-packed zig-zag [48,49]. Recent cryo E.M. studies of oligonucleosomes support an irregular zig-zag conformation of chromatin in solution [50]. In one model this zig-zag ribbon is coiled into a supercoil, i.e., a two-start helix [49]. Additional to being a one-start or a two-start helix these models differ markedly in the locations of histone H1 and linker DNA. In the former model, linker DNA and H1 are located in the hole along the axis of the supercoil of nucleosomes [46,47] whereas in the latter model H1 and linker DNA are located between adjacent nucleosomes in the coiled ribbon, i.e., in the "wall" of the nucleosome coil. A third type of model [51] is a two-start left-handed helix with linker DNA crossing from one side of the solenoid to the opposite side, similar to an earlier proposal [52]. Neutron scatter studies of long chromatin reconstituted with deuterated H1 have shown that the bulk of H1 is located in the hole along the axis of the 34 nm supercoil of nucleosomes [53]. This provides strong support for the supercoil [45] or solenoid [46,47] models for the "30 nm" filaments. More recently atomic probe microscopy studies of chromatin structure by van Holde and colleagues [54-57] have raised questions concerning the degree of order in the "30 nm fibril". One problem with the published models for the "30 nm fibril" [44-49] is that they are depicted as very regular structures extending over large distances. However, the scanning force microscopy studies [54-58] show that the regularity of the "30 nm filament" extends over short distances only. This accords with E.M. data [50] and with previous x-ray [47] and neutron [45] diffraction that give only a small number of diffuse diffraction peaks consistent with a low degree of order in the structure.

4. DNA Packing Ratio of Active Chromatin

In the E.M. active transcriptional units have been observed for the ribosomal RNA genes in the embryo of Oncopeltus fusciatus [59] and the Balbiani rings of the salivary glands of Chironomus tentaus [60,61]. For Balbiani rings a comparison of the length of the transcription product i.e., the 7SRNA with the length of the transcribing chromatin fiber gave a DNA packing ratio of 3 to 4:1. From E.M. tomographic studies, packing ratios of 4 to 8:1 have been obtained for different regions of the Balbiani ring transcription unit [62,63]. The latter value is comparable with the neutron scatter determined packing ratio of the extended 10 nm filament of 6 to 7:1 [44]. Thus, the unfolding of the 34 nm supercoil of nucleosomes to the extended form and beyond is probably a major step in the structural transition from inactive to active chromatin.

5. Factors Involved in Active Chromatin

The changes involved in chromatin structure and stability which precede the passage of RNA polymerase remain poorly defined. Correlations have been found of histone composition and subtypes, histone modifications and non-histone proteins with active chromatin. These include: i) full or partial depletion of histone H1 [64,65], ii) hyper acetylation of the core histones, particularly H3 and H4 [see 10-12]; iii) ubiquitination of histones H2A and H2B [14,16,17]; iv) "active" core particles which selectively bind RNA polymerase II are depleted in one (H2A,H2B) dimer [66] and v) the binding of high mobility group (HMG) proteins [67]. We have shown, however, that full or partial dissociation of the histone octamer is not required for transcript elongation although arrays of nucleosome cores by phage T7 RNA polymerase (see later) [68]. Concerning chromatin structure/function relationships we have shown that histone acetylations, in particular the acetylation of histones H3 and H4 [69,70] cause a reduction in the nucleosome DNA linking number change, ΔL_k, from -1.04 ± 0.08 to -0.82 ± 0.05 thus releasing negative DNA supercoiling from acetylated nucleosomes into a constrained chromatin domain which would facilitate chromatin domain unfolding. Very recently those studies have been extended to the effects of the fully acetylated forms of either histones H3 or H4. It was found that the full acetylation of H4 but not H3 caused the change in the linking number [71]. The sites of reversible ubiquitinations of H2A and H2B are located in their basic C-terminal tails [14,17,18,72]. It has been proposed that ubiquitin labels potentially active chromatin containing heat shock [14] and stress genes [17,18]. Thus, knowing the location of ubiquition and how ubiquitination of H2A and H2B modifies chromatin structure and stability are essential to understanding the functions of histone ubiquitination.

6. Nucleosome Positioning

Virtually all of the DNA in eukaryotic genomes is packaged by histones into nucleosomes and higher order chromatin structures. Some of these nucleosomes have been shown to be precisely positioned on the underlying DNA sequence, most probably for functional requirements [see 72,73]. The factors involved in the precise positioning of nucleosomes are not well-understood. Additional to the identification of nucleosome positioning sequences [see 73] statistical analyses of DNA sequences contained in native core particles suggest the involvement of more general sequence properties. Following an earlier proposal that DNA flexibility or bendability might be a factor in nucleosome positioning [74], sequence constraints as determinants of nucleosome core particle positioning have been identified [75,76]. Based on these findings, earlier observations that long stretches of some homo-nucleotide sequences cannot be assembled into nucleosomes [77,78] can be explained by their increased stiffness relative to native DNA sequences. However, from sequence engineering experiments [79,80] it has been concluded that although DNA bendability should be considered as a general property of DNA sequences, other more specific factors in nucleosome positioning cannot be excluded. Very lysine rich (VLR) histones that are required for the stability of the 30 nm supercoil or solenoid of nucleosomes [41], are thought also to be involved in nucleosome positioning [81].

To study the protein factors involved in nucleosome positioning my laboratory has used DNA substrates containing tandem repeats of nucleosome length DNA from the sea urchin lytechins 5S RNA gene. This DNA repeat has been shown to contain a unique nucleosome positioning sequence [82,83]. Head to tail tandem repeats of nucleosome length DNA have been constructed by Simpson's group [83]; these are 18 repeats of 207 bp DNA $(207)_{18}$ and 45 repeats of 172 bp DNA $(172)_{45}$. The 172 bp sequence contains 7 unique restriction sites and the longer 207 bp sequence contains 8 unique restriction sites. To determine the precise positions of nucleosome cores formed by the histone octamer, the assembled 172_{45} and 207_{18} chromatins were trimmed back to 146 bp core particles by micrococcal nuclease and 5' end-labeled. The 146±2 bp of DNA extracted from these core particles was digested with up to eight restriction enzymes. Analysis on denaturing polyacrylamide gels allowed the core particle boundaries to be mapped relative to the unique restriction sites. This analysis showed that most but not all of the histone octamers assemble on one dominant but not strictly unique position from nucleotide 6 to 153 bp on both the 172_{45} and 207_{18} DNA head-to-tail multimers. The unique position reported for the histone octamer assembled on the 260 bp monomer from 5SrDNA [82] lies 10 to 15 bp downstream from the above major site identified for the 172 bp and 207 bp DNA multimers. This could imply strongly that regions of DNA external to the 172 bp and 207 bp DNA influence the final position of the histone octamer on the 260 bp DNA monomer.

In addition to the dominant position from 6 to 153 bp occupied by a large proportion of the histone octamers, minor positions were identified that flanked the major position. Relative to this major position one of the minor positions in the 207 bp DNA was 10 bp upstream; two minor positions were 10 and 20 bp downstream and two other minor positions were further away, 40 bp downstream and 50 bp upstream. The 172 bp multimer gave the same octamer positions except that the more distant locations of 40 bp downstream and 50 bp upstream were absent. It is to be noted that all of the minor positions on the 172 bp and 207 bp DNA multimers are located in multiples of ±10 bp away from the dominant position. This is significant because 10 bp is the helical repeat of B-form DNA coiled around the histone octamer in the nucleosome core particle [36].

These observations, for nucleosome cores assembled on the tandemly repeated nucleosome positioning DNA sequences, of a major position flanked by minor positions spaced by multiples of 10 bp DNA most probably result from the dynamic nature of the primarily electrostatic interactions between the basic histones and the DNA coil. They raise the possibility that nucleosome cores have the ability to move between the major and minor positions depending on solution ionic conditions and temperature.

7.   Effects of VLR Histones on Nucleosome Positions

Mirococcal nuclease digestion of $(207)_{18}$ and $(172)_{45}$ chromatins assembled with histone H5 showed regularly spaced nucleosomes. Both the core particle 146 bp and, importantly, the chromatosome 168 bp nuclease digestion stops were well-defined suggesting that H5 (and H1) provide protection similar to that of native chromatin. Of considerable interest was the finding that the complexity of the 207 bp DNA nucleosome bands discussed above was reduced by H5. The addition of histone H5 appeared therefore to reposition many of the nucleosome cores between the minor position and dominant positions. DNA was extracted from the 208 bp and 172 bp chromatosome bands and digested with up to eight restriction enzymes as described above for the 146 bp DNA from the core particle. The digested DNA from the 207 bp chromatosome corresponded to two major positions with comparable populations 10 bp apart. Surprisingly, the 172 bp chromatosome was more complex and gave digestion bands corresponding to four major positions of similar abundance all spaced by differences of 10 bp. The boundaries of these 172 bp and 207 bp chromatosomes were in the same "10 bp phase" as found for the 146 bp core particles suggesting identical rotational settings of the DNA for both types of particles. Clearly, the interactions of the outer regions of the chromatosome DNA with the globular domain of H1 and H5 influences the probability of some nucleosome core locations. Thus on the 5S rDNA 207 bp and 172 bp nucleosome positioning sequences the chromatosome is a real positioning entity defined by both the octamer/DNA interactions and chromatosome/H1 or H5 interactions.

8. Nucleosome Mobility

The observation of a cluster of nucleosome core positions flanking a dominant position on a strongly positioning DNA sequence suggests that the nucleosome cores may be able to exchange between positions in the cluster depending on ionic solution conditions and temperature. Two-dimensional gel electrophoresis has been used to investigate the effects of buffer and temperature on the positions of the nucleosome cores [85]. Mononucleosomes excised from the long 207_{18} chromatin by the restriction enzyme AvaI migrate as three bands in a nondenaturing nucleoprotein particle polyacrylamide gel. This AvaI digest was divided into two aliquots for two nucleoprotein polyacrylamide gels and run in parallel at 4°C. One gel was incubated at 37°C for one hour and the other gel kept at 4°C. Both gels were run in parallel in the second dimensions under the same conditions as the first dimension. For the gel which was incubated at 4°C between the first and second dimensions the three nucleosome core bands run on the diagonal as expected. For the gel that was incubated at 37°C between the first and second dimension each of the three bands from the first dimension redistributed into three bands in the second dimension. The two-dimensional gel assay experiment was repeated with an AvaI digest that had been incubated at 37°C for one hour in buffer prior to gel loading. When this gel was incubated between the first and second dimensions, the 207 bp nucleosomes migrated as a square of spots, indicating that each of the three original bands had again redistributed into three bands. Taken together these results show that 207 bp nucleosome cores excised from $(207)_{18}$ chromatin assembled with histone octamers have the ability to redistribute at 37°C. Nucleosome positioning therefore appears to have a dynamic character. The mobility was observed in low salt at 37°C but not at 4°C. At 4°C the mobility may be too slow to detect. The positions in the cluster have the same coiling of DNA around the nucleosome, but the boundaries of the nucleosome cores lie at 10 bp increments, i.e., the B-form DNA helical repeat, along the path of the DNA coil. Because the dominant position of the octamer on the repetitive sequence is flanked by weaker positions spaced by 10 bp intervals, it would appear that the positioning signal has at least some rotationally defined character to it, such as bendability. A purely translational signal (requiring alignment at defined points in the nucleosome) would not allow this type of fluctuation around an energetically favored nucleosome binding site. The mobility of nucleosome cores on the strongly positioning 5S gene DNA sequence suggests that mobility is a general property of H1 depleted chromatin. This was tested for long H1 depleted chromatin digested with restriction enzymes to avoid the trimming of overhanging DNA ends [86]. This was necessary because a range of mononucleosome lengths greater than 170 bp is required to distinguish between differently positioned nucleosomes in gel electrophoresis. The slowest migrating nucleosomes contained DNA lengths ranging from about 220 bp to greater than 300 bp.

The two dimensional gel electrophoresis was carried out as described above with gel stripes containing the bands of interest from the first dimension gel. Both first and second dimensions were at 4°C. Between the two runs the gel strips were incubated for 1 hour at 4°C for the control and for 37°C for the mobility experiment. The control incubation of mononucleosomes at 4°C shows the diagonal line expected of immobile nucleosomes whereas for the incubation at 37°C many but not all the mononucleosomes have become mobile on the underlying DNA sequences as shown by an off-diagonal "fan" of DNA intensity. There is a marked bias for an increase in electrophoretic velocity indicating a preference for nucleosome cores to occupy end positions. Thus at low ionic strengths the mobility of histone octamers is potentially a general behavior of a large proportion of native nucleosomes [86].

An important determinant of nucleosome positioning is the DNA anisotropy of flexibility required to accomodate the DNA tight bending around the histone octamer [76]. Because the binding affinity is the cummulative effect of many small bends positioning is often rotationally unique but translationally degenerate [87]. There are several examples of multiple positions spaced by 10 bp [88,89]. The mobility of a nucleosome core appears to depend on the sequences flanking its position. The histone octamer would be mobile if the DNA coil continued beyond the immediate location of the nucleosome core. Within this extended DNA coil the histone octamer would be able to jump through units of a B-form DNA helical repeat. This mobility would be limited by the same elements that act as boundaries to nucleosome positioning [reviewed in 73]. The significance of this nucleosome core mobility is that unlike the nucleosome sliding observed at higher non-physiological ionic strengths, which suppresses histone DNA interaction, all of the histone/DNA contacts are maintained at the low ionic strengths used in these experiments. It would appear that chromatin is a more dynamic structure than is widely assumed [86].

In vivo, local ionic conditions differ and a number of other factors such as histone modifications, binding of very lysine rich histones, DNA binding proteins or interactions with adjacent nucleosomes may suppress or enhance nucleosome mobility. Nucleosome cores may be fixed or free to move depending on functional requirements. Binding sites for transacting factors could become exposed to the factors involved in the control of gene expression.

9. Very Lysine Rich Histones Suppress Nucleosome Mobility

The hypothesis that nucleosome core mobility may be suppressed for functional requirements was tested by the binding of the very lysine rich histones H1 and H5. Very lysine rich histones have been identified as general repressors of transcription

[90,91]. Both H1 and H5 can be faithfully reconstituted into 5SrDNA chromatin [84]. Using the two dimensional gel electrophoresis, the mobility of histone octamers positioned on constructs of sea urchin 5SrDNA was shown to be efficiently suppressed by the binding of H1 or H5 to nucleosomes [92]. This implies that if nucleosome mobility is required for access to the underlying DNA sequences then the very lysine rich histones could function as general gene repressors through the immobilization of nucleosome cores. This function would be additional to the role of very lysine rich histones in stabilizing the "30 nm" supercoil of nucleosomes. Histone H5 was found to be a stronger inhibitor of nucleosome core mobility which correlates with its stronger binding to chromatin [92]. These results have been reproduced using the Xenopus 5SrDNA sequence repeat. Very lysine rich histones were found to inhibit nucleosome core mobility and repress transcription, demonstrating that stable states of gene repression can be established even at the nucleosome level [93]. All the above studies raise the possibility that during development the redistribution of very lysine rich histone subtypes provide another mechanism for suppressing the activities of genes not required at particular stages of development.

In vivo mechanisms for controlling nucleosome organization and functions have recently been identified that involve complex cofactors. Nucleosome rearrangements have been reported for H1 containing chromatin assembled in a cell free Drosophila embryo extract. On the addition of transcription factors it was found that nucleosomal arrays at the promoters of hsp70 and hsp26 genes were disrupted but only in the presence of ATP [94-96]. Thus this nucleosome rearrangement was dependent on ATP hydrolysis. A nucleosome remodeling cofactor (NURF), a 500 kDa protein complex, was identified in these extracts [97]. The NURF is thought to function with transcription factors in an ATP dependent manner to rearrange H1 containing nucleosomes positioned on the promoter regions prior to transcription. NURF is distinct from the previously reported SWI/SWF transcription activator comples [reviewed in 98]. The SWI/SNF complex is a 2 MDa protein complex that relieves the constraints of nucleosomal packaging of DNA prior to transcription through a mechanism that is different from that of NURF. These *in vivo* mechanisms for the disruption or reorganization of nucleosomal arrays are presumably to allow access of transacting factors to their specific DNA binding sites. They are more complex than nucleosome core mobility in the absence of histone H1 that has also been shown to allow access of transcription factors to their DNA binding sites [93].

10. Transcription Through Nucleosomes

Chromatin structure presents a barrier to the efficient transcription of active genes. Chromatin changes associated with active genes have been listed above and include the hyperacetylation of the histone octamer, particularly histones H3 and H4. Presumably the nucleosome remodeling cofactor [93-97] binds to the active or accessible promoter regions of specific genes to disrupt nucleosomes for transacting factors to bind and initiate gene expression. Nucleosome mobility [85,86,92] also provides a mechanism for transacting factors to bind to gene regulatory sequences [93] and initiate gene expression. Following gene activation RNA polymerases are faced with transcribing the DNA packaged into "active" nucleosomes. The bacteriophage SP6 RNA polymerase and eukaryotic RNA polymerases II and III have been shown to transcribe through one nucleosome [99,100] or short stretches of nucleosomes [101,102]. The problems with these studies were that the nucleosome templates were not fully defined as regards protein composition, nucleosome spacing and positioning or they contained only one or a few nucleosomes.

Our studies of transcription through nucleosomes [68,103-104] have used the well-characterized tandemly repeated nucleosome positioning sequence $(207)_{18}$ described above for the nucleosome positioning and mobility studies. The DNA construct $(207)_{18}$ was inserted between the T7 and SP6 transcription promotors of pGEM-32. Nucleosome cores were assembled on supercoiled, closed circular $pT(207)_{18}$ and double label experiments were performed to determine the effect of nucleosome cores on both the initiation and elongation of transcripts by T7 RNA polymerase. Both transcript initiation and elongation were inhibited, the extent of the inhibition being directly proportional to the number of nucleosome cores assembled on the $pT(207)_{18}$ DNA template. Continuous regularly spared linear arrays of nucleosome cores were obtained by digesting the assembled $pT(207)_{18}$ chromatin with Dra1, for which a unique restriction site lies within the nucleosome positioning sequence of the 207 bp repeat. This site is protected from Dra1 by the formation of nucleosome cores. Dra1 will cut only naked DNA repeats and not the DNA repeats assembled into nucleosome cores. Thus the digestion of partially assembled $pT(207)_{18}$ with Dra1 gives the T7 promotor followed by continuous runs of assembled nucleosome cores of different lengths. *In vitro* transcription with T7 RNA polymerase gave an RNA ladder with a 207 nucleotide spacing demonstrating that transcription had proceeded through continuous arrays of positioned nucleosome cores. It was shown that nucleosome cores partially inhibit the elongation of transcripts by T7 RNA polymerase, while allowing passage of the polymerase through each nucleosome core at an upper efficiency of 85%. Hence, complete transcripts are produced with high efficiency from short nucleosomal templates, whereas the production of full length transcripts from large nucleosomal arrays is relatively ineffective. These results indicate that nucleosome cores have significant inhibitory effects *in vitro* not only in transcription initiation but also on transcription elongation and that special mechanisms probably exist to overcome those inhibitory effects *in vivo*. The question of whether histone octamer disociation was required for the process of transcription was addressed by the extensive cross-linking of the histone octamer prior to its assembly into $pT(207)_{18}$ chromatin [68]. Transcription studies of this heavily cross-linked $(207)_{18}$ assembled chromatin lacking H1 have demonstrated that there is no need for the disassociation of the histone octamer during elongation because transcription was not affected by the irreversible crosslinking of histones within the histone octamers.

As discussed above, histone H1 has a profound effect on the stability of nucleosomes and is required for the generation and stability of the 34 nm supercoil of nucleosomes. H1 has been strongly implicated in the formation of transcriptionally silent chromatin and has been demonstrated to be a general repressor of transcription initiation *in vitro* [106,107]. The effects of histone H1 on transcription of the $pT(207)_{18}$ assembled chromatin by T7 RNA polymerase have been investigated and both transcription initiation and elongation were found to be fully suppressed by histone H1. One interpretation of these results is that very lysine rich histones may inhibit transcription by stabilizing nucleosomal structures. This would suppress the mobility of nucleosome cores located on promotors, and thus reduce the accessibility of promoter regions to transacting factors and RNA polymerases. Further, stabilization of

122

nucleosome cores may provide resistance to the progress of RNA polymerase during transcription elongation. This raises the possibility that histone H1 provides another mechanism for regulation of transcription of nucleosomal templates by the suppressor of nucleosome mobilities.

Summary

All aspects of DNA processing are proving to be dauntingly complex and involve not only polymerases and many regulatory factors but also the functions of nucleosomes. Our understanding of chromatin structure/function relationships has advanced very slowly. It is now clear, however, that an understanding of histone and nucleosome functions are integral to understanding DNA processing. Histones are no longer thought of as passive structural components of chromosomes but through the reversible chemical modifications of histones are clearly involved in chromosome functions. Our view of nucleosome cores as static structures has changed drastically. Nucleosome cores are capable of short-range mobility and can exchange between a cluster of positions spaced by intervals of 10 bp DNA. This nucleosome mobility clearly has functional significance because the potential now exists for sequence specific DNA binding proteins to influence the position of nucleosome cores and allow access to their DNA binding sites. The ability of the very lysine rich histones to suppress nucleosome core mobility implies that they can function at the individual nucleosome level in addition to their role in generating and stabilizing higher order chromatin structures.

Acknowledgements

Research support is acknowledged to EMB from the Department of Energy (DE-FG03-88ER 60673), and the National Institutes of Health (PHS GM26901), and to SP from the Wellcome Trust .

References

1. Saitoh, Y., Laemmli, U.K. (1993) From the Chromosomal Loops and the Scaffold to the Classic Bands of Metaphase Chromosomes, *Cold Spring Harbor Symposia on Quantitative Biology* **58**, 755-765.
2. van Holde, K.E. (1988) *Chromatin*, (A. Rich, ed.) Springer-Verlag, New York, Berlin, Heidelberg, London, Paris, Tokyo.
3. Bradbury, E.M. (1992) Reversible Histone Modifications and the Chromosome Cell Cycle, *Bioassays* **14**, 9-16.
4. Moss, T., Cary, P.D., Abercrombie, B.D., Crane-Robinson, C., and Bradbury, E.M. (1976) A pH-Dependent Interaction Between Histones H2A and H2B Involving Secondary and Tertiary Folding, *Eur. J. Biochem.*, **71**, 337-350.
5. Arents, G., Burlingame, R.W., Wang, B.C., Love, W.E., and Moudrianakis, E.N. (1991) The Nucleosomal Core Histone Octamer at 3.1 A resolution: A Tripartite Protein Assembly and a Left-Handed Superhelix, *Proc. Natl. Acad. Sci. USA* **88**, 10148-10152.
6. Moss, T., Cary, P.D., Crane-Robinson, C., and Bradbury, E.M. (1976) Physical Studies on the H3/H4 Histone Tetramer, *Biochemistry* **15**, 2261-2267.
7. Bradbury, E.M., Chapman, G.E., Danby, S.E., Hartman, P.G., and Riches, P.L. (1976) Studies on the Role and Mode of Operation of the Very-Lysine-Rich Histone H1 (F1) in Eukaryote Chromatin. The Properties of the N-Terminal and C-Terminal Halves of Histone H1, *Eur. J. Biochem.* **57**, 521-528.
8. Hartman, P.G., Chapman, G.E., Moss, T., and Bradbury, E.M. (1977) Studies on the Role and Mode of Operation of the Very-Lysine-Rich Histone H1 in Eukaryote Chromatin, *Eur. J. Biochem.*, **77**, 456.

9. Chapman, G.F., Hartman, P.G., Cary, P.D., Bradbury, E.M., and Lee, D.R. (1978) A Nuclear Magnetic Resonance Study of the Globular Structure of the H1 Histone, *Eur. J. Biochem.*, **86**, 35.
10. Johnson, E.M., and Allfrey, V.G. (1978) In: *Biochemistry Actions of Hormones*", Vol. 5 (G. Litwac. ed.) Academic Press, New York.
11. Matthews, H.R. and Waterborg, J. (1985) in "The Enzymology of Post Translational Modifications of Proteins", Vol. 2, pp. 125-185.
12. Yasuda, H., Mueller, R.D., and Bradbury, E.M. (1986) Molecular Regulation of Nuclear Events in "Mitosis and Meiosis", (Schlegel, R.A., Halleck, M.S., and Rao, P.N., eds.), Academic Press, New York pp. 391-361.
13. Busch, H. and Goldknoph, I.L. (1981) Ubiquitin-Protein Conjugates, *Mol. Cell Biol.* **40**, 173-187.
14. Christensen, M.E. and Dixon, G.H. (1982) Hyperacetylation of Histone-H4 Correlates with the Terminal, Transcriptionally Inactive Stages of Spermatogenesis in Rainbow Trout, *Dev. Biol.* **93**, 404-415.
15. Goldknoph, I.L., Wilson, G., Ballard, N.R., and Busch, H. (1980) Chromatin Conjugate Protein A24 is Cleaved and Ubiquitin is Lost During Chicken Erythropoiesis, *J. Biol. Chem.* **255**, 10555-10558.
16. Levinger, L. and Varshavsky, A. (1982) Selective Arrangement of Ubiquitinated and D1 Protein-Containing Nucleosomes Within the Drosophila Genome, *Cell* **28**, 375-385.
17. Matsui, S.I., Seon, B.K., and Sandberg, A.A. (1979) Disappearance of a Structural Chromatin Protein A24 in Mitosis: Implications for Molecular Basis of Chromatin Condensation, *Proc. Natl. Acad. Sci., USA* **76**, 6386-6390.
18. Mueller, R.D., Yasuda, H., Hatch, C.L., Bonner, W.M., and Bradbury, E.M. (1985) Phosphorylation of Histone H1 Through the Cell Cycle of *Physarum polycephalum*, *J. Biol. Chem.* **260**, 5147-5153.
19. Bradbury, E.M., Inglis, R.J., and Matthews, H.R. (1974) Control of Cell Division by Very Lysine-Rich Histone F1 Phosphorylation, *Nature* **241**, 257-261.
20. Bradbury, E.M., Inglis, R.J., Matthews, H.R., and Langan, T.A. (1974) Molecular Basis of Mitotic Cell Division in Eukaryotes, *Nature* **249**, 553.
21. Gurley, L.R., D'Anna, J.A., Halleck, M.S., Barham, S.S., Walters, R.A., Jett, J.H., and Tobey, R.A. (1981) In: *Cold Spring Harbor Conferences on Cell Proliferation* **8**, 1073-1093.
22. McGhee, J.D. and Felsenfeld, G. (1980) Nucleosome Structure, *Ann. Rev. Biochem.* **40**, 1115-1156.
23. Bradbury, E.M. and Matthews, H.R. (1982) Chromatin Structure, Histone Modifications in the Cell Cycle, in "Cell Growth" (Nicolini, C. ed.) Plenum Press, NY, pp. 411-454.
24. Korberg, R.D. and Klug, A. (1981) The Nucleosome, *Sci. American* **244**, 48-60.
25. Simpson, R.T. (1978) Structure of the Chromosome, a Chromatin Particle Containing 160 base pairs of DNA and all the Histones, *Biochemistry* **17**, 5524-5531.
26. Bradbury, E.M., Baldwin, J.P., Carpenter, B.G., Hjelm, R.P., Hancock, R., and Ibel, K. (1975) Neutron-Scattering Studies of Chromatin, *Brookhaven Symp. Biol.* **27**, *IV* (Schoenborn, B.P., ed.) pp. 97-116.
27. Pardon, J.F., Worcester, D.C., Wooley, J.C., Tatchell, K., van Holde, K.E., and Richards, B.M. (1975) Low-Angle Neutron Scattering from Chromatin Subunit Particles, *Nucl. Acids Res.* **2**, 2163-2176.
28. Suau, P., Kneale, G.G., Braddock, G.W., Baldwin, J.P., and Bradbury, E.M. (1977) A Low Resolution Model for the Chromatin Core Particle by Neutron Scattering, *Nucleic Acids Research* **4**, 3769-3786.
29. Hjelm, R.P., Kneale, G.G., Suau, P., Baldwin, J.P., and Bradbury, E.M. (1977) Small Angle Neutron Scattering Studies of Chromatin Subunits in Solution, *Cell* **10**, 139-151.
30. Richards, B.M., Pardon, J., Lilley, D.M.J., Cotter, P., and Wooley, J. (1977) Sub-Structure of Nucleosomes, *Cell Biol. Int. Rep.* **1**, 107-116.
31. Braddock, G.W., Baldwin, J.P., and Bradbury, E.M. (1981) Neutron Scattering Studies of the Structure of the Chromatin Core Particle, *Biopolymers* **20**, 327-343.
32. Finch, J.T., Lutter, L.C., Rhodes, D., Brown, R.S., Rushton, B., Levitt, M., and Klug, A. (1977) Structure of Nucleosome Core Particles of Chromatin, *Nature* **269**, 29-36.
33. Finch, J.T., Brown, R.S., Rhodes, D., Richmond, T., Rushton, B., Lutter, L.C., and Klug, A. (1981) X-Ray-Diffraction Study of a New Crystal Form of the Nucleosome Core Showing Higher Resolution, *J. Mol. Biol.* **145**, 757-769.
34. Bentley, C.A., Finch, J.T., and Lewit-Bentley, A. (1981) Neutron Diffraction Studies on Crystals of Nucleosome Cores Using Contrast Variation, *J. Mol. Biol.* **145**, 771-784.
35. Richmond, T.J., Klug, A., Finch, J.T., and Lutter, L.C. (1981) In: *Proc. 2nd SUNY Conversation in Biomolecular Stereodynamics* (Sarma, R.H. ed.) Vol. II., Adenine Press, NY, pp. 109-123.
36. Richmond, T.J., Finch, J.T., Rushton, B., Rhodes, D., and Klug, A. (1984) Structure of the Nucleosome Core Particle at 7A Resolution, *Nature* **311**, 532-537.
37. Wood, M.J., Yau, P.M., Imai, B.S., Goldberg, M.W., Lambert, S.J., Fowler, A.L., Baldwin, J.P., Godfrey, J., Moudrianakis, E.N., Ibel, K., May, R.P., Koch, M., and Bradbury (1991) Neutron and X-Ray Scatter Studies of the Histone Octamer and Amino and Carboxyl Domain Trimmed Octamers, *J. Biol. Chem.* **266**, 5696-5702.
38. Schroth, G.P., Yau, P.M., Imai, B.S., Gatewood, J.M., and Bradbury, E.M. (1990) A NMR Study of Mobility in the Histone Octamer, *FEBS Lett.* **268**, 117-120.

124

39. Imai, B.S., Yau, P.M., Baldwin, J.P., Ibel, K., May, R.P., and Bradbury, E.M. (1986) Hyperacetylation of Core Histones Does not Cause Unfolding of Nucleosomes: Neutron Scatter Data Accords with Disc Structure of the Nucleosome, *J. Biol. Chem.* **261**, 8784-8792.

40. Usachenko, S.I., Barykin, S.G., Gavin, I.M., and Bradbury, E.M. (1994) Rearrangement of the Histone H2A C-Terminal Domain in the Nucleosome, *Proc. Natl. Acad. Sci. USA*, **91**, 6845-6849.

41. Thoma, F., Koller, T.H., and Klug, A. (1979) Involvement of Histone H1 in the Organization of the Nucleosome and of the Salt-Dependent Superstructures of Chromatin, *J. Cell Biol.* **83**, 403-427.

42. Crane-Robinson, C., Bohm, L., Puigdomenech, P., Cary, P.D., Hartman, P.G., and Bradbury, E.M. (1980) Structural Domains in Histones, *FEBS DANA-Recombination Interactions and Repair*, Pergamon Press, Oxford and New York.

43. Allan, J., Hartman, P.G., Crane-Robinson, C., and Aviles, F.X. (1980) The Structure of Histone H1 and its Location in Chromatin, *Nature* **288**, 675-679.

44. Suau, P., Bradbury, E.M., and Baldwin, J.P. (1979) Higher-Order Structures of Chromatin in Solution, *Eur. J. Biochem.* **97**, 593-602.

45. Carpenter, B.G., Baldwin, J.P., Bradbury, E.M., and Ibel, K. (1976) Organization of Subunits in Chromatin, *Nucl. Acids Res.* **3**, 1739-1746.

46. Finch, J.T. and Klug, A. (1976) Solenoidal Model for Superstructure in Chromatin, *Proc. Natl. Acad. Sci. USA* **73**, 1897.

47. Widom, J. and Klug, A. (1985) Structure of the 300A Chromatin Filament: X-Ray Diffraction from Oriented Samples, *Cell* **43**, 207-213.

48. Worcel, A., Strongatz, S., and Riley, D. (1981) Structure of the 300A Chromatin Filament: X-Ray-Diffraction from Oriented Samples, *Proc. Natl. Acad. Sci. USA* **78**, 1461-1465.

49. Woodcock, C.L.F., Frado, L.L.Y., and Rattner, J.B. (1984) The Higher-Order Structure of Chromatin: Evidence for a Helical Ribbon Arrangement, *J. Cell Biol.* **99**, 42-52.

50. Bednar, J. Horowitz, R.A., Dubochet, J., and Woodcock, C.L. (1995) Compaction: 3-Dimensional Structural Information from Cryoelectron Microscopy, *J. Cell Biol.* **131**, 1365-1376.

51. Williams, S.P., Athey, B.D., Muglia, L.J., Scheppe, R.S., Gough, A.H., and Langmore, J.P. (1986) Chromatin Fibers are Left-Handed Double Helices with Diameter and Mass per Unit Length that Depend on Linker Length, *Biophys. J.* **49**, 233-248.

52. Staynov, D.Z. (1983) Possible nucleosome arrangements in the higher-order structure of chromatin. *Int. J. Biol. Macromol.* **5**, 3-9.

53. Graziano, V., Gerchman, V., Schneider, D.K., and Ramakrishnan, V. (1994) Histone H1 is Located in the Interior of the Chromatin 30-nm Filament, *Nature* **368**, 351-354.

54. Yang, G., Leuba, S.H., Bustamante, C., Zlatanova, J., and van Holde, K. (1994) Linker DNA Accessibility in Chromatin Fibers of Different Conformations: A Re-Evaluation, *SPIE Proc.* **2384**, 13-21.

55. Leuba, S.H., Yang, G., Robert, C., Samori, B., van Holde, K., Zlatanova, J., and Bustamante, C. (1994) Three-Dimensional Structure of Extended Chromatin Fibers as Revealed by Tapping-Mode Scanning Force Microscopy, *Proc. Natl. Acad. Sci.* **91**, 11621-11625.

56. Zlatanova, J., Leuba, S.H., Yang, G., Bustamante, C., and van Holde, K. (1994) Linker DNA Accessibility in Chromatin Fibers of Different Conformations: A re-Evaluation, *Proc. Natl. Acad. Sci.* **91**, 5277-5280.

57. Leuba, S.H., Zlatanova, J., and van Holde, K. (1994) On the Location of Linker DNA in the Chromatin Fiber. Studies with Immobilized and Soluble Micrococcal Nuclease, *J. Mol. Biol.* **235**, 871-880.

58. Allen, M.J. (1995) Ph.D. Dissertation "Application of Atomic Force Microscopy to In Vitro and In Situ Investigations of Somatic and Sperm Chromaton Structure", University of California-Davis.

59. Foe, V.E. (1977) Modulation of Ribosomal RNA Synthesis in Oncopeltus Fasciatus: An Electron Microscopic Study of the Relationship Between Changes in Chromatin Structure and Transcriptional Activity, *Cold Spring Harbor Symp. Quant. Biol.* **42**, 723-740.

60. Andersson, K., Mahr, R., Bjorkroth, B., and Daneholt, B. (1982) Rapid Reformation of the Thick Chromosome Fiber Upon Completion of RNA Synthesis at the Balbiani Ring Genes in Chironomus Tentans, *Chromosoma* **87**, 33-84.

61. Daneholt, B. (1982) In: Insect Ultrastructure 1 (King and Akai, eds.) Plenum Publishing Corporation, pp. 382-401.

62. Olins, A.L., Olins, D.E., and Lezzi, M. (1982) Ultrastructural studies of Chironomus salivary-gland cells in different states of Balbiani ring activity. *Eur. J. Cell Biol..* **27**, 161-169.

63. Olins, D.E., Olins, A.L., Levy, H.A., Durfee, R.C., Margie, S.M., Timnel, E.P., and Dover, S.D. (1983) Electron-Microscope Tomography: Transcription in 3-Dimensions, *Science* **220**, 498-500.

64. Levy-Wilson, B. and Dixon, G.H. (1979) Limited Action of Micrococcal Nuclease on Trout Testis Nucleo Generates Two Mononucleosome Subsets Enriched in Transcribed DNA Sequences, *Proc. Natl. Acad. Sci. USA* **76**, 1682-1686.

65. Yasuda, H., Mueller, R.D., Logan, K.A., and Bradbury, E.M. (1986) Histone H1 in Physarum polycephalum; Its High Level in the Plasmodial State Increases in Amount and Phosphorylation in the Sclerotial Stage, *J. Biol. Chem.* **261**, 2349-2354.

66. Baer, B.W. and Rhodes, D. (1983) Eukaryotic RNA Polymerase-II Binds to Nucleosome Cores from Transcribed Genes, *Nature* **301**, 482-488.

67.   Weisbrod, S. and Weintraub, H. (1981) Isolation of Actively Transcribed Nucleosomes Using Immobilized HMG 14 and 17 and an Analysis of Alpha-Globin Chromatin, *Cell* **23**, 391-400.
68.   O'Neill, T.E., Smith, J.G., and Bradbury, E.M. (1993) Histone Octamer Dissociation is not Required for Transcript Elongation Through Arrays of Nucleosome Cores by Phage T7 RNA Polymerase in vitro, *Proc. Natl. Acad. Sci. USA* **90**, 6203-6207.
69.   Norton, V.G., Imai, B.S., Yau, P.M., and Bradbury, E.M. (1989) Histone Acetylation Reduces Nucleosome Core Particle Linking Number Change, *Cell* **57**, 449-457.
70.   Norton, V.G., Marvin, K.W., Yau, P.M., and Bradbury, E.M. (1990) Nucleosome Linking Number Change Controlled by Acetylation of Histones H3 and H4, *J. Biol. Chem.* **265**, 19,848-19,852.
71.   Cao, Y., Yau, P., and Bradbury, E.M., manuscript in preparation.
72.   West, M.H.P. and Bonner, W.M. (1980) Histone 2B can be Modified by the Attachment of Ubiquitin, *Nucl. Acids Res.* **8**, 4671.
73.   Simpson, R.T. (1986) Nucleosome Positioning in vivo and in vitro, *Bioessays* **4**, 172-176.
74.   Trifonov, E.N. (1980) Helical Model of Nucleosome Core, *Nucl. Acids Res.* **8**, 4041-4053.
75.   Drew, H.R. and Travers, A.A. (1985) DNA Bending and its Relation to Nucleosome Positioning, *J. Mol. Biol.* **186**, 773-790.
76.   Satchwell, S.C., Drew, H.R. and Travers, A.A. (1986) Sequence Periodicities in Chicken Nucleosome Core DNA, *J. Mol. Biol.* **191**, 659-675.
77.   Rhodes, D. (1979) Nucleosome Cores Reconstituted from poly(dA-dT) and the Octamer of Histones, *Nucl. Acids Res.* **6**, 1805-1816.
78.   Prunell, A. (1982) Nucleosome reconstitution on Plasmid-Inserted Poly(dA)•poly(dT), *EMBO J.* **1**, 173-179.
79.   Neubauer, B., Linxweiler, W., and Horz, W. (1986) DNA Engineering shows that Nucleosome Phasing on the African-Green Monkey Alpha-Satellite is the Result of Multiple Additive Histon, *J. Mol. Biol.* **190**, 639-645.
80.   Thoma, F. and Zatchei, M. (1988) Chromatin Folding Modulates Nucleosome Positioning in Yeast Minichromosomes, *Cell* **55**, 945-953.
81.   Stein, A. and Mitchell, M. (1988) Generation of Different Nucleosome Spacing Periodicities in vitro: Possible Origin of Cell Type Specifficity, *J. Mol. Biol.* **203**, 1029-1043.
82.   Simpson, R.T. and Stafford, D.W. (1983) Structural Features of a Phased Nucleosome Core Particle, *Proc. Natl. Acad. Sci., USA* **50**, 51-55.
83.   Simpson, R.T., Thoma, F., and Brubaker, J.M. (1985) Chromatin Reconstituted from Tandemly Repeated Cloned DNA Fragments and Core Histones: A Model System for Study of Higher-Order St, *Cell* **42**, 799-808.
84.   Meersseman, G., Pennings, S., and Bradbury, E.M. (1991) Chromatosome Positioning on Assembled Long Chromatin: Linker Histones Affect Nucleosome Placement on 5S rDNA, *J. Mol. Biol.* **220**, 89-100.
85.   Pennings, S., Meersseman, G., and Bradbury, E.M. (1991) Mobility of Positioned Nucleosomes on 5S rDNA, *J. Mol. Biol.* **220**, 101-110.
86.   Meersseman, G., Pennings, S., and Bradbury, E.M. (1992) Mobile Nucleosomes - A General Behavior, *EMBO J.* **11**, 2951-2959.
87.   Shrader, T.E. and Crothers, D.M. (1989) Artificial Nucleosome Positioning Sequences, *Proc. Natl. Acad. Sci., USA*, **86**, 7418-7422.
88.   Lowman, H. and Bina, M. (1990) Correlation Between Dinucleotide Periodicities and Nucleosome Positioning on Mouse Satellite DNA, *Biopolymers*, **30**, 861-876.
89.   Simpson, R.T. (1991) Nucleosome Positioning: Occurrence, Mechanisms, and Functional Consequences, *Prog. Nucl. Acids Res. Mol. Biol.*, **40**, 143-184.
90.   Wolffe, A.P. (1989) Dominant and Specific Repression of Xenopus Oocyte 5S RNA Genes and Satellite I DNA by Histone H1, *EMBO J.*, **8**, 527-537.
91.   Laybourn, P.J. and Kadonaga, J.T. (1991) Role of Nucleosomal Cores and Histone H1 in the Regulation of Transcription by RNA Polymerase II, *Science*, **254**, 238-245.
92.   Pennings, S., Meersseman, G., and Bradbury, E.M. (1994) Linker Histones H1 and H5 Prevent the Mobility of Positioned Nucleosomes, *Proc. Natl. Acad. Sci. USA*, **91**, 10275-10279.
93.   Ura, K., Hayes, J.J., and Wolffe, A.P. (1995) A Positive Role for Nucleosome Mobility in the Transcriptional Activity of Chromatin Templates: Restriction by Linker Histones, *EMBO J.*, **14**, 3725-3765.
94.   Tsukiyama, T., Becker, P.B., and Wu, C. (1994) ATP-Dependent Nucleosome Disruption at a Heat-Shock Promoter Mediated by Binding of GAGA Transcription Factor, *Nature*, **367**, 525-532.
95.   Wall, G., Varga-Weisz, zp.D., Sandaltzopoulos, R., and Becker, P.B. (1995) Chromatin Remodeling by GAGA Factor and Heat Shock Factor at the Hypersensitive Drosophila hsp26 Promoter in vitro, *EMBO J.*, **14**, 1727-1736.
96.   Pazin, M.J., Kamakaka, R.T., and Kadonaga, J.T. (1994) ATP-Dependent Nucleosome Reconfiguration and Transcriptional Activation from Preassembled Chromatin Templates, *Science*, **266**, 2007-2011.
97.   Tsukiyama, T. and Wu, C. (1995) Purification and Properties of an ATP-Dependent Nucleosome Remodeling Factor, *Cell*, **83**, 1011-1020.

126

98.   Peterson, C.L. and Tamkun, J.W. (1995) The SWI-SNF Complex: A Chromatin Remodeling Machine, *Trends. Biochem. Sci.*, **20**, 143-146.
99.   Lorch, Y., LaPointe, J.W., and Kornberg, R.D. (1987) Nucleosomes Inhibit the Initiation of Transcription but Allow Chain Elongation with the Displacement of Histones, *Cell,* **49**, 203-210.
100.   Losa, R. and Brown, D.D. (1987) A Bacteriophage RNA Polymerase Transcribes *in vitro* Through a Nucleosome Core Without Displacing it, *Cell,* **50**, 801-808.
101.   Morse, R.H. (1989) Nucleosomes Inhibit Both Transcriptional Initiation and Elongation by RNA Polymerase III in vitro, *EMBO J.,* **8**, 2343-2351.
102.   Felts, S.J., Weil, P.A., and Chalkley, R. (1990) Transcription Factor Requirements for in vitro Formation of Transcriptionally Competent 5S rDNA Gene Chromatin, Mol. Cell. Biol., **10**, 2390-2401.
103.   O'Neill, T.E., Roberge, M., and Bradbury, E.M. (1992) Nucleosome Arrays Inhibit both Initiation and Elongation of Transcripts by T7 RNA Polymerase, *J. Mol. Biol.,* **223**, 67-78.
104.   O'Neill, T.E., Pennings, S., Meersseman, G., and Bradbury, E.M. (1995) Deposition of Histone H1 Onto Reconstituted Nucleosome Arrays Inhibits Both Initiation and Elongation of Transcripts by T7 RNA Polymerase, *Nucleic Acids Res.,* **23**, 1075-1082.

CHROMATIN STRUCTURE AND GENE REGULATION BY STEROID HORMONES

MIGUEL BEATO[1], SEBASTIÁN CHÁVEZ[2], KARIN EISFELD[1], CHRISTIAN SPANGENBERG[1] and MATHIAS TRUSS[1]

[1] I.M.T, E.-Mannkopff-Str. 2, D35037 Marburg, Germany
[2] Departamento de Genética, Universidad de Sevilla, E41080 Sevilla, Spain

Hormonal induction of the Mouse Mammary Tumor Virus (MMTV) is mediated by a complex hormone responsive region (HRR) composed of 5 hormone receptor binding sites, upstream of a binding site for the transcription factor NFI, two octamer motifs, and the TATA box. Optimal induction requires the integrity of all these *cis*-acting elements, but the corresponding factors can not bind to free DNA simultaneously. In cells carrying chromosomal MMTV sequences, the HRR is organized into a phased nucleosome which allows binding of the hormone receptors to two of their five cognate sites, while precluding access to the NFI site and to the octamer motifs. Receptor binding is determined by the rotational orientation of the relevant major grooves, but the NFI site is inaccessible, no matter the helical orientation, as long as it is included within a positioned nucleosome. Hormone treatment leads to a rapid alteration in chromatin structure that makes the dyad axis of the regulatory nucleosome more accessible to digestion by DNaseI and restriction enzymes. Concomitantly, all five receptors binding sites, the NFI, and the octamer motifs are occupied, while the nucleosome remains in place, suggesting that it may facilitate full loading of the promoter with transcription factors. This notion is supported by studies on the influence of nucleosome depletion on basal and induced expression of MMTV promoter in yeast. The MMTV promoter exhibits a positioned nucleosome in *S. cerevisiae* with similar location as in metazoan cells. Hormonal induction of MMTV transcription in yeast depends on a functional synergism between the glucocorticoid receptor (GR) and NFI. Following depletion of nucleosomes, independent transactivation by NFI or by GR as well as binding of the individual proteins to the promoter are enhanced, whereas the NFI-dependent hormone induction of the promoter and simultaneous binding of receptor and NFI are compromised. Thus, positioned nucleosomes do not only account for constitutive repression but participate in induction by mediating cooperative binding and functional synergism between GR and NFI. However, given the starting nucleosomal organization, occupancy of all sites on the promoter would require remodeling the chromatin. This could involved a change in stoichiometry of the core histones, since the NFI site and the octamer motifs are more accessible upon removal of histone H2A/H2B dimers, or a modification of the histone tails, since moderate hyperacetylation enhances promoter activity even in the absence of hormone. Elucidating the biochemistry of the hormone induced nucleosome remodeling is the next goal.

127

C. Nicolini (ed.), Genome Structure and Function, 127–144.
© 1997 Kluwer Academic Publishers. Printed in the Netherlands.

1. Introduction

Gene expression in eukaryotes is orchestrated by precise arrays of *cis* elements in the promoter and/or enhancer regions of genes, which specify the combinatorial interaction among sequence-specific and general transcription factors. Recently it has been found that the implementation of these interactions often involves the action of adaptor factors or co-activators, which do not bind DNA and can act as integrators of various signal transduction pathways [1]. The assembly of large multiprotein complexes on promoters and enhancers is influenced by the chromatin packaging of the corresponding DNA sequences. Their nucleosomal organization modulates the accessibility of various *cis* elements in ways that are not fully understood. Evidence is accumulating that local chromatin remodeling is an important prerequisite for efficient transcription factor binding and induction of transcription. Large protein complexes have been discovered that counteract the repressing function of chromatin. One such assembly, the SWI/SNF complex, originally found in yeast genetic screens, was characterized by suppressor mutations which affect chromatin proteins [2]. Homologues to the SWI2/SNF2 component of the complex have been identified in *Drosophila*, mouse and human. These genes encode DNA-dependent ATPases whose function is essential for the remodeling activity of the complex. The isolated SWI/SNF complex is able to influence the structure of nucleosomes and the binding of transcription factors to DNA sequences wrapped around histone octamers [3, 4]. Other activities related to, but distinct from the SWI/SNF complex, have been recently identified in *Drosophila* [5], suggesting that the eukaryotic cell devotes a large number of genes to deal with chromatin structure in the context of transcriptional gene activation [6].

The behavior of nucleosomes is influenced by the covalent modification of the core histones, in particular the acetylation of lysine ε-amino groups in their N-terminal domains [7]. Analysis of natural and artificial DNA sequences has shown that their organization in nucleosomes restricts access of transcription factors, and that acetylation of the histone tails facilitates factor access to DNA [8]. In addition to covalent modifications, the structure of the nucleosome can also be influenced by selective dissociation of its constituent histones. For instance, dissociation of the histone H2A/H2B dimers, which may be promoted by the SWI/SNF complex or by nucleoplasmin [4], apparently leaves the basic structure of the core particle unchanged but leads to a higher accessibility of DNA sequences for nucleases, and transcription factors.

To approach the question of how chromatin structure influences the interaction of transcription factors with their target sequences, we have studied the regulated transcription of the Mouse Mammary Tumor Virus (MMTV) promoter which is silent in the absence of hormones but highly active in response to glucocorticoids or progestins [see [9] and references therein]. Events leading to hormonal activation of the MMTV promoter are initiated by the interaction of the corresponding hormone receptors with the promoter in chromatin. Before going into a detail description of this system, we will summarize our present knowledge of the structure and function of steroid hormone receptors.

2. Steroid Hormone Receptors Are Modular Ligand Activated Transcription Factors

The physiological and pharmacological effects of steroid hormones are mediated by their interaction with specific receptors, which bind the corresponding hormone with high affinity and specificity. The hormone receptors are ligand activated transcriptional modulators, which regulate genes by virtue of their binding to imperfect palindromic sequences on DNA called hormone responsive elements (HRE). The exact mechanism by which this regulation takes place is still unclear, but in the last years a large number of new proteins interacting with the hormone receptors has been identified and it is becoming progressively clear that the mechanism of action of the receptors is more complicated than originally envisaged [10]. Mutational analysis demonstrated that steroid hormone receptors have a modular structure with a DNA binding domain, nuclear localization signals, a ligand binding domain, as well as several transactivation domains. Each of these modules has been shown to be functional when placed out of context by fusion to a different protein domain.

The three dimensional structure of the DNA binding domains of the glucocorticoid receptor (GR) and of the estrogen receptor (ER) have been solved in solution and as complexes with DNA, and the mechanism of DNA sequence recognition is well understood [11]. The domain is structure in form of two C_2-C_2 zinc fingers, and the α-helices involved in sequence specific DNA binding and in homodimerization are located at the knuckle of the first and the second zinc fingers, respectively.

The crystal structure of the unoccupied ligand binding domain of RXRα has also been solved, allowing a first modeling of the ligand binding cavity, which has been probably conserved among various members of the nuclear receptor family [12]. The structure is highly α-helical with only two antiparallel ß-strands. One of the α-helices, helix 12, encompasses the activation function AF2 conserved in all members of the nuclear receptor family [13]. AF2 protrudes out of the globular structure of the unliganded receptor. A comparison with the structure of RARγ complexed with ligand shows that upon ligand binding the C-terminal helix folds back towards the ligand contributing to seal the hydrophobic pocket. This mouse trap mechanism repositions the amphipatic α-helix of the AF-2 activation region exposing it for interaction with co-activators [14]. The structure of the ligand binding domain of the thyroid hormone receptor (T3R) has also been reported, and a comparison with the structure of RAR and RXR confirms the general fold of the domain and suggest an active participation of the ligand in stabilizing the three dimensional structure [15].

The structure of the N-terminal half of the receptors with the strong transactivation function is still unknown, but studies with a short transactivation region from the GR, called τ, suggest a relatively flexible structure which could be dramatically influenced by protein-protein interactions [16].

3. The Receptors Can Interact Directly Or Indirectly With General Transcription Factors

Steroid hormone receptors are able to interact with several components of the general transcriptional machinery including the TATA box binding protein TBP [17] and TBP associated factors, TAFIIs [18, 19]. Human $TAF_{II}30$ is required for transactivation by ER and both proteins interact [20], and *Drosophila* $TAF_{II}110$ interacts with the DNA binding domain of PR [21]. An interaction between ER and PR with $TF_{II}B$ has also been reported [22] though its significance remains to be established [21].

Recently a number of potential Transcription Intermediary Factors (TIFs) have been identified which are supposed to mediate the interaction of steroid hormone receptors with the general transcription factors. Several proteins interact with the ligand binding domain of ER in a ligand dependent manner [23-27]. Using a yeast two-hybrid screen with RXRγ, a protein interacting with its AF2 has been identified and cloned, which also interacts with ER and PR in a ligand dependent way [28]. This protein, called TIF1, does not interact with components of the general transcriptional machinery and could mediate transactivation through interaction with factors involved in chromatin dynamics (see below). Interestingly, TIF1 belongs to the family of RING finger proteins with two B boxes, which also includes PML. PML, the transcription factor to which RARα is fused in acute promyelocytic leukemia, has been independently shown to enhance transactivation by PR, without influencing its ligand binding nor its affinity for DNA [29]. To the same family of proteins belongs Efp, which is encoded by an estrogen responsive gene [30]. Therefore, a whole family of proteins exists that facilitates chromatin involving steps in gene induction by steroid hormones.

Another interesting auxiliary factor, namely SUG1, was identified in yeast genetic screens with thyroid hormone receptor (T3R) [31] and RAR [32] and has been reported to be a component of the mediator [33], a subcomplex of the RNA polymerase-II holoenzyme, but also of the 26S proteasome [31]. SUG1 also interacts with the transactivation domains of GAL4 and VP16, as well as with TBP [34] and with TAFII30 [32]. Intriguingly, although binding of both SUG1 and TIF1 requires the AF2 amphipathic helix of nuclear receptors, the details of the interaction are different and the relative affinities of the two proteins for various members of the superfamily are also different [32].

Additional proteins have been identified in genetic screens for intermediary factors [35]. Many of them belong to a family which prototype is SRC-1, an ubiquitous factor isolated by the two hybrid system which interacts with PR *in vitro* in a ligand-dependent fashion [36]. SRC-1 enhances transactivation by PR, ER, GR, T3R and RXR, but also by VP16 and Sp1. The N-terminal truncated SRC-1 is a dominant negative repressor of steroid hormone transactivation. Recently two related proteins with similar properties have been identified, TIF-2 [37] and GRIP1 [38]. Some of these co-activators may act by bridging to the general transcription factors but an interaction with more general integrators, such as the CREB (cyclic AMP responsive element binding protein) binding protein CBP, has been postulated [1]. CBP itself binds the nuclear receptors, but also members of the SRC-1 family as well as several other transcription factors, such as CREB an members of the AP1 family. The elucidation of the functional significance of TIFs and co-activators under physiological conditions is one of the main challenges for the near future and will require the inactivation of the corresponding genes.

4. Steroid Hormone Receptors Participate In Transcriptional Repression

Steroid hormones can repress transcription by virtue of a receptor-dependent inhibition of the activity of other transcription factors. Often the inhibitory effect is reciprocal, in that the activity of the receptors on HREs containing promoters can be inhibited by interaction with the corresponding factors. Inhibitory effects of glucocorticoids on transcription can be mediated by so-called "composite HREs" [39]. like the one found in the proliferin promoter [40, 41]. This 25 bp long element can bind both GR and AP1. Different members of the AP1 family behave similarly in the absence of GR, but have opposite effects on GR action [42]. Four amino acids within the DNA recognition α-helix in the basic region of the AP1 are crucial for this effect. The four residues are solvent accessible on the helical surface away from DNA and could be involved in direct interactions with GR [42]. Under conditions in which GR represses AP1-stimulation, the mineralocorticoid receptor (MR) is inactive [43]. Using GR/MR chimeras a region of the N-terminus of rat GR (amino acids 150-440) has been shown to be required for repression and, therefore, has been called a composite specificity domain [43].

According to recent findings the receptors can exists in at least two different conformations, and the equilibrium between these conformations is influenced by the nucleotide sequence of the HREs. Point mutations in the composite HRE of proliferin promoter lead to transactivation under conditions where the wild type HRE is inactive or represses, for instance in F9 cells which do not have AP1 [44]. A mutant GR, K461A, which exhibits decreased binding to and transactivation from a single canonical GRE, activates transcription in all GRE contexts and even when tethered to the collagenase A promoter via AP1 [44]. In a yeast screen of 60,000 GR mutants for strong activation from the composite HRE of proliferin promoter, all 13 isolated mutants contain a K461 mutation [44]. When this residue is mutated in the mineralocorticoid receptor or in RARß it leads to receptors which activate at response elements from which the wild type receptors repress or are inactive. Thus, this residue serves as an allosteric "lock" that restricts receptors to inactive or repressing configurations except in response to GRE contexts which convert them to activators [44].

A different mechanism of repression is operative in the glucocorticoid inhibition of AP1-mediated induction in genes lacking HREs, such as the collagenase A gene (for a review see [45]). The DNA binding domain of GR, but not the binding of GR to DNA, is required for collagenase transrepression. GR seems to inhibit the activity of the Fos/Jun complex by a direct interaction with the AP1 complex [46, 47]. Although *in vitro* GR can disturb binding of Jun homodimers to a phorbol ester responsive element, it does not influence binding of Jun/Fos heterodimers, under conditions where it completely inhibits phorbol ester induction [48]. These effects may be mediated by a competition between receptors and AP1 for binding to the integrator CBP [1], but a direct interaction between GR and AP1 factors is supported by the reverse effect [46, 49, 50]. The outcome of the cross-talk between steroid hormones and the AP1 signaling pathway can be a reciprocal inhibition or a potentiation depending on the receptor, the AP1 factors, the cell type, and the promoter used [51].

Glucocorticoids exert an inhibitory effect on NF-κB mediated transcription by virtue of two different mechanisms. On the one hand, the NF-κB mediated activation of the IL-6 and the IL-8 genes is inhibited by glucocorticoids through an interaction of GR with the p65 subunit of NF-κB [52, 53]. Conversely, overexpression of p65 inhibits glucocorticoid induction of MMTV [53]. There is also evidence for a direct physical interaction between p65 and GR [53, 54]. Like with AP1, the DNA binding domain of GR is essential, even if no direct DNA binding is necessary [55]. However, there is also evidence for an interaction of GR with the p50 subunit of NF-κB [56]. Apparently, GR blocks the DNA binding activity of NF-κB not only when bound to agonistic ligands but also when complexed to the antiglucocorticoid RU486 [56]. On the other hand glucocorticoids induce expression of I-κBα and thus maintain NF-κB in an inactive cytoplasmic complex [57, 58]. These effects could be important for the antiinflamatory and immunosuppressive effects of glucocorticoids, as the expression of many interleukins are induced by mechanisms involving NF-κB sites.

A connection between glucocorticoids and erythroid differentiation has been postulated in mouse erythroleukemia cells, in which glucocorticoids block globin synthesis. The N-terminal 106 amino acids of GR bind the transcription factor GATA-1 and inhibit its binding to DNA. Several globin genes exhibit GREs close to the GATA regulatory elements and, on these promoters, the inhibitory action of glucocorticoids is more evident. Since GATA-1 is involved in a positive autoregulatory loop, glucocorticoids could inhibit erythroid differentiation by blocking GATA-1 action [59]. Estrogens have been reported to induce apoptosis in erythroid cells by virtue of their inhibitory effect on GATA-1 [60], suggesting a more general relevance for this interaction.

5. Hormone Receptors Can Be Activated By Signals Acting Through Other Transduction Pathways

One unexpected finding has been the discovery that hormone receptors can be activated in response to signals interacting with membrane receptors, even in the absence of the corresponding steroid hormone [61, 62]. Induction of HRE containing reporter genes in response to these signals can be inhibited by antagonistic steroid hormone receptor ligands and is therefore mediated by the corresponding receptor [63, 64]. The mechanism of this cross-talk between membrane signaling and nuclear receptors is unclear. Recently it has been shown that the ER acts as substrate for MAP kinase *in vitro* and in response activation of c-Ha-*Ras in vivo* [65]. A permissive role for the cAMP signaling pathway in glucocorticoid action has been described, but the molecular mechanism was unclear [66, 67]. In hepatoma cells GR is required for recruitment of the transcription factor HNF3 to the enhancer of the tyrosine aminotransferase gene. However, no footprint over the GRE is visible *in vivo* unless the protein kinase A pathway is activated [68]. The potential role of CBP in this cross-talk remains to be established. This kind of connection among different signal transduction pathways could be of great physiological significance, as it would allow the cell to integrate and coordinate its response to various metabolic and endocrine signals.

Very recently another kind of cross-talk, in a sense symmetric to the one describe above, has been reported, namely the steroid-hormone dependent activation of signaling pathways normally activated by membrane receptors. This connection could be essential for understanding the effects of estrogens on cell proliferation in cancer cells. In addition to their effects on transcription of the early genes *fos* and *jun* [69], ovarian hormones also influence the Ras/MAP kinase pathway, which is activated by estrogens in mammary cells probably through a rapid and direct effect on a c-Src-like tyrosine kinase [70]. On the other hand, activation of the Ras/MAP kinase cascade enhances the AF1 function of ER, probably by phosphorylation of the receptor at a specific serine [65]. In that respect it is interesting that the growth effects of EGF in uterine cells may be mediated by ER [71], and that estrogens activate the tyrosine kinase activity of ErbB2 [72]. The elucidation of the molecular mechanisms involved in this cross-talk would be facilitated by the availability of mice lacking GR [73], ER [74] or PR [75].

6. The Function Of The Hormone Responsive Region Of The MMTV Promoter Is Modulated By Its Assembly In Nucleosomes

The MMTV promoter is transcriptionally controlled by steroid hormones, in particular glucocorticoids and progestins. The hormone receptors bind to a hormone responsive region (HRR) and facilitate the interaction of other transcription factors, including Nuclear Factor I (NFI, CTF1) and the octamer transcription factor OTF1 (Oct-1), with the MMTV promoter (for a review see [9]). A role for chromatin structure in MMTV regulation has been postulated based on functional studies with cell lines carrying stable minichromosomes [76, 77] and on *in vitro* nucleosome assembly studies [78-80]. The combined results from these studies suggest that while hormone receptors recognize the HRR of MMTV in chromatin, access of the transcription factor NFI to the MMTV promoter is hindered by its nucleosomal organization [9]. Though the exact positioning of the nucleosome over the MMTV promoter has been debated [81, 82], it seems that under a variety of conditions a dominant nucleosome phase is found which precludes binding of NFI [80, 83]. Hormone induction was believed to cause a displacement or rearrangement of the nucleosome over the HRR, as manifested by the hypersensitivity of this region to DNaseI [76, 84], which enables free access of NFI to its binding site and leads to its transcriptional activation [77].

However, in cells carrying a single copy of the MMTV promoter integrated in chromosomal DNA and organized into a precisely positioned nucleosome, hormone induction does not lead to displacement of this nucleosome, but to a rearrangement which enables simultaneous binding of receptors, NFI and OTF1 [83]. These findings suggest that the basal repression of the MMTV promoter prior to hormone induction is due to its organization in chromatin, and that hormone induction results in a remodeling of chromatin allowing the recruitment of relevant transcription factors to the promoter. Hormone receptors and NFI can not bind simultaneously to the MMTV promoter as free DNA [85], and NFI binding prevents access of OTF1 to the promoter distal octamer motif [86]. Therefore, the observation that all these sites are occupied simultaneously on the surface of a positioned nucleosome, suggests the interesting possibility that the organization in chromatin may be a prerequisite for optimal induction of the MMTV promoter [83].

Positioned nucleosome can fulfill a structural role which facilitates the function of conventional *cis*-acting elements. Examples are the heat shock elements in the *Drosophila hsp26* gene [87] and the vitellogenin B1 promoter of *Xenopus*, where a positioned nucleosome may bring an HRE in apposition to other promoter elements and thus enable interaction between hormone receptors and other transcription factors [88]. Chromatin structure can also facilitate the interaction among more distant DNA sequence as suggested by the "nuclear ligation assay" [89]. Estrogen treatment stimulates 2-3 fold the ligation efficiency between the distal enhancer and the proximal promoter regions of the rat prolactin gene, suggesting that these two regions are juxtaposed in minichromosomes only after hormonal induction.

6.1 THE MMTV PROMOTER IN YEAST

When the MMTV promoter is introduced in *Saccharomyces cerevisiae* and analyzed by genomic footprinting it is precisely organized in a nucleosome which covers the region between -43 and -195. The DNA adopts the same rotational orientation on the surface of this nucleosome as found in animal cells and in chromatin assembly reactions [90]. Therefore a short region of the MMTV LTR is able to impose over the yeast chromatin assembly machinery the positioning characteristic of animal cells in the absence of linker histones and sequence specific transcription factors. This implies that in addition to coding and regulatory information, there is conformational or topological information in DNA that is implemented in chromatin, modulates the accessibility of regulatory information, and is therefore, critical for the realisation of the genetic program.

The MMTV promoter is silent in yeast in the absence of cotransfected transactivators. Addition of NFI expression vectors results in negligible activity of either the wild type or a truncated MMTV promoter lacking the HRR. However, upon expression of a chimeric NFI-VP16 transactivator the truncated MMTV promoter is activated to a much higher extent than the wild type promoter containing the HRR [91]. These results support the hypothesis that the HRR represses access to the NFI site by positioning a nucleosome over the MMTV promoter. Upon expression of GR and in the presence of an agonistic ligand, the MMTV promoter is activated in an NFI-dependent fashion. Following hormone induction there is no indication of displacement or disruption of the nucleosome over the MMTV promoter. The only change observed is a nuclease hypersensitivity over the TATA box region located in the linker between two nucleosomes [90]. In yeast strains carrying an inducible histone H4 gene have shown that nucleosome depletion leads not only to increase constitutive activity of the MMTV promoter in the absence of receptor but also to a higher accessibility of the HRR for the receptors . However, the functional synergism between receptors and NFI which is essential for full induction in response to hormone does not take place when the MMTV promoter is depleted of nucleosomes [92]. These findings support the concept that the nucleosomal organization of the MMTV promoter is involved in both constitutive repression and induced transcription.

6.2 NFI BINDING TO RECONSTITUTED MMTV NUCLEOSOMES

The repressive role of the nucleosomal assembly can be reproduced in vitro in nucleosome assembly reactions with the MMTV promoter. Under no conditions did we detect binding of NFI to its cognate site on the promoter assembled into a nucleosomal core particle [93, 94]. Even when the rotational phasing of the NFI site was changed to the opposite of that found in the wild type configuration, there was no binding of NFI, as long as the cognate site was encompassed within the limits of the core particle [95] . This is probably due to the extensive nature of the contacts between the NFI and DNA, which embrace the whole circumference of the double helix [95]. On the contrary the hormone receptors are known to contact the DNA double helix through a narrow sector of its circumference [96, 97]. These findings may explain the difference in affinity of NFI and receptors for free DNA and nucleosomes. While NFI binds free MMTV promoter fragments with a much higher affinity than hormone receptors, the relationship is reversed when these fragments are assembled on a nucleosome core particle. It is likely that this observations can be generalized, and that proteins binding free DNA with high affinity, e.g. interacting with many atoms of the double helix, are unable to bind to their cognate sites when wrapped around a histone octamer, whereas proteins with relatively low affinity and few contacts to free DNA, may be able to bind to nucleosomally organized targets, provided the double helix is properly oriented.

7. Hormone Induction Involves Remodeling Of The Chromatin Structure Of The MMTV Promoter

7.1 MULTIPROTEIN COMPLEXES INVOLVED IN CHROMATIN REMODELING

Experiments with GR expressed in yeast pointed for the first time to a potentially relevant set of receptor interacting proteins, namely components of the SWI/SNF complex. The SWI/SNF complex, composed of 11 polypeptides, includes a set of pleiotropic transactivators which appear to be important for transcription of inducible genes, probably by mechanism involving remodeling of chromatin [98]. Yeast strains with mutations in genes encoding components of the SWI/SNF complex are defective in transactivation by GR, and GR interacts with SWI3 *in vivo* and *in vitro* [99]. One homologue of SWI2 in *Drosophila* is an helicase with DNA dependent ATPase activity, called *brahma* (*brm*), which if mutated suppress *polycomb* mutations and results in low homeotic gene activity [100]. Curiously, a suppressor of the *swi/snf* phenotype, *Spt6*, potentiates transactivation by ER in yeast, and has been postulated to act as co-activator [101].

A human homologue of *Drosophila brm*, *hbrm*, is a 180 kDa nuclear factor which can act as a transcription factor when fused to an heterologous DNA binding domain. The mouse *brm* is expressed ubiquitously, whereas *hbrm* is not expressed in all cells. In cells lacking *hbrm*, such as C33 and SW13, GR induction of a transfected reporter is weak and can be selectively enhanced by cotransfection of a hbrm expression vector [102]. Cooperation requires the DBD of GR and two regions of *hbrm*, including the helicase domain. A second human homologue to yeast SWI2 and *Drosophila brm*, named BRG1, is a nuclear protein of 205 kDa which ATPase domain can, as a chimeric protein, restore GR-dependent transcription in yeast strains lacking SWI2 [103]. Like SWI2 in yeast, *hbrm* is part of a large multiprotein complex, which could be involved in the remodeling of chromatin required for transcription and/or act as a bridge between activators and the basal transcriptional machinery. The human SWI/SNF complex isolated from HeLa cells mediates ATP-dependent disruption of a nucleosome containing GAL4 binding sites, and enables binding of transactivators linked to the GAL4 DNA binding domain to the nucleosome core [3]. The mechanism of this nucleosome disruption is not completely clear but could involve a displacement of histone H2A/H2B dimers [4]. Moreover, BRG1 binds specifically the retinoblastoma product Rb in a yeast two-hybrid system, and exhibits tumor suppressor activity in SW13 cells. This activity is lost in BRG1 mutants unable to bind Rb, or in the presence of Adeno E1A [104]. Moreover, Rb upregulates GR-mediated transactivation, and this effect is dependent on hBrm [105]. Thus, components of the SWI/SNF complex could also act as a connection between gene regulation and the cell cycle.

Very recently a novel chromatin remodeling complex of 500 kDa, NURF, has been characterized and purified from *Drosophila* embryo extracts [5]. NURF is composed of 4 polypeptides, one of which is ISWI, is a member of the SWI/SNF2 family, with ATPase activity which is stimulated not by free DNA or histones but by nucleosomes [106]. NURF and hydrolyzable ATP are required to facilitate the GAGA factor-induced remodeling of the nucleosome containing the promoter of the *Hsp70* gene in *Drosophila* embryo extracts [5]. Thus, very likely several chromatin remodeling machineries exist in the cell nucleus, which can be selectively recruited to the genetic control regions by mechanisms still to be uncovered.

7.2 DNASE I HYPERSENSITIVITY OVER THE MMTV NUCLEOSOME

Evidence for a participation of chromatin structure in the regulated expression of steroid inducible genes was initially derived from an analysis of DNaseI hypersensitive sites in chromatin after hormone induction. In chicken liver, the vitellogenin gene exhibits an estrogen dependent DNaseI hypersensitive site in the upstream promoter region, which has been the subject of detailed studies [107, 108]. Estrogens induce the demethylation of CpG sites in the 5′-flanking region of the gene [109], by mechanisms which may involved other DNA binding proteins [110]. However, the exact nucleosomal organization of the relevant region has not been defined. Hormone induced DNaseI hypersensitive sites have been found in many other inducible genes including the rat tyrosine aminotransferase gene in liver [111], the chicken lysozyme gene in oviduct cells [112] and the rabbit uteroglobin gene in endometrium [113]. In all these cases it was assumed that the DNaseI hypersensitive regions reflect the hormone induced removal or disruption of a nucleosome [114]. Evidence for this kind of mechanism has been presented for the enhancer of the rat tyrosine aminotransferase gene in hepatocytes. This enhancer located around 2.5 kb upstream of the transcription initiation site, is regularly organized in nucleosomes [115], five of which appear to be precisely positioned. Upon induction, the chromatin structure is altered in a reversible way, in that DNaseI hypersensitive sites appear 20 min after hormone treatment and disappear following removal of the hormone or addition of the antiglucocorticoid RU486 [116]. There is some controversy about the nature of the chromatin changes, one group finds no evidence for disruption of nucleosomes over the enhancer [116], another group claims that two out of the five positioned nucleosomes are disrupted upon induction [117]. This is reminiscent of the situation in the MMTV promoter, where gentle genomic footprint analysis does not reveal a long region of DNaseI hypersensitivity, but rather a narrow region over the pseudo-dyad axis of a nucleosome [83]. However, using more extensive experimental manipulations a long region of nuclease hypersensitivity was observed over the MMTV promoter ([84] and our own unpublished results), the hormone induced chromatin remodeling may be similar in both cases.

7.3 MODIFICATIONS OF THE CORE HISTONES

The hormone induced conformational change of chromatin is not stable and generates a nucleosome, which does not resist manipulation of cell nuclei. The nature of this conformational change remains obscure, but experiments with inhibitors of histone deacetylases suggest a role for modification of the histone tails. Among the changes in nucleosomes that have been associated with transcriptionally active chromatin, are hyperacetylation of lysine residues in the N-terminal tails of all four core histones [118]. Recently it has been found that restriction of GAL4 binding to reconstituted nucleosomes containing GAL4 binding sites, can be alleviated by proteolytic digestion of the histone tails, suggesting a general repressive role for this highly charged domains of the core histones [119, 120]. Similarly, binding of the transcription factor TFIIIA to a reconstituted nucleosome carrying a 5S RNA gene, is enhanced by acetylation of the histone tails [8]. In the case of MMTV, complete inhibition of histone deacetylase with 5-10 mM sodium butyrate inhibits hormone induction and nucleosome remodeling [121]. However, lower concentrations of butyrate activate hormone-independent transcription from single copy integrated MMTV-reporters [122]. A similar response is observed with a more selective inhibitor of histone deacetylase activity, trichostatin A, which acts at nanomolar concentrations [123]. Moreover, inducing concentrations of butyrate or trichostatin A generate the same type of DNaseI hypersensitivity over the pseudo-dyad axis of the regulatory nucleosome that we observed following hormone induction [92, 122]. Though we do not know whether hormone treatment alters the acetylation of core histones, these results suggest that a nucleosome remodeling similar to that induced by receptor binding can be generated by changes in the core histone tails. However, alternative changes in nucleosome structure, such as removal of histone H2A/H2B [124], have to be considered as additional possibilities for facilitating factor binding to nucleosomally organized DNA sequences.

7.4 DISSOCIATION OF DIMERS OF HISTONES H2A AND H2B

It is intriguing that the hormone-induced structural alteration in chromatin structure takes place at the nucleosome pseudo dyad axis, as this region has been shown to be crucial for control of chromatin-mediated gene expression involving the SWI/SNF complex in yeast [98, 125]. Mutations in histones H3 and H4 located near the nucleosome dyad axis suppress the phenotype of the *swi/snf* mutations [98], suggesting that the architecture of this region of the nucleosome is an important determinant of transcriptional activity. Recently, a soluble complex of all SWI/SNF gene products has been identified, which may play an important role in facilitating chromatin transcription [126, 127]. As mentioned above, this complex enhances binding of GAL4 derivatives to nucleosomally organized GAL 4 binding sites in a ATP-dependent manner. This process is accompanied by a depletion of histone H2A/H2B dimers and is facilitated by nucleoplasmin [4]. It is conceivable that hormone induction of chromatin remodeling is accomplished by a receptor-dependent recruitment of the SWI/SNF complex to MMTV promoter, followed by destabilization of the histone octamer, which would then facilitate access to masked HREs, to the NFI binding site and to the octamer distal motif. Support for this notion comes from our recent studies on the binding of NFI and OTF1 to the MMTV promoter assembled with either a histone octamer or a tetramer of recombinant histones H3 and H4 [128]. With the core histone octamer particle we

138

confirm previous results, showing that even at very high concentrations of NF1, sufficient to completely shift the free DNA fragment, no binding could be detected to a nucleosomally organized MMTV promoter DNA [80, 93, 129]. In contrast, the MMTV promoter assembled with a H3/H4 tetramer is accessible to NF1 binding, as demonstrated by the appearance of a ternary complex of the tetrameric particle and NF1 in band shift experiments. This finding was unexpected since exonuclease III cleavage and hydroxyl radical footprinting show that the NF1 binding site between -75 and -62 is confined within the limits of the H3/H4 tetramer particle and the rotational orientation of the DNA in the tetramer particle, which is the same as in the octamer particle, should preclude NF1 binding [93]. Possibly, NF1 binding to the tetramer particle reflects the higher accessibility of the DNA over the NF1 binding site as demonstrated in the DNaseI digestion pattern.

7.5 ADDITIONAL MECHANISMS

Other mechanisms may also modulate the binding of factors to nucleosomally organized DNA. One possibility is a direct interaction with core histones, as has been recently reported between the SIR3 and SIR4 and the N-terminal regions of histones H3 and H4 in the context telomere silencing in yeast [130]. Recently, an interaction between NFI and the globular domain of histone H3 has been described [131]. However, this interaction is unlikely to play an important role for binding of NFI to the MMTV nucleosome, as the relevant region of the protein, is not present in the variant of NFI, namely CTF-2, used for the majority of our studies.

As the linker histones are important for condensation of the 10 nm chromatin fibre [132], they could also influence the accessibility to DNA sequences in chromatin. Moreover, important cis-elements of the MMTV promoter are located at the edge of the positioned nucleosome or in the linker DNA, which are regions contacted by histone H1 [133]. In fact, it has been claimed that upon hormone induction there is a depletion of histone H1 from the MMTV promoter, and this could influence the accessibility of promoter sequences [134]. Although this remains an attractive possibility, it can not be an essential prerequisite for hormonal induction of MMTV transcription, since, the MMTV promoter is perfectly regulated in budding yeast, which lacks linker histones [90]. Nevertheless, it is possible that in animal cells linker histones play a role in modulating promoter accessibility, possibly through their effect on nucleosome mobility [135].

Acknowledgments

The experimental work reported in this chapter was supported by grants from the Deutsche Forschungsgemeinschaft, the European Union, the Schering Foundation, and the Fond der Chemischen Industrie.

References

1. Kamei Y., Xu L., Heinzel T., Torchia J., Kurokawa R., Gloss B., Lin S.C., Heyman R.A., Rose D.W., Glass C.K. et al. (1996) A CBP integrator complex mediates transcriptional activation and AP-1 inhibition by nuclear receptors, *Cell* **85**, 403-414

2. Peterson C.L.and Tamkun J.W. (1995) The SWI-SNF complex: a chromatin remodeling machine, *Trends Biochem Sci* **20**, 143-146

3. Kwon H., Imbalzano A.N., Khavari P.A., Kingston R.E.and Green M.R. (1994) Nucleosome disruption and enhancement of activator binding by a human SWI/SNF complex, *Nature* **370**, 477-481

4. Côte J., Quinn J., Workmann J.L.and Peterson C.L. (1994) Stimulation of GAL4 derivative binding to nucleosomal DNA by the yeast SWI/SNF complex, *Science* **265**, 53-59

5. Tsukiyama T.and Wu C. (1995) Purification and properties of an ATP-dependent nucleosome remodelling factor, *Cell* **83**, 1011-1020

6. Kingston R.E., Bunker C.A.and Imbalzano A.N. (1996) Repression and activation by multiprotein complexes that alter chromatin structure, *Genes Dev* **10**, 905-920

7. Turner B.M. (1993) Decoding the nucleosome, *Cell* **75**, 5-8

8. Lee D.Y., Hayes J.J., Pruss D.and Wolffe A.P. (1993) A positive role for histone acetylation in transcription factor access to nucleosomal DNA, *Cell* **72**, 73-84

9. Beato M. (1991) Transcriptional regulation of mouse mammary tumor virus by steroid hormones, *Crit Rev Oncogen* **2**, 195-210

10. Beato M., Herrlich P.and Schütz G. (1995) Steroid hormone receptors: Many actors in search of a plot., *Cell* **83**, 851-857

11. Schwabe J.W.R., Chapman L., Finch J.T.and Rhodes D. (1993) The crystal structure of the estrogen receptor DNA-binding domain bound to DNA: How receptors discriminate between their response elements, *Cell* **75**, 567-578

12. Bourguet W., Ruff M., Chambon P., Gronemeyer H.and Moras D. (1995) Crystal structure of the ligand-binding domain of the human nuclear receptor RXR-alpha, *Nature* **375**, 377-382

13. Danielian P.S., White R., Lees J.A.and Parker M.G. (1992) Identification of a conserved region required for hormone dependent transcriptional activation by steroid hormone receptors, *EMBO J* **11**, 1025-1033

14. Renaud J.P., Rochel N., Ruff M., Vivat V., Chambon P., Gronemeyer H.and Moras D. (1995) Crystal structure of the RAR-γ ligand-binding domain bound to all-*trans* retinoic acid, *Nature* **378**, 681-689

15. Wagner R.L., Apriletti J.W., Mcgrath M.E., West B.L., Baxter J.D.and Fletterick R.J. (1995) A structural role for hormone in the thyroid hormone receptor, *Nature* **378**, 690-697

16. Dahlman-Wright K., Baumann H., McEwan I.J., Almlof T., Wright A.P.H., Gustafsson J.A.and Härd T. (1995) Structural characterization of a minimal functional transactivation domain from the human glucocorticoid receptor, *Proc Natl Acad Sci USA* **92**, 1699-1703

17. Sadovsky Y., Webb P., Lopez G., Baxter J.D., Fitzpatrick P.M., Gizang-Ginsberg E., Cavailles V., Parker M.G.and Kushner P.J. (1995) Transcriptional activators differ in their responses to overexpression of TATA-box-binding protein, *Mol Cell Biol* **15**, 1554-1563

18. Tjian R.and Maniatis T. (1994) Transcriptional activation: a complex puzzle with few easy pieces, *Cell* **77**, 5-8

19. Verrijzer C.P., Chen J.L., Yokomori K.and Tjian R. (1995) Binding of TAFs to core elements directs promoter selectivity by RNA polymerase II, *Cell* **81**, 1115-1125

20. Jacq X., Brou C., Lutz Y., Davidson I., Chambon P.and Tora L. (1994) Human TAFII30 is present in a distinct TFIID complex and is required for transcriptional activation by the estrogen receptor, *Cell* **79**, 107-117

21. Schwerk C., Klotzbücher M., Sachs M., Ulber V.and Klein-Hitpass L. (1995) Identification of a transactivation function in the progesterone receptor that interacts with the TAFII110 subunit of the TFIID complex, *J Biol Chem* **270**, in press

22. Ing N.H., Beekman J.M., Tsai S.Y., Tsai M.J.and O'Malley B.W. (1992) Members of the steroid hormone receptor superfamily interact with TFIIB (S300-II), *J Biol Chem* **267**, 17617-17623

23. Halachmi S., Marden E., Martin G., Mackay H., Abbondanza C.and Brown M. (1994) Estrogen receptor-associated proteins: possible mediators of hormone-induced transcription, *Science* **264**, 1455-1458

24. Cavaillès V., Dauvois S., Danielian P.S.and Parker M.G. (1994) Interaction of proteins with transcriptionally active estrogen receptors, *Proc Natl Acad Sci USA* **91**, 10009-10013

25. Cavailles V., Dauvois S., L'Horset F., Lopez G., Hoare S., Kushner P.J.and Parker M.G. (1995) Nuclear factor RIP140 modulates transcriptional activation by the estrogen receptor, *EMBO J* **14**, 3741-3751

26. Kurokawa R., Söderström M., Hörlein A., Halachmi S., Brown M., Rosenfeld M.G.and Glass C.K. (1995) Polarity-specific activities of retinoic acid receptors determined by a co-repressor, *Nature* **377**, 451-454

27. Eggert M., Möws C.C., Tripier D., Arnold R., Michel J., Nickel J., Schmidt S., Beato M.and Renkawitz R. (1995) A fraction enriched in a novel glucocorticoid receptor-interacting protein stimulates receptor-dependent transcription in vitro, *J Biol Chem* **270**, 30755-30759

140

28. LeDouarin B., Zechel C., Garnier J.M., Lutz Y., Tora L., Pierrat B., Heery D., Gronemeyer H., Chambon P.and Losson R. (1995) The N-terminal part of TIF1, a putative mediator of the ligand-dependent activation function (AF-2) of nuclear receptors, is fused to B-raf in the oncogenic protein T18, *EMBO J* **14**, 2020-2033

29. Guiochon-Mantel A., Savouret J.F., Qignon F., Delabre K., Milgrom E.and De The H. (1995) Effect of PML and PML-RAR on the transcription properties and subcellular distribution of steroid hormone receptors, *Mol Endocrin* **9**, 1791-1803

30. Inoue S., Orimo A., Hosoi T., Kondo S., Toyoshima H., Kondo T., Ikegami A., Ouchi Y., Orimo H.and Muramatsu M. (1993) Genomic binding-site cloning reveals an estrogen-responsive gene that encodes a RING finger protein, *Proc Natl Acad Sci USA* **90**, 11117-11121

31. Lee J.W., Ryan F., Swaffield J.C., Johnston S.A.and Moore D.D. (1995) Interaction of thyroid-hormone receptor with a conserved transcriptional mediator, *Nature* **374**, 91-94

32. vom Baur E., Zechel C., Heery D., Heine M., Garnier J.M., Vivat V., LeDouarin B., Gronemeyer H., Chambon P.and Losson R. (1995) Differential ligand-dependent interactions between the AF-2 activation domain of nuclear receptors and the putaive transcriptional intermediary factors mSUG1 and TIF1, *EMBO J* **15**, 110-124

33. Kim Y.J., Bjorklund S., Li Y., Sayre M.H.and Kornberg R.D. (1994) A multiprotein mediator of transcriptional activation and its interaction with the C-terminal repeat domain of RNA polymerase II, *Cell* **77**, 599-608

34. Swaffield J.C., Melcher K.and Johnston S.A. (1995) A highly conserved ATPase protein as a mediator between acidic activation domains and the TATA-binding protein, *Nature* **374**, 88-91

35. Lee J.W., Choi H.S., Gyuris J., Brent R.and Moore D.D. (1995) Two classes of proteins dependent on either the presence or absence of thyroid hormone for interaction with the thyroid hormone receptor, *Mol Endocrinol* **9**, 243-254

36. Oñate S.A., Tsai S.Y., Tsai M.-J.and O´Malley B.W. (1995) Sequence and characterization of a coactivator for the steroid hormone receptor superfamily, *Science* **270**, 1354-1357

37. Voegel J.J., Heine M.J.S., Zechel C., Chambon P.and Gronemeyer H. (1996) TIF2, a 160 kDa transcriptional mediator for the ligand-dependent activation function AF-2 of nuclear receptors, *EMBO J* **15**, 3667-3675

38. Hong H., Kohli K., Trivedi A., Johnson D.L.and Stallcup M.R. (1996) GRIP1, a novel mouse protein that serves as a transcriptional coactivator in yeast for the hormone binding domains of steroid receptors, *Proc Natl Acad Sci USA* **93**, 4948-4952

39. Miner J.N.and Yamamoto K.R. (1991) Regulatory Crosstalk at Composite Response Elements, *Trends Biochem Sci* **16**, 423-426

40. Mordacq J.C.and Linzer D.I.H. (1989) Co-localization of elements required for phorbol ester stimulation and glucocorticoid repression of proliferin gene expression , *Genes&Develop* **3**, 760-769

41. Diamond M.I., Miner J.N., Yoshinaga S.K.and Yamamoto K.R. (1990) Transcription factor interactions: selectors of positive or negative regulation from a single DNA element, *Science* **249**, 1266-1272

42. Miner J.N.and Yamamoto K.R. (1992) The basic region of AP-1 specifies glucocorticoid receptor activity at a composite response element, *Gene Develop* **6**, 2491-2501

43. Pearce D.and Yamamoto K.R. (1993) Mineralocorticoid and glucocorticoid receptor activities distinguished by nonreceptor factors at a composite response element, *Science* **259**, 1161-1165

44. Starr D.B., Matsui W., Thomas J.R.and Yamamoto K.R. (1996) Intracellular receptors use a common mechanism to interpret signaling information at response elements, *Genes Dev* **10**, 1271-1283

45. Schüle R.and Evans R.M. (1991) Cross-Coupling of Signal Transduction Pathways - Zinc Finger Meets Leucine Zipper, *Trends Genet* **7**, 377-381

46. Schüle R., Rangarajan P., Kliewer S., Ransone L.J., Bolado J., Yang N., Verma I.M.and Evans R.M. (1990) Functional antagonism between oncoprotein c-Jun and the glucocorticoid receptor., *Cell* **62**, 1217-1226

47. Jonat C., Rahmsdorf H., Park K., Cato A., Gebel S., Ponta H.and Herrlich P. (1990) Antitumor promotion and antiinflamation: Down-regulation of AP-1 (Fos/Jun) activity by glucocorticoid hormone, *Cell* **62**, 1189-1204

48. König H., Ponta H., Rahmsdorf H.J.and Herrlich P. (1992) Interference between pathway-specific transcription factors: glucocorticoids antagonize phorbol ester-induced AP-1 activity without altering AP-1 site occupation *in vivo*, *EMBO J* **11**, 2241-2246

49. Lucibello F.C., Slater E.P., Joos K.U., Beato M.and Müller R. (1990) Mutual transrepression of Fos and the glucocorticoid receptor: involvement of a functional domain in Fos which is absent in FosB, *EMBO J* **9**, 2827-2834

50. Yang-Yen H.-F., Chambard J.-C., Sun Y.-L., Smeal T., Schmidt T.J., Drouin J.and Karin M. (1990) Transcriptional interference between c-Jun and the glucocorticoid receptor: Mutual inhibition of DNA binding due to direct protein-protein interaction, *Cell* **62**, 1205-1215

51. Shemshedini L., Knauthe R., Sassone-Corsi P., Pornon A.and Gronemeyer H. (1991) Cell-specific inhibitory and stimulatory effects of Fos and Jun on transcription activation by nuclear receptors, *EMBO J* **10**, 3839-3849

52. Mukaida N., Morita M., Ishikawa Y., Rice N., Okamoto S., Kasahara T.and Matsushima K. (1994) Novel mechanism of glucocorticoid-mediated gene repression: nuclear factor-kB is target for glucocorticoid-mediated interleukin 8 gene repression, *J Biol Chem* **269**, 13289-13295

53. Ray A.and Prefontaine K.E. (1994) Physical association and functional antagonism between the p65 subunit of transcription factor NF-κB and the glucocorticoid receptor, *Proc Natl Acad Sci USA* **91**, 752-756

54. Ray A., Siegel M.D., Prefontaine K.E.and Ray P. (1995) Anti-inflammation: direct physical association and functional antagonism between transcription factor NF-κB and the glucocorticoid receptor, *Chest* **107**, S139

55. Caldenhoven E., Liden J., Wissink S., Van de Stolpe A., Raaijmakers J., Koenderman L., Okret S., Gustafsson J.A.and Van der Saag P.T. (1995) Negative cross-talk between RelA and the glucocorticoid receptor: a possible mechanism for the antiinflammatory action of glucocorticoids, *Mol Endocrinol* **9**, 401-412

56. Scheinman R.I., Gualberto A., Jewell C.M., Cidlowski J.A.and Baldwin A.S. (1995) Characterization of mechanisms involved in transrepression of NF-κB by activated glucocorticoid receptors, *Mol Cell Biol* **15**, 943-953

57. Scheinman R.I., Cogswell P.C., Lofquist A.K.and Baldwin A.S. (1995) Role of transcriptional activation of I-kBα in mediation of immunosuppression by glucocorticoids, *Science* **270**, 283-286

58. Auphan N., DiDonato J.A., Rosette C., Helmberg A.and Karin M. (1995) Immunosuppression by glucocorticoids: inhibition of NF-kB activity through induction of IkB synthesis, *Science* **270**, 286-290^

59. Chang T.J., Scher B.M., Waxman S.and Scher W. (1993) Inhibition of mouse GATA-1 function by the glucocorticoid receptor: possible mechanism of steroid inhibition of erythroleukemia cell differentiation, *Mol Endocrinol* **7**, 528-542

60. Blobel G.A.and Orkin S.H. (1996) Estrogen-induced apoptosis by inhibition of the erythroid transcription factor GATA-1, *Mol Cell Biol* **16**, 1687-1694

61. Power R.F., Mani S.K., Codina J., Conneely O.M.and O'Malley B.W. (1991) Dopaminergic and ligand-independent activation of steroid hormone receptors, *Science* **254**, 1636-1639

62. O'Malley B.W., Schrader W.T., Mani S., Smith C., Weigel N.L., Conneely O.M.and Clark J.H. An alternative ligand-independent pathway for activation of steroid receptors. In: CW Bardin, eds. Book. 525 B Street, Suite 1900, San Diego, CA 92101-4495: Academic Press Inc, 1995:333-347.

63. Smith C.L., Conneely O.M.and O'Malley B.W. (1993) Modulation of the ligand-independent activation of the human estrogen receptor by hormone and antihormone, *Proc Natl Acad Sci USA* **90**, 6120-6124

64. Aronica S.M., Kraus W.L.and Katzenellenbogen B.S. (1994) Estrogen action via the cAMP signaling pathway: Stimulation of adenylate cyclase and cAMP-regulated gene transcription, *Proc Natl Acad Sci USA* **91**, 8517-8521

65. Kato S., Endoh H., Masuhiro Y., Kitamoto T., Uchiyama S., Sasaki H., Masushige S., Gotoh Y., Nishida E., Kawashima H. et al. (1995) Activation of the estrogen receptor by Ras through phosphorylation by mitogen-activated protein kinase, *Science* **270**, 1491-1494

66. Schmid E., Schmid W., Jantzen M., Mayer D., Jastorff B.and Schütz G. (1987) Transcription activation of the tyrosine aminotransferase gene by glucocorticoids and cAMP in primary hepatocytes, *Eur J Biochem* **165**, 499-506

67. Ruppert S., Boshart M., Bosch F.X., Schmid W., Fournier R.E.K.and Schütz G. (1990) Two genetically defined trans-acting loci coordinately regulate overlapping sets of liver-specific genes, *Cell* **61**, 895-904

68. Espinás M.L., Roux J., Pictet R.and Grange T. (1995) Glucocorticoids and protein kinase A coordinately modulate transcription factor recruitment at a glucocorticoid-responsive unit, *Mol Cell Biol* **15**, 5346-5354

69. Weisz A.and Bresciani F. (1993) Estrogen regulation of proto-oncogenes coding for nuclear proteins, *Critical Rev Oncogen* **4**, 361-388

70. Migliaccio A., DiDomenico M., Castoria G., de Falco A., Nola E.and Auricchio F. (1995) Tyrosine kinase/p21ras/MAP-kinase pathway activation by estradiol-receptor complex im MCF-7 cells, *EMBO J* **15**, 1292-1300

71. Ignar-Trowbridge D.M., Nelson K.G., Bidwell M.C., Curtis S.W., Washburn T.F., McLachlan J.A.and Korach K.S. (1992) Coupling of dual signaling pathways. Epidermal growth factor action involves the estrogen receptor, *Proc Natl Acad Sci USA* **89**, 4658-4662

72. Matsuda S., Kadowaki Y., Ichino M., Akiyama T., Toyoshima K.and Yamamoto T. (1993) 17ß-estradiol mimics ligand activity of the c-erb B2 protoöncogen product, *Proc Natl Acad Sci, USA* **90**, 10803-10807

73. Cole T.J., Blendy J.A., Monaghan A.P., Krieglstein K., Schmid W., Aguzzi A., Fantuzzi G., Hummler E., Unsicker K.and Schütz G. (1995) Targeted disruption of the glucocorticoid receptor gene blocks adrenergic chromaffin cell development and severely retards lung maturation, *Genes & Dev* **9**, 1608-1621

74. Lubahn D.B., Moyer J.S., Golding T.S., Couse J.F., Korach K.S.and Smithies O. (1993) Alteration of reproductive function but not prenatal sexual development after insertional disruption of the mouse estrogen receptor gene, *Proc Natl Acad Sci USA* **90**, 11162-11166

142

75. Lydon J.P., DeMayo F.J., Funk C.R., Mani S.K., Hughes A.R., Montgomery C.A., Shyamala G., Conneeely O.M.and O´Malley B.W. (1995) Mice lacking progesterone receptor exhibit pleiotropic reproductive abnormalities, *Genes & Dev* **9**, 2266-2278

76. Richard-Foy H.and Hager G.L. (1987) Sequence-specific positioning of nucleosomes over the steroid-inducible MMTV promoter, *EMBO J* **6**, 2321-2328

77. Cordingley M.G., Riegel A.T.and Hager G.L. (1987) Steroid-dependent interaction of transcription factors with the inducible promoter of mouse mammary tumor virus in vivo, *Cell* **48**, 261-270

78. Perlmann T.and Wrange Ö. (1988) Specific glucocorticoid receptor binding to DNA reconstituted in a nucleosome, *EMBO J* **7**, 3073-3079

79. Piña B., Truss M., Ohlenbusch H., Postma J.and Beato M. (1990) DNA rotational positioning in a regulatory nucleosome is determined by base sequence. An algorithm to model the preferred superhelix, *Nucleic Acids Res* **18**, 6981-8987

80. Archer T.K., Cordingley M.G., Wolford R.G.and Hager G.L. (1991) Transcription factor access is mediated by accurately positioned nucleosomes on the mouse mammary tumor virus promoter, *Mol Cell Biol* **11**, 688-698

81. Fragoso G., John S., Roberts M.S.and Hager G.L. (1995) Nucleosome positioning on the MMTV LTR results from the frequency-biased occupancy of multiple frames, *Genes & Dev* **9**, 1933-1947

82. Roberts M.S., Fragoso G.and Hager G.L. (1995) Nucleosomes reconstituted in vitro on Mouse Mammary Tumor Virus B region DNA occupy multiple translational and rotational frames, *Biochemistry* **34**, 12470-12480

83. Truss M., Bartsch J., Schelbert A., Haché R.J.G.and Beato M. (1995) Hormone induces binding of receptors and transcription factors to a rearranged nucleosome on the MMTV promoter *in vivo*, *EMBO J* **14**, 1737-1751

84. Zaret K.S.and Yamamoto K.R. (1984) Reversible and persistent changes in chromatin structure accompanying activation of a glucocorticoid-dependent enhancer element , *Cell* **38**, 29-38

85. Brüggemeier U., Rogge L., Winnacker E.L.and Beato M. (1990) Nuclear factor I acts as a transcription factor on the MMTV promoter but competes with steroid hormone receptors for DNA binding, *EMBO J* **9**, 2233-2239

86. Möws C., Preiss T., Slater E.P., Cao X., Verrijzer C.P., van der Vliet P.and Beato M. (1994) Two independent pathways for transcription from the MMTV promoter, *J Steroid Biochem Molec Biol* **51**, 21-32

87. Thomas G.H.and Elgin S.C.R. (1988) Protein/DNA architecture of the DNaseI hypersensitive region of the *Drosophila* hsp26 promoter, *EMBO J* **7**, 2191-2201

88. Schild C., Claret F.X., Wahli W.and Wolffe A.P. (1993) A nucleosome-dependent static loop potentiates estrogen-regulated transcription from the Xenopus vitellogenin-B1 promoter invitro, *EMBO J* **12**, 423-433

89. Cullen K.E., Kladde M.P.and Seyfred M.A. (1993) Interaction between transcription regulatory regions of prolactin chromatin, *Science* **261**, 203-206

90. Chávez S., Candau R., Truss M.and Beato M. (1995) Constitutive repression and nuclear factor I-dependent hormone activation of the Mouse Mammary Tumor Virus promoter in yeast, *Mol Cell Biol* **15**, 6987-6998

91. Candau R., Chávez S.and Beato M. (1996) The hormone responsive region of Mouse Mammary Tumor Virus positions a nucleosome and precludes access of nuclear factor I to the promoter, *J Steroid Biochem Molec Biol* **57**, 19-31

92. Truss M., Candau R., Chávez S.and Beato M. (1995) Transcriptional control by steroid hormones: the role of chromatin, *Ciba Found Symp* **191**, 7-23

93. Piña B., Brüggemeier U.and Beato M. (1990) Nucleosome positionining modulates accessibility of regulatory proteins to the mouse mammary tumor virus promoter, *Cell* **60**, 719-731

94. Piña B., Barettino D., Truss M.and Beato M. (1990) Structural features of a regulatory nucleosome, *J Mol Biol* **216**, 975-990

95. Eisfeld K., Candau R., Truss M.and Beato M. (1996) Binding of NF1 to the MMTV promoter in nucleosomes: Influence of rotational phasing, translational positioning and histone H1, *J Mol Biol* **submitted**,

96. Scheidereit C.and Beato M. (1984) Contacts between receptor and DNA double helix within a glucocorticoid regulatory element of mouse mammary tumor , *Proc Natl Acda Sci USA* **81**, 3029-3033

97. Truss M., Chalepakis G.and Beato M. (1990) Contacts between steroid hormone receptors and thymines in DNA: An interference method, *Proc Natl Acad Sci USA* **87**, 7180-7184

98. Winston F.and Carlson M. (1992) Yeast SNF/SWI transcriptional activators and the SPT/SIN chromatin connection, *Trends Genet* **8**, 387-391

99. Yoshinaga S.K., Peterson C.L., Herskowitz I.and Yamamoto K.R. (1992) Roles of SWI1, SWI2, and SWI3 proteins for transcriptional enhancement by steroid receptors, *Science* **258**, 1598-1604

100. Tamkun J.W., Deuring R., Scott M.P., Kissinger M., Pattatucci A.M., Kaufman T.C.and Kennison J.A. (1992) brahma: A regulator of Drosophila homeotic genes structurally related to the yeast transcriptional activator SNF2/SWI2, *Cell* **68**, 561-572

101. Baniahmad C., Nawaz Z., Baniahmad A., Gleeson M.A.G., Tsai M.J.and O´Malley B.W. (1995) Enhancement of human estrogen receptor activity by SPT6: A potential coactivator, *Mol Endocrinol* **9**, 34-43

102. Muchardt C.and Yaniv M. (1993) A human homologue of *Saccharomyces cerevisiae SNF2/SWI2* and *Drosophila-brm* genes potentiates transcriptional activation by the glucocorticoid receptor, *EMBO J* **12**, 4279-4290

103. Khavari P.A., Peterson C.L., Tamkun J.W., Mendel D.B.and Crabtree G.R. (1993) BRG1 contains a conserved domain of the SWI2/SNF2 family necessary for normal mitotic growth and transcription, *Nature* **366**, 170-174

104. Dunaief J.L., Strober B.E., Guha S., Khavari P.A., Alin K., Luban J., Begemann M., Crabtree G.R.and Goff S.P. (1994) The retinoblastoma protein and BRG1 form a complex and cooperate to induce cell cycle arrest, *Cell* **79**, 119-130

105. Singh P., Coe J.and Hong W.J. (1995) A role for retinoblastoma protein in potentiating transcriptional activation by the glucocorticoid receptor, *Nature* **374**, 562-565

106. Tsukiyama T., Daniel C., Tamkun J.and Wu C. (1995) ISWI, a member of the SWI2/SNF2 ATPase family, encodes the 140 kD subunit of the nucleosome remodelling factor, *Cell* **83**, 1021-1026

107. Burch J.B.E.and H. W. (1983) Temporal order of chromatin structural changes associated with activation of the major chicken vitellogenin gene, *Cell* **33**, 65-76

108. Jost J.P., Seldran M.and Geiser M. (1984) Preferential binding of estrogen-receptor complex to a region containing the estrogen-dependent hypomethylation site preceding the chicken vitellogenin II gene, *Proc Natl Acad Sci USA* **81**, 429-433

109. Wilks A.F., Cozens P.J., Mattaj I.W.and Jost J.P. (1982) Estrogen induces a demethylation at the 5′ end region of the chicken vitellogenin gene, *Proc Natl Acad Sci USA* **79**, 4252-4255

110. Bruhat A.and Jost J.P. (1995) In vivo estradiol-dependent dephosphorylation of the repressor MDBP-2-h1 correlates with the loss of in vitro preferential binding to methylated DNA, *Proc Natl Acad Sci USA* **92**, 3678-3682

111. Nitsch D., Stewart A.F., Boshart M., Mestril R., Weih F.and Schütz G. (1990) Chromatin structure of the rat tyrosine aminotransferase gene relate to the function of its cis-acting elements, *Mol Cell Biol* **10**, 3334-3342

112. Fritton H.P., Sippel A.E.and Igo-Kemenes T. (1983) Nuclease-hypersensitive sites in the chromatin domain of the chicken lysozyme gene, *Nucl Acids Res* **11**, 3467-3485

113. Jantzen C., Fritton H.P., Igo-Kemenes T., Espel E., Janich S., Cato A.C.B., Mugele K.and Beato M. (1987) Partial overlapping of binding sequences for steroid hormone receptors and DNaseI hypersensitive sites in the rabbit uteroglobin gene region, *Nucl Acid Res* **15**, 4535-4552

114. Elgin S.C.R. (1988) The formation and function of DNaseI hypersensitive sites in the process of gene activation, *J Biol Chem* **263**, 19259-19262

115. Weih F., Nitsch D., Reik A., Schütz G.and Becker P.B. (1991) Analysis of CpG methylation and genomic footprinting at the tyrosine aminotransferase gene: DNA methylation alone is not sufficient to prevent protein binding *in vivo*, *EMBO J* **10**, 2559-2567

116. Reik A., Schütz G.and Stewart A.F. (1991) Glucocorticoids are required for establishment and maintenance of an alteration in chromatin structure: induction leads to a reversible disruption of nucleosomes over an enhancer, *EMBO J* **10**, 2569-2576

117. Carr K.D.and Richard-Foy H. (1990) Glucocorticoids locally disrupt and array of positioned nucleosomes on the rat amino transferase promoter in hepatoma cells, *Proc Natl Acad Sci USA* **87**, 9300-9304

118. Hebbes T.R., Thorne A.W.and Crane-Robinson C. (1988) A direct link between core histone acetylation and transcriptionally active chromatin, *EMBO J* **7**, 1395-1402

119. Vettese-Dadey M., Walter P., Chen H., Juan L.-J.and Workman J.L. (1994) Role of the histone amino termini in facilitated binding of a transcription factor, GAL4-AH, to nulceosomal cores, *Mol Cell Biol* **14**, 970-981

120. Juan L.-J., Utley R.T., Adams C.C., Vettese-Dadey M.and Workman J.L. (1994) Differential repression of transcription factor binding by histone H1 is regulated by the core histone amino termini, *EMBO J* **13**, 6031-6040

121. Bresnick E.H., John S., Berard D.S., LeFebvre P.and Hager G.L. (1990) Glucocorticoid receptor-dependent disruption of a specific nucleosome on the mouse mammary tumor virus promoter is prevented by sodium butyrate, *Proc Natl Acad Sci USA* **87**, 3977-3981

122. Bartsch J., Truss M., Bode J.and Beato M. (1996) Moderate histone acetylation activates the MMTV promoter and remodells its ncleosomal organization, *Proc Natl Acad Sci USA* **in press**,

123. Yoshida M., Kijima M., Akita M.and Beppu T. (1990) Potent and specific inhibition of mammalian histone deacetylase both in vivo and in vitro by trichostatin A, *J Biol Chem* **265**, 17174-17179

124. Hayes J.J.and Wolffe A.P. (1992) Histones H2A/H2B inhibit the interaction of transcription factor-IIIA with the *Xenopus borealis* somatic 5S RNA gene in a nucleosome, *Proc Natl Acad Sci USA* **89**, 1229-1233

125. Peterson C.L.and Herskowitz I. (1992) Characterization of the yeast SWI1, SWI2, and SWI3 genes, which encode a global activator of transcription, *Cell* **68**, 573-583

126. Cairns L.A., Crotta S., Minuzzo M., Moroni E., Granucci F., Nicolis S., Schiró R., Pozzi L., Giglioni B., Ricciardi-Castagnoli P. et al. (1994) Immortalization of multipotent growth-factor dependent hemopoietic progenitors from mice transgenic for GATA-1 driven SV40 tsA58 gene, *EMBO J* **13**, 4577-4586

127. Peterson C.L., Dingwall A.and Scott M.P. (1994) Five SWI/SNF gene products are components of a large multisubunit complex required for transcriptional enhancement, *Proc Natl Acad Sci USA* **91**, 2905-2908

128. Spangenberg C., Eisfeld K., Luger K., Richmonds T.J., Truss M.and Beato M. (1996) The MMTV promoter positioned on a tetramer of histones H3 and H4 binds nuclear factor 1 and OTF1^, *J Mol Biol* **submitted,**

129. Blomquist P., Li Q.and Wrange Ö. (1996) The affinity of nuclear factor 1 for its DNA site is drastically reduced by nucleosome organization irrespective of its rotational or translational position, *J Biol Chem* **271**, 153-159

130. Hecht A., Laroche T., Strahl-Bolsinger S., Gasser S.M.and Grunstein M. (1995) Histone H3 and H4 N-termini interact with SIR3 and SIR4 proteins: a molecular model for the formation of heterochromatin in yeast, *Cell* **80**, 583-592

131. Alevizopoulos A., Dusserre Y., Tsai-Pflugfelder M., von der Weid T., Wahli W.and Mermod N. (1995) A proline-rich TGFß-responsive transcriptional activator interacts with histone H3, *Genes Dev* **9**, 3051-3066

132. Van Holde K.E. (1993) The omnipotent nucleosome, *Nature* **362**, 111-112

133. Hayes J.J.and Wolffe A.P. (1993) Preferential and asymmetric interaction of linker histones with 5S DNA in the nucleosome, *Proc Natl Acad Sci USA* **90**, 6415-6419

134. Bresnick E.H., Bustin M., Marsaud V., Richard-Foy H.and Hager G.L. (1992) The transcriptionally-active MMTV promoter is depeleted of histone H1, *Nucl Acids Res* **20**, 273-278

135. Ura K., Hayes J.J.and Wolffe A.P. (1995) A positive role for nucleosome mobility in the transcriptional activity of chromatin templates: restriction by linker histones, *EMBO J* **14**, 3752-3765

THE NUCLEAR PI CYCLE: ITS RELEVANCE TO NUCLEAR STRUCTURE AND FUNCTION

R Stewart Gilmour[1], Alberto M Martelli[2], Lucia Manzoli[3], Anna M Billi[4] and Lucio Cocco[4].

[1] Department of Cellular Physiology, Babraham Institute, Cambridge, England.

[2] Department of Morphology, University of Trieste, Via Manzoni, 16, 3413 8 Trieste, Italy.

[3] Institute of Human Morphology, University of Chieti, Via dei Vestini, 66013 Chieti, Italy.

[4] Institute of Anatomy, University of Bologna, Via Irnerio 48, 40126 Bologna, Italy.

A typical diploid mammalian cell contains about two meters of double-stranded DNA packed into a nucleus of about 10 μm diameter. To achieve this the DNA is coiled and folded through a heirarchy of defined higher order structures. Intuitively it is reasonable to think that only a highly structured nucleus can deal with the topological difficulties associated with, for example, DNA replication or RNA transcription. This degree of spatial organisation might apply not only to DNA but also to the molecular complexes which interact with it. Indeed the nucleus has been shown to contain a heterogeneous group of inclusions such as interchromatin granules [1-3], nuclear bodies [4] and coiled bodies [5] and subsequent studies using antibodies against molecules involved in DNA replication [6-9] and transcription [10, 11], DNA repair [12], MRNA processing and transport [13-16] and steroid binding [17] show that these processes take place in well-defined domains some of which correspond to the above inclusions. These studies demonstrate that compartmentalisation exists in the nucleus. Although the mechanism for its maintainance is unclear at present, it provides a potential explanation of how functions are structurally integrated within the highly condensed nuclear interior. A particular situation exists with the passage of second messengers from the cytoplasm to the nucleus and indeed a case can be argued for the unambiguous segregation within the nucleus of those signals that are destined to initiate proliferation from those found in quiescent cells. Whether signal transduction involves the translocation of a regulatory factor into the nucleus or the modification of a preexisting nuclear protein the problem of the propagation and functional integration of events within the nucleus arises. The discovery of an autonomous polyphosphotidylinositol (PI) signalling pathway in the nucleus [18] which is quite distict from its plasma membrane counterpart adds a new dimension to the control of nuclear activities by cytoplasmic signals. This chapter reviews our current understanding of this so-called nuclear PI cycle and examines its relevance to cell proliferation.

C. Nicolini (ed.), Genome Structure and Function, 145–154.
© 1997 Kluwer Academic Publishers. Printed in the Netherlands.

1. Discovery Of The Nuclear Pi Cycle

Nuclear PIs were first recognised by Smith and Wells [19] who reported the existence of PI kinases in isolated rat liver nuclei. Some years later Cocco et al. [20] showed that PIs are synthesised in isolated mouse erythroleukeia (MEL) cell lines and that, as the cells undergo terminal differentiation, there is an accumulation of phosphatidylinositol 4,5-phosphate (PIP_2) formed by the phosphorylation of phosphatidylinositol 4-phosphate (PIP). When it became clear that the PI intermediates and their interconversions were virtually identical to the PI signalling mechanism at the plasma membrane, a crucial issue became the unequivocal demonstration that the two PI pools are indeed distinct from one another and not simply due to adventitious contamination of nuclei with cytoplasmic membranes during the isolation procedures. In these experiments nuclei were isolated with a non-ionic detergent which strips off both inner and outer nuclear membranes and accompanying endoplasmic reticulum. The absence of cytoplasmic contamination as judged by electron microscopy and the absence of cytosolic enzymes and tubulin attest to the purity of the isolated nuclei.

An important subsequent finding was that PIP and PIP_2 labelling in nuclei isolated from Swiss 3T3 fibroblasts underwent a 50% decrease within minutes of treatment of the cells with mitogenic concentrations of insulin-like growth factor-1 (IGF-1) and thereafter returned to pre-stimulatory levels about 30 mins. later [21]. These data, which measured the turnover of isotopic phosphate on the inositol moiety of nuclear PIs, were shown to be due to changes in the actual mass of PIs present. Cells labelled to equlibrium with tritiated myo-inositol prior to stimulation with IGF-1 showed the same rapid decrease in PI labelling on treatment with IGF-1 thus suggesting a transient fall in the nuclear PI mass. A detailed description of the PI lipid changes that accompany IGF-1 stimulation of 3T3 cells was later given by Divecha et al.[23] using picomole-sensitive mass assays for the various components of the nuclear PI cycle. As before, within minutes of IGF-1 treatment the nuclear mass of PIP and PIP_2 halved while that of nuclear diacylglycerol (DAG) doubled. They also demonstrated an inverse relation between the magnitude of the response and nuclear purity suggesting that these phenomena are truly nuclear and not due to cytoplasmic contamination. Although there was a close correlation between DAG appearance and PIP and PIP_2 disappearance the hydrolysis of other lipids by phospholipase C (PLC) cannot be ruled out. These data however highlighted an important point: that nuclei are capable of self-generating the potent second messenger, DAG.

Subsequent work [24] has shown that IGF-1 rapidly induces a four-fold activation of PLC in 3T3 cell nuclei and that this is due to a pool of PLCβ-1 isoform which is present in the nucleus. No evidence was found for differences in the amounts of PLCβ protein before and after stimulation as judged by Western blotting, therefore activation does not involve translocation of the enzyme into the nucleus. In the 3T3 cell PLCγ is specific for the cytosol and is unaffected by IGF-1 treatment.

2. Nuclear PI Cycle Responses

A number of external stimuli have been shown to affect the nuclear PI cycle. The IGF-1 response is one of the most researched. IGF acting through its type I receptor does not lead to detectable breakdown of PIP_2 or DAG production at the plasma membrane, however these effects are seen a few minutes later in the nucleus. Conversely, bombesin acting through its receptor causes a marked hydrolysis of PIP_2 (catalysed by cytosolic PLCβ) at the plasma membrane and not in the nucleus [23], thus demonstrating that the plasma membrane and nuclear PI cycles are experimentally separable. Additionally, the isolation of a 3T3 clone that bound IGF-1 to its receptor but failed to respond immediately in terms of nuclear PI turnover or later by DNA synthesis suggested that the IGF-1 nuclear signal is involved in the onset of the mitogenic response [25].

Treatment of Daudi lymphoma cells with interferon (INFα) caused an activation of nuclear PLC and the hydrolysis of PIP_2 resulting in a peak of DAG production 90 mins. later. At earlier times, (2 mins.), INFα caused an increase in cytosolic DAG by hydrolysis of plasma membrane phosphatidylcholine [26,27]. This latter activity can be inhibited without affecting the nuclear response.

Interleukin 1α stimulates the activity of nuclear PLCβ in human osteosarcoma SaOS-2 cells causing a rapid breakdown of PIP and PIP_2 within 2 mins. of treatment [28].

The cellular proliferation associated with regenerating rat liver has been characterised by an increase in nuclear DAG [29] and ascribed to an enhanced nuclear PLC activity previously found by Kuriki et al. [30].

Finally, the terminal differentiation of NEL cells [31] is accompanied by an inhibition of nuclear PLCβ activity due to a down-regulation of its mRNA expression, leading to an accumulation of nuclear PIP and PIP_2 [32].

A recurrent feature of these examples is a positive correlation of the PI cycle with cell division and the involvement of an exclusively nuclear PLC which in a few cases has been identified as the β-1 isoform. More examples are however required to establish the general validity of these conclusions.

3. Downstream Effects Of The Nuclear PI Cycle

3.1 DAG AND PROTEIN KINASE C (PKC)

By analogy with signalling at the plasma membrane, PI hydrolysis in the nucleus should yield two potential second messengers namely DAG and IP_3. Although nuclear DAG has been verified following IGF-1 stimulation of 3T3 cells [23], the prevalence of IP_3 is less clear. A possible consequence of nuclear DAG accumulation is the activation of protein kinase C in the nucleus. PKC has been shown to exist in the nuclei of rat liver hepatocytes [33,34] and in a variety of cell types treated with phorbol esters [35 -40].

In FDCP1 haemopoietic cells, IL-3 and bryostatin induce a rapid translocation of PKC to the nuclear envelope [41] as did thrombin in 11C9 fibroblasts [42]. However an association with the production of nuclear DAG was not established in these cases. Circumstantial evidence for this comes from experiments with IGF-1 stimulated Swiss 3T3 cells in which the translocation of PKC to the nucleus is temporally related to an increase in nuclear DAG concentrations [23, 25, 43, 44].

Several potential substrates for nuclear PKC have been identified. Lamin B phosphorylation is observed after PKC activation by bryostatin in HL60 nuclear envelopes [38] and by phorbol ester in mouse lymphocyte nuclei [45]. In NIH 3T3 cells PKC activation by phorbol ester and by bryostatin induced phoshorylation of lamins A, B and C while PDGF only stimulated PKC-dependent phosphorylation of lamins A and C [46]. PKC is reported to phosphorylate DNA topoisomerase I [47] and RNA polymerase II *in vitro* [48]. A number of nuclear proteins, including histone H1 are substrates for PKC during rat liver generation [49,50]. Also identified as substrates are a 110 kD polyA binding protein of the nuclear envelope and a putative IP_3 receptor [52]. Mitogenic stimulation of Swiss 3T3 cells with a combination of IGF-1 and bombesin enhanced the phosphorylation of histone H1 and a 21 kD histone-like protein both *in vivo* and in an *in vitro* PKC dependent nuclear assay [43].

The presence of PKC in the nucleus of the quiescent cell and its massive translocation after the application of an appropriate stimulus might suggest that PKC is continuously cycling in and out of the nucleus and that the increase in nuclear DAG might act as a sink, fixing it inside the nucleus at the same time as activating it. However no consensus nuclear-targetting sequences have been identified in the various PKC isoforms know to exist in the cytosol [34] and therefore the mechanism involved in the accumulation of nuclear PKC remains unclear for the present.

3.2 NUCLEAR CALCIUM

The nuclear envelope [53] consists of an inner and outer membrane traversed frequently by nuclear pores [54]. The outer membrane is continuous with the endoplasmic reticulum [55] while the inner membrane envelopes the nucleoplasm and is lined by the nuclear lamina to which are attached the nuclear lamins [56]. The outer membrane contains an endoplasmic reticulum-type calcium pump ATPase [57] which accounts for an ATP mediated nuclear calcium uptake [58]. However the outer membrane also contains IP_4 receptors [59,60]. Both ATP and IP_4 can induce nuclear calcium uptake. Apparently the free calcium concentration in the nuclear environment determines the choice of mechanism for calcium transport [60].

The inner membrane has IP_3 receptors [59-62] and the nucleus itself has the potential to produce IP_3 from the hydrolysis of PIP and PIP_2 [22,23]. Thus, in principle, there are mechanisms which, given the appropriate signals, transport calcium from the cytoplasm into the lumen of the nuclear envelope in an ATP- and IP_4-dependent manner and a mechanism to transport calcium from the lumen into the nucleus in response to endogenous IP_3 production from the nuclear PI cycle. The mechanism of regulation of nuclear calcium however is a subject of considerable debate at present [63]. Nuclear calcium signals have been implicated in a number of functions including DNA synthesis, transcription and repair and nuclear envelope breakdown [64-68]. There is evidence that nuclear and cytosolic calcium signals are differently regulated and independent of one another [69-75]. Although the source of nuclear calcium is thought to be cytosolic [76] there is good evidence that there is a nuclear barrier to the free entry of cytosolic calcium [77] and that the nucleus is insulated from cytosolic calcium fluxes [75].

The evidence that links the activation of the nuclear PI cycle with calcium is circumstantial at present. *In vitro* experiments which looked at the activation of translocated PKC in IGF-1 stimulated Swiss 3T3 cells concluded that PKC-dependent phosphorylation of nuclear proteins required activation of PKC by both DAG and calcium [43]. These studies were not designed to reveal the source of the calcium and hence it was not possible to link the activation of the PI cycle to an influx of calcium into the nucleus. In this regard, it is enigmatic that attempts to image by confocal microscopy nuclear calcium transients in IGF-1 stimulated Swiss 3T3 cells using calcium binding fluorescent dyes has so far proved unsuccessful. [R. S. Gilmour, unpublished data].

4. Immunocytochemical Evidence For The Nuclear PI Cycle

Biochemical studies on the nuclear PI cycle have defined in broad terms the distribution of the components of the cycle within the intracellular compartments of the cell. Using immunocytochemistry these observations have been confirmed at a much finer level of discrimination and in some cases indications of structural associations have been revealed.

4.1 NUCLEAR PHOSPOLIPIDS

A puzzling feature of the original observations was that the phospholipids remain associated with the nucleus after purification using non-ionic detergents like NP-40 and Triton X-100 which strip off the nuclear envelope. The fine localisation of inositol phospholipids within the nucleus has been demonstrated by cytochemical methods using colloiodal gold at the electron microscope level. Using an anti-PIP monoclonal antibody [78], the nuclear phospholipids have been identffied with the interchromatin domains, ie.within the inner nucleus, mainly in association with the ribonucleoprotein particles and the nuclear matrix but not with the peripheral lamina [79].

4.2 PLC ISOFORMS: WHOLE CELL STUDIES

The published immunological analyses of PLCs in different cell types do not indicate the precise localisation of the enzymes. Immunofluorescence examination of whole cells by confocal microscopy provides at least a partial answer [80]. In Swiss 3T3 cells, the fluorescence associated with an anti-PLCβ-1 antibody is almost entirely located in the nucleus and appears as a fine network of fluorescent filaments which are diffused throughout the nucleoplasm except for the nucleolar regions. By contrast, using an anti-PLCγ antibody, the nucleus appears virtually negative while an intense cytoplasmic fluorescence is seen to be associated with filament bundles. These structures resemble cytoskeletal actin filaments when visualised with an anti-actin antibody. Taken together these results suggest that in 3T3 cells the specific isoforms are not randomly diffused in either the nucleoplasm or cytosol but are associated with the seletal structures of the two compartments. It is worth noting that these experiments detected very small amounts of PLCβ-1 associated with the plasma membrane. This is significant because it is known that 3T3 cells respond to agonists like bombesin which are known to generate cytoplasmic second messengers via the hydrolysis of plasma membrane inositol phosphoplipids. Together with the biochemical evidence, the importance of these findings is not so much the demonstration of an absolute compartmentalisation of the two PLC isoforms within the 3T3 cell but the partitioning of the PLCβ-1 isoform into two operationally distinct areas of activity. Indeed, in the MEL cell nuclei both PLCβ and γ isoforms are present however during the process of differentiation when the nuclear PI cycle is shut down only the PLCβ isoform is down-regulated [32].

At the electron microscope level, the distribution of the PLC isoforms within the nucleus and cytoplasm can be determined more precisely. In 3T3 cell nuclei, immunogold labelling with an anti-PLCβ-1 monoclonal antibody is mainly concentrated in the interchromatin regions while the heterochromatin and the nuclear envelope are virtually unlabelled. An anti-PLCγ antibody localises to cytoplasmic regions that are devoid of organelles [80]. Quantitative evaluation of the gold particle distribution indicates that the PLCβ-1 is eight-fold more abundant in the nucleus compared to the the cytoplasm while PLCγ is five-fold more concentrated in the cytoplasm than the nucleus. These results agree in general with immunological analyses of isolated cell fractions except that the preferential localisation of the isoforms appears almost clear-cut [24]. This may reflect differences in the sensitivity and background noise of the individual methods employed.

4.3 PLC ISOFORMS: LOCALISATION IN SUBCELLULAR FRACTIONS

Attempts have been made to identify the specific cellular structures associated with PLCs by analysis of the proteins in isolated and *in situ* cell subtractions. For example, immunoblotting of the proteins obtained from the isolated nuclear matrix of 3T3 cells indicates that about 60% of the PLCβ-1 protein is present [80]. This would suggest that a substantial fraction of nuclear PLCβ-1 is not in the soluble nucleoplasm but nuclear matrix bound. This figure may in fact be much higher since it was observed that the high salt extraction involved in matrix preparation did not contain solubilised PLCβ-1 *per se,* but fragments of nuclear matrix containing the bound enzyme which were generated by chromatin breakage during the isolation procedure. In support of this is the additional observation using immunoelectron microscopy that there is only a slight decrease in the density distribution of gold labelling with anti-PLCβ antibody in the isolated matrices compared with intact, isolated nuclei [81].

A more precise localisation of PLCβ-1 on the nuclear matrix has been obtained from *in situ* matrix preparations in which the integrity of the matrix and skeletal structure is preserved [82]. Since the extraction procedures were performed on adherent cells the nuclei could be examined at different stages of the preparation. With these structures immunogold detection incontrovertibly confirmed the absence of the enzyme in the nuclear pore-lamina complex and an association with the filaments of the inner nuclear matrix in close proximity to clusters of nucleoprotein particles [81]. By contrast, the PLCγ isoform was confined to the cytoplasm and appeared closely linked with structures which were identified as actin filaments.

5. Concluding Remarks

There is now a growing body of evidence for the existence of a nuclear PI cycle with a unique structural and functional identity. A number of cell types have been shown to possess distinct pools of PLC, in some cases more than one isoform per pool exists [32], however in all cases so far investigated PLCβ-1 is present and it is the activity of this isofonn that changes in parallel with the turnover of nuclear PIs. Another common characteristic is the association of an active PI cycle with the proliferative response to a selective number of growth factors and cytokines. The differentiated cell on the other hand is characterised by an inactive nuclear PI cycle and a progressive loss of PLCβ-1 protein. It may be that there are exceptions to this rule, hence it is important that further cell types and their responses are investigated to establish the general validity of these conclusions.

Many receptor-coupled signalling mechanisms at the plasma membrane however activate a cytosolic PLCβ-1 via the Gαq class of G proteins. The demonstartion that bombesin can activate this isoform in the cytoplasm and not in the nucleus [23] gives additional support for the idea of independent PI signalling mechanisms in these compartments. This also accommodates the finding that IGF-1 only activates the nuclear PLCβ-1 [24] and PDGF activates both cytosolic and nuclear PLCβs [32]. The important conclusion is not the absolute partitioning of PLC isoforms within the cell but the existence of independently regulated pools of a specific isoform in the cytosol and nucleus. The fact that this appears to be the β-1 isoform is intriguing. Comparison of the protein sequence of the known PLC isoforms shows that the C-terminus of PLCβ-1 is poorly conserved and contains a number of candidate nuclear localisation motifs [83]. In addition there is a G protein binding region which has been shown to be essential for G protein-dependent activation of PLCβ-1 and which may account for receptor-coupled activation at the plasma membrane [84]. This region also contains a potential MAP kinase phosphorylation site and it is tempting to speculate that this may be involved in the activation of nuclear PLCβ-1.

Localisation studies on nuclear subfractions have provided separate evidence in support of the existence of components of the nuclear PI cycle. The presence of PI, PIP, PIP_2, DAG kinase and PLC activities following isolation of purified nuclei using non-ionic detergents and subsequent robust extraction procedures to prepare nuclear subtractions are consistent with an association of these components with the inner proteinaceous framework of the nucleus. The existence of these enzymes in the nucleus is wholly consistent with the presence of their substrates and the considerable body of evidence demonstrating intra-nuclear PI metabolism. Thus the puzzling feature that membrane-free nuclei nevertheless contain inositol phospholipids is reconciled by the conclusion that they do not appear to be part of the nuclear envelope but are incorporated into a structure in the inner, interchromatin regions of the nucleus. The exact nature of this structure has yet to be identified, if indeed it forms one as discrete as the previously mentioned interchromatin granules. The identification of a PI-binding consensus sequence $K(X)_nKXKK$ (where n=3-7) in proteins which are known to associate with PIs [83] provides a potential mechanism whereby PIs in the nuclear interior remain nuclear-bound after detergent treatment. The PI-binding consensus sequence appears in a large number of cytoplasmic as well as many nuclear proteins. The latter include, in addition to PLCβs, lamin B, histones, HMG proteins, DNA and RNA polymerases, nucleolin, nucleoplasmin, Sn and Hn ribonucleoprotein, steroid hormone receptors, DNA topoisomerase and helicase and the transcription factors AP-1, AP-2 and jun. These interactions could be anchoring mechanisms for specific proteins which could be released from matrix sites by PLC hydrolysis of the PIP or PIP_2 phosphodiester bond at the same time liberating DAG which might fulfil a role in the activation of other nuclear proteins via PKC activation. Alternatively, as outlined in the accompanying diagram, the interchromatin filament structures might act as repositories for nuclear PIs possibly also involving an interaction with a PI binding sequence as well as providing a nearby anchorage for PLCβ-1 by, as yet, undefined interactions. This complex might also be predicted to contain the kinases and phosphatases involved in the metabolism of nuclear PI. Further detailed analyses of these structures by electron microscopic techniques and using specific antibodies for the individual components of

152

the nuclear PI cycle wffi help to elucidate the nature of this complex which appears to play a vital role in transmitting mitogenic signals to their targets within the nucleus.

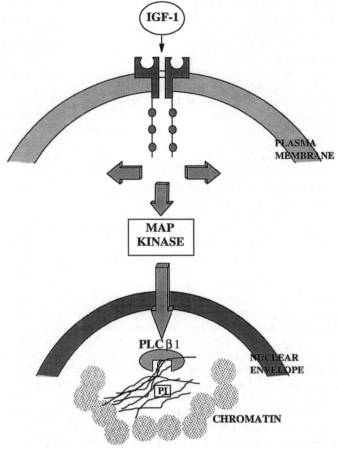

A model for the activation of the nuclear PI cycle by cytoplasmic second messengers proposes that growth factor signalling is propagated to the nuclear interior by MAP kinase where it in turn activates PLCβ-1 which is bound within the fibrillar, interchromatin region of the nucleus. The nuclear inositol phosphofipids are also sequestered in this structure in close proximity to the PLC and subsequent hydrolysis of PI by the activated form of the enzyme releases DAG into the nucleus.

References

1. Clevenger C. V. and Epstein A. L. (1984). *Exp. Cell Res.* **151**, 194-207.
2. Thiry M. (1993). *Cell Biol.* **62**, 259-269.
3. Ferreira J. A., Carmo-Fonseca M. and Lamond A. 1. (1994). *J. Cell Biol.* **126**, 11-23.
4. Brasch K. and Ochs R.L. (1992). *Exp. Cell Res.* **202**, 211-223.
5. Thiry M. (1994). *Chromosoma* **103**, 268-276.
6   Nakayasu H. and Berezney R. (1989). *Cell Biol.* **108**, 1-11.
7. Mazzotti G., Rizzoli R., Galanzi A., Papa S., Vitale M., Falconi M., Neri L M., Zini N. and Maraldi N.M. (1990). *J. Histochem. Cytochem.* **38**, 13-22.
8. Neri L. M., Mazzotti G., Capitani S., Maraldi N.M., Cinti C., Baldini N., Rana R. and Martelli A. M. (1992). *Histochemistry* **98**, 19-32.
9. Hassan A. B. and Cook P.R. (1993) *J. Cell Sci.* **105**, 541-550.
10. Lawrence J. B., Singer R.H. and Marselle L.M, (1989) *Cell* **57**, 493-502.
11. Wansink D. G., Schul W., van der Kraan I., van Steensel B., van Driel R. and de Jong L. (1993). *J. Cell Biol.* **122**, 283-293.
12. Jackson D. A., Balajee A. S., Mullenders L. and Cook P.R. (1994) *J.Cell Sci.* **107**, 1745-1752.
13. Carter K.C., Taneja K.L. and Lawrence J.B. (1991). *J. Cell Biol.* **115**, 1191-1202.
14. Zhang G., Taneja K.L., Singer R.H. and Green M.R. (1994). *Nature* **372**, 809-812.
15. Blencowe B. J., Nickerson J. A., Issner R., Penman S. and Sharp P.A. (1994). *J. Cell. Biol.* **1275** 593-607.
16. Bisotto S., Lauriault P., Duval M. and Vincent M. (1995). *J. Cell Sci.* **108**, 1873-1882.
17. van Steensel B., Brink M., van der Meulen K., van Binnendijk E. P., Wansink D. G., de Jong L., de Kloet E.R. and van Driel R. (1995). *J. Cell Sci.* **108**, 3003-3011.
18. Cocco L., Martelli A. M. and Gilmour R.S. (1994). *Cell. Signal.* **6**, 481-485.
19. Smith C.D.andWells W.W. (1983). *J. Biol. Chem.* **258**, 9368-9373.
20. Cocco L., Gilmour R. S., Ognibene A., Letcher A. J., Manzoli F. A. and Irvine R. F. (1987). *Biochem. J.* **248**, 765-770.
21. Cocco L., Martelh A. M., Gilmour R. S., Ognibene A., Manzoli F. A. and Irvine R.F. (1988), *Biochem. Biophys. Res. Comm.* **154**, 1266-1278.
22. Cocco L., Martelfi A. M., Gilmour R. S., Ognibene A., Manzoli F. A. and Irvine R. F. (1989). *Biochem. Biophys. Res. Comm.* **159**, 720-725.
23. Divecha N., Banfic H. and Irvine R.F. (1991). *EMBO J.* **10**, 3207-3214.
24. Martelli A. M., Gilmour R. S., Bertagnolo V., Neri L. M., Manzoli L. and Cocco L. (1992). *Nature* **358**, 242-245.
25. Martelli A. M., Gilmour R. S., Neri L. M., Manzoli L., Corps A. N. and Cocco L. (1991). *FEBS Lett.* **283**, 243-246.
26. Cataldi A., Di Primio R., Lisio R., Rana R. A., Robuffo I., Bosco D. and Miscia S. (1992) *Cell. Signal.* **5**, 331-336.
27. Cataidi A., Rana R., Bareggi R., Lisio R., Robuffo I., di Valerio V. and Miscia S. (1993) *Cytokine* **5**, 235-239.
28. Marmiroli S., Ognibene A., Bavelloni A., Cinti C., Cocco L. and Maraldi N.M. (1994). *J. Biol. Chem.* **269**, 13-16.
29. Banfic H., Zizak M., Divecha N. and Irvine R.F. (1993). *Biochem J.* **290**, 633-636.
30. Kuriki H., Tauiya-Koizumi K., Asano M., Yoshida S., Kojima K. and Nimura Y. (1992). *J. Biochem.* **111**, 283-286.
31. Billi A. M., Cocco L., Mártelli A. M., Gilmour R. S., and Weber G. (1993). *Biochem. Biophys. Res. Comm.* **195**, 8-12.
32. Martelli A. M., Billi A. M., Gilmour R. S., Neri L. M., Manzoli L., Ognibene A. and Cocco L. (1994). *Cancer Res.* **54**, 2536-2540.
33. Capitani S., Girard P. R., Mazzei G. J., Kuo J.F., Berezney R. and Manzoli F. A. (1987). *Biochem. Biophys. Res. Commun.* **142**, 367-375.
34. Rogue P., Labourdette G., Masmoudi A., Yoshida Y., Huang F. L., Huang K.-P., Zwiller J., Vincedon G. and Malviya A.N. (1990). *J. Biol. Chem.* **265**, 4161-4165.
35. Zoltan K., Deli E. and Kuo J.F. (1988). *FEBS Lett.* **231**, 41-46.
36. Thomas T. P., Talwar H. S. and Anderson W.B. (1988). *Cancer Res.* **48**, 1910-1919.
37. Beny N., Ase K., Kikkawa U., Kishimoto A. and Nishizuka Y. (1989). *J. Immun.* **143**, 1407-1413.
38. Fields A. P., Pettit G. R. and Stratford May W. (1988). *J. Biol. Chem.* **263**, 8253-8260.
39. James G. and Olson E. (1992). *J. Cell. Biol.* **116**, 863-874.
40. Eldar H., Ben-Chaim J. and Livneh E. (1992). *Expl. Cell Res.* **202**, 259-266.
41. Fields A. P., Pincus S. M., Kraft A. S. and Stratford May W. (1989). *J. Biol. Chem.* **264**, 21895-215901.
42. Leach K. L., Ruff V. A., Jarpe M. B., Adams L. D., Fabbro D. and Raben D.M. (1992) *J. Biol. Chem.* **267**, 21816-21822.
43. Martelli A. M.. Gilmour R. S., Falcieri E., Manzoli F. A. and Cocco, L. (1989). *Expl. Cell Res.* **185**, 191-202.

154

44. Martelli A. M., Neri L. M., Gilmour R. S., Barker P. J., Huskisson N. J., Manzoli F. A. and Cocco L. (1991) *Biochem. Biophys. Res. Commun.* **177**, 480-487.
45. Hombeck P., Huang K.-P., and Paul W.E. (1988). *Proc. Natn. Acad. Sci. U.S.A.* **85**, 2279-2283.
46. Fields A. P., Tyler G., Kraft A. and Stratford May W. (1990) *J. Cell Sci.* **96**, 107-114.
47. Samuels D. S., Shimizu Y. and Shimizu N. (1989) *FEBS Lett.* **259**, 57-60.
48. Chuang L. F., Zhao F.-K. and Chuang R. Y. (1989). *Biochem. Biophys. Acta* **992**, 87-95.
49. Martelli A. M., Carini C., Marmiroli S., Mazzoni M., Barker P. J., Gilmour R. S. and Capitani S. (1991). *Expl. Cell Res.* **195**, 255-262.
50. Mazzoni M., Carini C., Matteucci A., Martelli A. M., Bertagnolo V., Previati M., Rana R., Cataldi A. and Capitani S. (1992). *Cell. Signal.* **4**, 313-319.
51. Schafer P., Aitken S. J. M., Bachmann M., Agutter P. S., Muller W. E. G. and Prochnow D. (1993). *Cell. Molec. Biol.* **39**, 703-714.
52. Matter N., Ritz M.-F., Freemuth S., Rogue P. and Malviya A. N. (1993). *J. Biol. Chem.* **268**, 732-738.
53. Newport J. W. and Forbes D. J. (1987). *Ann. Rev. Biochem.* **56**, 535-565.
54. Dingwall C. and Laskey R. (1992). *Science* **258**, 942-947.
55. Dale B., Defelice L. J., Kyozuka K., Santella L. and Tosti E. (1994). *Proc. R. Soc. Lond. Ser. B Biol. Sci.* **255**, 119-124.
56. Gerace L. and Foisner R. (1994). *Trends Cell Biol.* **4**, 127-131.
57. Lanini L., Bachs O. and Carafoli E. (1992). *J. Biol. Chem.* **267**, 11548-11552.
58. Nicotera P., McConkey D. J., Jones D. P. and Orrenius S. (1989).
59. Koppler P., Mersel M. and Malviya A. N. (1993). *Biochemistry* **33**, 14707-14710.
60. Humbert J.-P., Matter N., Artault J.-C., Koppler P. and Malviya A. N. (1996). *J. Biol. Chem.* **271**, 478-485.
61. Malviya A. N., Rogue P. and Vincendon G. (1990). *Proc. Natl. Acad. Sci. U. S.A.* **87**, 9270-9274.
62. Matter N., Ritz M. F., Freyermuth S., Rogue P. and Malviya A. N. (1993). *J. Biol. Chem.* **268**, 732-736.
63. Hock J. B. and Thomas A. P. (1994). *Cell Calcium* **16**, 237-338.
64. Karin M. (1992). *FASEB J.* **6**, 2581-2590.
65. Tombes R. M., Simerly C., Borisy G. G. and Schatten G. (1992). *J. Cell Biol.* **117**, 799-811.
66. Peunova N. and Enikolopov G. (1993). *Nature* **364**, 450-453.
67. Wegner M., Coa Z. and Rosenfeld M. G. (1992) *Science* **256**, 370-373.
68. Steinhardt R.A. and Alderton J. (1988). *Nature* **332**, 364-366.
69. Hernandez-Cruz A., Sala F. and Adams P. R. (1990). *Science* **247**, 858-862.
70. Waybill M. M., Yelamarty R. V., Zhang Y. L., Scaduto R.C., LaNoue K. F., Hsu C. J., Smith B. C., Tillotson D. L., Yu F. T. and Cheung J. Y. (1991). *Am. J. Physiol.* **261**, E49-E57.
71. Przywara D. A., Bhave S. V., Bhave A., Wakade T. D. and Wakade A. R. (1991). *FASEB J.* **51** 217-222.
72. Birch B. D., Eng D. L. and Koesis J. D. (1992). *Proc. Natl. Acad. Sci. U.S.A.* **89**, 7978-7982.
73. Nakato K., Furuno T., Inagaki K., Terao R. and Nakamishi M. (1992). *Eur. J. Biochem.* **209**, 745-749.
74. Williams D. A., Becker P. L. and Fay F. S. (1987). *Science* 1644-1648.
75. Al-Mohanna F.A., Caddy K. W. T. and Bolsover S. R. (1994). *Nature* **367**, 745-750.
76. Allbritton N. L., Oancea E., Kuhn M. A. and Meyer T. (1994). *Proc. Natl. Acad. Sci. U.S.A.* **91**, 12458-12462.
77. Badminton M. N., Kendall J. M., Sala-Newby G. and Campbell A. K. (1995). *Expl. Cell Res.* **2165**, 236-243.
78. Voorhout W. F., van Genderen I. L., Yoshioka T., Fukami K., Geuze H. J. and van Meer G. (1992). *Trends Glycosci. Glycosechn* **4**, 533-546.
79. Maraldi N.M., Zini N., Squarzoni S., del Coco R., Sabatelli P. and Manzoli F.A. (1992). *J. Hystochem.Cytochem.* **40**, 1383-1392.
80. Zini N., Martelli A. M., Cocco L., Manzoli F. A. and Maraldi N. M. (1993) *Exptl. Cell Res.* **2085** 257-269.
81. Maraldi N. M., Cocco L., Capitani S., Mazzotti G., Barnabei O. and Manzoli F. A. (1994). *Advan. Enzyme Regul.* **34**, 129-143.
82. Fey E. G., Wan K. M. and Penmann S. (1984). *J. Cell Biol.* **98**, 1973-1984.
83. Yu F.-X., Sun H.-Q., Janmey P. A. and Yin H. L. (1992). *J. Biol. Chem.* 264616-14621.
84. Park D., Jhon D.-Y., Lee C.-W., Ryu S. H. and Rhee S.G. (1993). *J. Biol. Chem.* **268**, 3710-3714.

HISTONE ACETYLATION
A global regulator of chromatin function

BRYAN M.TURNER
Chromatin and Gene Expression Group
Department of Anatomy
University of Birmingham Medical School
Edgbaston, Birmingham B15 2TT, U.K.

1. Chromatin Is A Major Determinant Of Gene Expression In Eukaryotes

In prokaryotic organisms (eg. bacteria such as E.Coli), genes are switched on and off in response to environmental signals, such as the availability, or otherwise, of nutrients. E.coli has only a single RNA polymerase that is directed towards particular genes by association with ancilliary proteins, called sigma factors, which confer upon it the ability to recognise DNA sequences in the promotors of these genes. By changing the availability of these factors and of other regulators in response to environmental cues, bacteria are able to maintain an appropriate pattern of gene expression. In recent years it has been shown that eukaryotic organisms also use ancilliary, DNA-binding proteins, generally called Transcription Factors, to regulate transcription of specific genes. However, it has also become clear that the transcription of eukaryotic genes always requires complex assemblies of several different proteins in addition to RNA polymerase. Protein-protein as well as protein-DNA interactions are of crucial importance in transcription.

Regulation of gene expression in eukaryotes must accomodate one further complicating factor not present in bacteria, namely that eukaryotes package their DNA by complexing it with small basic proteins, the histones, to form chromatin. The basic unit of chromatin structure, the nucleosome core particle, consists of eight histones (two each of H2A, H2B, H3 and H4) around which are wrapped 146 base pairs of DNA. It is described in detail elsewhere in this volume and is shown, in a very diagrammatic form, in Figure 1. Virtually all DNA, including coding DNA, in a typical eukaryotic cell is packaged as nucleosomes and the close association of histones with DNA will inevitably effect the ability of transcription factors and other DNA binding proteins to recognise and bind to specific DNA sequences. In addition, the need for the transcribing polymerase somehow to circumvent the nucleosomes that are known to be retained on actively transcribed genes, gives a further possible stage at which control might be

C. Nicolini (ed.), Genome Structure and Function, 155–171.
© 1997 *Kluwer Academic Publishers. Printed in the Netherlands.*

156

exercised through chromatin. So chromatin, and the nucleosome in particular, must be major regulators of transcription in eukaryotes.

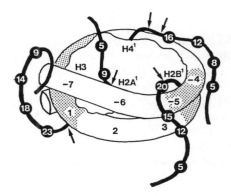

Figure 1. A diagram of the nucleosome core particle showing the sizes and likely locations of the amino-terminal domains of the core histones. Lysine residues that can be acetylated in vivo are shown as numbered circles. Accessible trypsin cutting sites are indicated by arrows. Only one amino-terminal domain is shown for each pair of histones. (From Turner, J.Cell Sci. 99, 13-20)

The role of chromatin in regulating gene expression, though important, could be regarded as essentially passive. That is, nucleosomes influence transcription by blocking access of transcription factors or polymerases to DNA or impeding polymerase movement. If that were so, then understanding the role of chromatin in transcription would be a matter of defining the mechanisms by which nucleosomes were positioned or displaced in order to obscure or reveal different parts of the genome. These processes are undoubtedly important and are considered elsewhere in this volume. But recent results suggest that the nucleosome can regulate gene expression in even more subtle ways.

2. Histones Are Highly Conserved But Extensively Modified

The histones of the nucleosome core particle have been highly conserved through evolution, as has the core particle itself. Using physical or microscopical criteria, it would be hard to distinguish nucleosomes from humans and plants. However, this highly conserved structure is subject to a bewildering array of enzyme-catalyzed, post-translational modifications, most of which (though not quite all) are made on specific amino acids in the N-terminal domains of the four core histones. Possible modifications are detailed in Table 1. Not surprisingly, these modifications have attracted considerable interest. The intimate association of histones with cellular DNA makes it likely that any change, particularly one which involves a change in net charge, will have an effect on chromatin function. But, unravelling the possible effects of these changes has not been helped by the fact that the role(s) of the histone regions within which most changes occur, namely the N-terminal domains (or "tails"), is itself unclear. The following properties of the N-terminal tails are well established:

Table 1. Modification of histone N-terminal domains in higher eukaryotes

Histone	Modification			
	Acetylation	Phosphoryl'n	Methylation	ADPRibosyl'n
H2A	K5 (12)[8]	S1[1]	-	-
H2B	K5, 10, 13, 18	- [2]	-	E2[3]
H3	K9, 14, 18, 23	R2,T3,S10[4]	K 9,27 [6,7]	-
H4	K5, 8, 12,16	S1[1,5]	K20 [7]	-

Where: K = lysine, S = serine, R = arginine, T = threonine and E = glutamic acid

Notes and References
(1)   Detected by sequencing N-terminal peptides (*Sung & Dixon, PNAS 67, 1616*).
(2)   The N-term domain of H2B is phosphorylated in sea urchin sperm. Sperm H2B has a more extended N-terminal domain than that in somatic cells (*Green & Poccia, Dev Biol 108, 235*)
(3)   All core histones can be ADPRibosylated on glutamic acid, but, as shown in Figure 2, only H2B has a glutamic acid residue in its *N-term* domain (*Boulikas PNAS 86, 3499 and EMBO J 7, 57*)
(4)   Kinases able to phosphorylate H3 on R2 or T3 have been isolated, respectively, from mouse leukaemia cells and bovine thymus (*Wakim & Aswad, JBC 269, 2722, Shoemaker & Chalkley, JBC 255, 11048*)
       S10 is phosphorylated in vivo in HeLa cells (*Whitlock et al. JBC 258, 1299*) and is the *mitosis specific* site (*Taylor, JBC 257, 6056; Paulson & Taylor JBC 257, 6064*). A cAMP-dependent H3 kinase phosphorylates S10 on DNA-bound H3 in vitro (*Shibata et al, Eur.J.Biochem. 192, 87*)
(5)   Two kinases have been identified in murine lymphosarcoma cells using synthetic peptides as substrates, one specific for S1 and one for S47 (*Masaracchia et al. JBC 252, 7109*)
(6)   Early sequencing results indicated that lys27 is 100% methylated (*DeLange et al PNAS 69, 882; FEBS Lett 23, 357*)
(7)   Early sequencing results identified sites of methylation (and acetylation) on H4 (*Ogawa et al JBC 244, 4387*)
       Radiolabelling confirmed that only H3 and H4 are methylated in vivo in chick erythrocytes (*Hendzel & Davie, JBC 264, 19208*)
(8)   Tetrahymena H2A can be acetylated at lysines 5 and 12 (*Fusauchi & Iwai J.Biochem. 95, 147*).
(9)   See refs. 1 and 2 in main list for further details and for C-term. modifications

1.   They are accessible to proteases and antibodies and so must be located on the surface of the core particle (ie outside the DNA)
2.   They are not detected by X-ray crystallographic analysis of core particle crystals and so cannot have a unique structure or location in the core particle. (They are usually designated as having a "random coil" structure).
3.   They have been particularly well conserved through evolution. For example, the H4 N-terminal domains of humans, yeast and fruit flies are identical. The N-terminal domains of H2A and H2B in the same species show differences, but the family resemblence is still clear. The amino acid sequences of the N-terminal tails of each of the four core histones are shown in Figure 2.

Of the various post-translational changes to which the histone tails are subject, acetylation has been the most intensively studied. Acetylation offers some experimental advantages, such as chemical stability and relative ease of detection, but perhaps the most significant reason for its popularity is that it has consistently been associated with

158

the fundamental cellular processes of differentiation and development. Some specific examples are listed in Table 2.

3. Histones Undergo Cyclical Acetylation And Deacetylation

Post-translational acetylation of the histones which organise the nucleosome core particle has been found in all animal and plant species so far examined. Acetylation occurs at specific lysine residues, all of which occur in the amino-terminal domains of the core histones (Figures 1 and 2). Note that the α-amino groups of the amino-terminal serine residues on H2A and H4 are acetylated ("blocked") during or immediately after synthesis. This modification occurs on many proteins other than histones and is irreversible. It does not, as far as is known, play any role in the specific functions of H2A and H4.

The transfer of acetate groups from Acetyl-CoA to histones and their subsequent removal is catalyzed by specific enzymes, the histone acetyl-transferases and deacetylases (Figure 3). The net level of acetylation of a histone at a particular genomic site is the result of an equilibrium between these two types of enzyme. The steady state level of acetylation and the rates at which acetate groups are turned over vary between different cell types, with half-lives ranging from a few minutes to several hours. The dynamic nature of the histone acetylation/deacetylation cycle is readily seen by treating cells with inhibitors of histone deacetylating enzymes, such as butyric or propionic acid or more specific inhibitors such as the fungal antibiotic Trichostatin A. This results in the progressive accumulation of more acetylated isoforms (see below).

H2A
S.G.R.G.*K.Q.G.G.K.A.R.A.K.A.K.T.R.S.S.R....

H2B
P.E.P.A.*K.S.A.P.A.P.K.*K.G.S.*K.K.A.V.T.*K.A.Q.K.K.D....

H3
A.R.T.K.Q.T.A.R.*K.S.T.G.G.*K.A.P.R.*K.Q.L.A.T.*K.A.A.R.K.S.A.P..

H4
S.G.R.G.*K.G.G.*K.G.L.G.*K.G.G.A.*K.R.H.R.K....

Figure 2. N-terminal sequences of the four core histones in mammals

Notes
Acetyl-lysine is shown as *K. Two examples where identical motifs of 2 or 3 amino acids are differentially acetylated are highlighted by double underlining. Three of the four acetylatable lysines in the H4 N-terminal domain are embedded in GKG motifs (underlined). See Van Holde (1989) for details.

Table 2. Examples of changes in patterns of histone acetylation through differentiation and development

Example	*What changes*	*Ref*
Maturation of *Trout* testis	Steady state acetylation level goes up	1
Macronuclear differentiation in *Tetrahymena*	Acetylation level in macronuclei (high) and micronuclei (low)	2
Chemically-induced Differentiation of *Human* promyeloid cultured cells	Heterochromatin acetylation fluctuates during progression down granulocyte and macrophage pathways	3
Development of *Sea Urchin* embryos (2-64 cell stage)	Level of diacetylated H4 decreases in line with cell doubling rate	4
Xenopus embryos pre & post Mid Blastula Transition	Acetate turnover increases at MBT	5
Mouse embryos (1-4 cell stage)	Intranuclear distribution of specific isoforms of acetylated H4	6
Mouse Embryonic Stem cells developing *in vitro*	H4 acetylation is reduced in constitutive and facultative (Xi) heterochromatin	7

References:
1. Louie & Dixon. Proc.Natl.Acad.Sci. 69, 1975 (1972);
2. Chicoine & Allis. Dev.Biol. 116, 477 (1986);
3. O'Neill and Turner. EMBO J. 14, 3946 (1995);
4. Chambers & Ramsay-Shaw. J.Biol.Chem. 259, 13458 (1984);
5. Dimitrov et al. Dev.Biol. 160, 214 (1993);
6. Worrad et al. Development 121, 2949 (1995);
7. Keohane and Turner, unpublished.

4. Individual Lysine Residues Are Selectively Acetylated

Because each acetate group added to a histone reduces its net positive charge by one the acetylated isoforms differ in net charge and so can be separated by electrophoresis on the appropriate gel system (an example is shown in Figure 3). However, each electrophoretic isoform is not necessarily homogeneous but may itself be a mixture of molecules in which the total number of acetate groups is the same, but the sites acetylated are different. For example, in the mono-acetylated isoform of H4, the single acetate could be attached to lysine 5, 8, 12 or 16. The electrophoretic mobility would be the same in each case. It is important to recognise that the sites which are acetylated may be just as significant, in a functional sense, as the change in net charge. This consideration has encouraged attempts to establish whether the different acetylation sites are used in a controlled or a random fashion. This has been tested with histone H4 by isolating the mono-, di-, tri- and tetra-acetylated isoforms and sequencing the amino-terminal region. In each of the several species tested so far, site usage has been shown to be non-random, though the frequency with which sites are used differs between species. Thus, in mono-acetylated H4 ($H4Ac_1$) from human or bovine cells, lysine 16 is the predominant acetylated site, while in cuttlefish testis lysine 12 is used exclusively. In human or bovine cells there is some flexibility in site usage, particularly in the more-acetylated isoforms, while in cuttlefish lysines are acetylated in a fixed and invariant sequence. The sequencing approach has recently been extended to

160

include histones H3 and H2B from human and bovine cells and here too site usage is non-random but with a degree of flexibility reminiscent of that seen with H4.

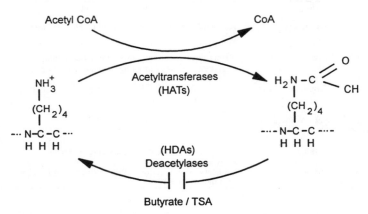

Figure 3. The enzyme-catalyzed acetylation and deacetylation of lysine residues on core histones

It should be emphasised that this type of analysis shows only the steady state level of acetylation at each site for each of the acetylated isoforms. For each isoform this level may be the net result of several different acetylation-deacetylation cycles, possibly occurring at different rates, in different parts of the genome and catalyzed by different enzymes.

5. Antibodies To Acetylated Histones Are Useful Experimental Tools

The fact that histone acetylation procedes in a controlled, non-random manner emphasizes that, in order to understand the roles that acetylation can play in chromatin function, it will be necessary to distinguish not only the effects of acetylation of individual histones, but also of acetylation of different lysines on the same histone. One way of addressing this problem is through the use of antibodies specific for particular histones acetylated at particular lysine residues. Over the past several years a panel of such antibodies has been prepared by immunizing rabbits with synthetic peptides corresponding to regions within the N-terminal domains of each of the core histones and containing acetyl-lysine at one or more of the sites acetylated in vivo. Such antibodies can be used in a variety of biochemical and microscopical techniques, some of which are referred to below.

In initial experiments, antisera were prepared that were specific for H4 acetylated at lysines 5, 8, 12 or 16. Such antisera have been used to immunostain Western blots from acid/urea/Triton gels (a gel system that separates on the basis of net charge and therefore can resolve mono-, di-, tri- and tetra-acetylated isoforms of the core histones). Typical results are shown in Figure 4. They show that the staining patterns are exactly what one would expect if the order of acetylation of H4 lysines is 16, followed by 12 or 8, followed by 5. That is, *all* acetylated isoforms (including the mono-acetylated, H4Ac$_1$) are acetylated at lysine 16, the di-, tri- and tetra-acetylated isoforms (H4Ac$_2$, H4Ac$_3$ and H4Ac$_4$ respectively) are also acetylated at lysines 8 and 12, and only the most highly acetylated isoforms (H4Ac$_3$ and H4Ac$_4$) are acetylated at lysine 5. There is therefore excellent agreement between the sequencing results and the results obtained by Western blotting and immunostaining.

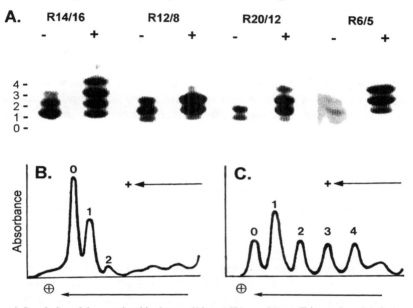

Figure 4. Resolution of the acetylated isoforms of histone H4 on acid/urea/Triton gels and their detection by (A) Western blotting and immunolabelling with antibodies to acetylated H4 and (B, C) staining with Coomassie Blue followed by laser densitometry. Histones were prepared from cells grown in the presence (+) or absence (-) of the deacetylase inhibitor sodium butyrate. Antibodies recognise H4 isoforms acetylated at specific lysine residues, namely 5 (R6/5), 8 (R12/8), 12 (R20/12) and 16 (R14/16). While the labelling pattern with antibodies to H4Ac16 resembles that seen with Coomassie Blue, those of the other antisera do not. This reflects the order in which the H4 lysines are acetylated. (From Belyaev et al. Exp.Cell Res. 225, 277-285, 1996).

6. Patterns Of Histone Acetylation Are Likely To Be Determined Primarily By The Specificities Of The Acetylating And Deacetylating Enzymes

The fact that the pattern of H4 acetylation differs between species, while the sequence of the amino terminal region of H4 and, presumably, the structure of the nucleosome, are identical, suggests that differences in acetylation reflect the specificities and relative activities of the acetylating and deacetylating enzymes. Information on these important enzymes has been limited until recently. Over the past year papers have appeared describing the purification of histone acetyltransferases and deacetylases and their characterization at the protein and DNA level. Some recent findings are summarized in Table 3. The results are revealing in that both types of enzyme show homology (or even complete identity) with known transcriptional regulators. It seems likely that at least some of these enzymes are integral components of the transcription complex, though whether their roles are to facilitate initiation (ie. binding of the complex to the promoter) or elongation (ie. nucleosome displacement during polymerase transit) or both, remains to be seen. It is also apparent that some of the enzymes form part of multi-subunit complexes, something that helps explain the difficulties encountered in their purification.

7. Histone Acetylation Is Associated With Various Basic Cellular Functions

As noted in Table 2, histone acetylation has been associated with progression along various pathways of differentiation and development. However, in most cases it is difficult to decide whether the changes observed are relevant to differentiation *per se*, or to changes in basic cellular processes that are themselves altered during differentiation and are now known to be intimately associated with shifts in the pattern of histone acetylation. These processes are:

1.   DNA replication and chromatin assembly
2.   Cell cycle progression
3.   Transcription
4.   Genomic marking

This is not to say that all changes in histone acetylation listed in Table 2. are contingent upon changes in these basic cellular processes, just that these processes must be taken into account when trying to define the role of acetylation in differentiation and development. Perhaps the most compelling example of a role for histone acetylation beyond the four processes listed above, is that of histone hyper- acetylation during the later stages of spermatogenesis. The use of hyperacetylation to weaken histone-DNA binding and thereby facilitate histone displacement by protamine, the major packaging agent for sperm DNA, seems entirely logical.

8. Histone Acetylation, DNA Replication And Chromatin Assembly

The traditional role of histones is in the packaging of DNA into chromatin and its organization in interphase and mitotic chromosomes. It has been known for many years that chromatin assembled on newly replicated DNA is enriched in acetylated histones. Most early work in the area was carried out using early amphibian embryos as a model system. In the fertilized amphibian egg, the single zygotic nucleus goes through several rounds of DNA replication and division before zygotic transcription begins. All the histones needed for this process are stored in the cytoplasm and transported into the nuclei for assembly into chromatin. It has been known for several years that the stored histones are acetylated. In the case of H4 the bulk of the stored histone is in the form of the di-acetylated isoform. Two proteins, nucleoplasmin and N1, have been identified in oocytes and fertilised eggs of the amphibian *X.laevis,* and their role has been shown to be the transport of histones from the cytoplasm to the nucleus and their assembly into chromatin. N1 deals with the transport and assembly of H3 and H4 and nucleoplasmin with that of H2A and H2B. It remains unclear whether the role of acetylation is to facilitate histone transport or assembly or both. It has recently been shown that the role of acetylation in chromatin assembly is not confined to the rather specific requirements of the developing frog embryo. Chromatin assembly coupled to DNA replication mediated by the Origin of Replication Complex (ORC) from mammalian cells has also been shown to require acetylated H4.

Table 3. The enzymology of histone acetylation and deacetylation

Enzyme (species, cloned subunit)	Properties	Homologies	Ref
Acetyltransferases			
HAT1 (yeast, Hat1p)	H4 lys12 specific	Various acetyltransferases	1
HAT A (Tetrahymena, p55)	Nuclear, single subunit active, complexes with other proteins	Yeast Gcn5p (transcriptional regulator) Bromodomain proteins Various acetyltransferases	2
HAT B (Drosophila)	Cytoplasmic, H4 lys 11/12 specific	Related enzymes studied in tetrahymena, pea and yeast	3
Deacetylases			
HD1 (human, p55)	Nuclear. Binds RbAp48 (human p50)	Yeast Rpd3p (transcription regulator). 2 related human genes.	4
HDA (yeast, HDA1)	350kD complex (3 sub-units)TSA-sensitive	Yeast Rpd3p	5
HDB (yeast, Rpd3)	600kD complex TSA insensitive	Rpd3 is a subunit	5

References:
1. *Kleff et al. J.Biol.Chem. 270, 24674 (1995);*
2. *Brownell et al. Cell 84, 843 (1996)*
3. *Sobel et al. J.Biol.Chem. 269, 18576 (1994);*
4. *Taunton et al. Science 272, 408 (1996);*
5. *Carmen et al. J.Biol.Chem. 271, 15837 (1996).*

It was shown some time ago that in the ciliated protozoan *Tetrahymena,* H4 deposited on newly-replicated DNA in the transcriptionally inactive micronucleus was selectively acetylated at lysines 4 and 11 (equivalent to 5 and 12 in other species) and that a cytosolic histone acetyltransferase (HAT B) activity showed just such a site specificity in vitro. *Tetrahymena* H4 is unusual in that its N-terminal residue is not blocked by acetylation, so direct sequencing of newly deposited H4 (marked by pulse-labelling with ^{3}H-lysine), can be used to locate acetyl-lysine residues. A chemical deblocking procedure has recently been used to enable the same sequencing approach to be applied to newly synthesized H4 in chromatin from other species. In both *Drosophila* and human cells, H4 is acetylated at lysines 5 and 12 prior to deposition. In contrast, the behaviour of H3 is more variable. *Drosophila* and *Tetrahymena* H3 are both di-acetylated prior to deposition, but at different residues, while human H3 is not acetylated at all. It is interesting to consider what factors regulate the transition from deposition-related H3/H4 acetylation to patterns appropriate for different genomic regions (see below). Apparently the deposition-related acetates turn over completely before the "transcription related" acetylation patterns are set.

9. Histone Acetylation And Cell Cycle Progression

Early studies on changes in histone acetylation through the cell cycle took advantage of the natural synchrony of nuclear divisions in the myxomycete *Physarum polycephalum*. Although the details of the findings and the interpretation of the results differed somewhat, there were clearly coordinated shifts in acetylation level as the cells progressed through the cycle. Cell cycle related changes in both the overall level of acetylation and in acetylation of particular lysine residues have also been found in mammalian cultured cells. However, such analyses are difficult to interpret. The changes observed will certainly be influenced by ongoing processes of chromatin assembly, transcription and changes in DNA packaging, processes that will vary not only through the cell cycle but from one part of the genome to another. To complicate the picture still further, recent results with mutants of the yeast *S.cerevisiae* in which the N-terminal tail of histone H4 has been altered in specific ways, indicate that histone acetylation may have a role in enabling progression through a specific cell cycle checkpoint, namely progression from G_2 to mitosis. Once again, interpretation of the phenotype observed is complicated by the fact that changes to the H4 tail can, potentially, influence various cellular functions.

10. Histone Acetylation And Transcription

Over the years there has been a steady accumulation of circumstantial evidence linking increased histone acetylation and transcriptionally active chromatin. The experiments reported usually demonstrate that chromatin fractions enriched in actively transcribed genes are also enriched in the more highly acetylated isoforms of the core histones. While these results are certainly suggestive, they do not prove that histone acetylation has any direct role in transcription, or indeed that the highly acetylated histones are in fact located on the transcribing genes themselves. They may simply co-purify. Indeed, a lack of any direct and general correlation between transcription and histone acetylation in mammalian cells was noted by Perry and Chalkley in 1982. In recent years three experimental approaches have thrown some light on what has proved to be a contentious issue. These are:
1.   immunoprecipitation with antibodies specific for acetylated histones
2.   the construction of yeast mutants with modified histones
3.   the analysis of transcription factor binding to nucleosomes in vitro

They will be considered in turn.

10.1   IMMUNOPRECIPITATION

Antibodies to acetyl-lysine have been used to immunoprecipitate chromatin fragments enriched in acetylated proteins, predominantly histones. These experiments have shown that in chicken erythrocytes, the highly acetylated (precipitated) chromatin fraction is enriched in the β-globin gene (expressed in these cells) but not in the ovalbumin gene, which is not expressed in the erythroid lineage. However, the β-globin gene is also enriched in the acetylated fraction when the starting chromatin is prepared from the erythrocytes of 5 day embryos, at which developmental stage the β-globin gene is not expressed. When the same approach was applied to human cultured K562 (erythroid) cells, the actively transcribed α-globin and inducible Platelet Derived Growth Factor B (PDGF-B) genes were both acetylated whereas the Human Growth Hormone gene was not. So histone acetylation seems not to be a general marker of transcription *per se* but rather of the potential for transcription. Because of the specificity of the antibody used for these experiments, it is not possible to say which core histones are hyperacetylated along (potentially) active genes. It could be just one or all four. Studies on the role of individual histones and their acetylation in transcription have so far been made only on H4. Immunoprecipitation of chromatin fragments from human cultured HL-60 cells with antibodies to acetylated H4, shows that a selection of transcriptionally active and quiescent genes both contain significant levels of acetylated H4, but this level of acetylation does not alter when transcription is induced or suppressed. Nor were significant differences noted between a selection of transcriptionally active, transcribable and "permanently" inactive genes. These same experiments have shown that the H4 associated with DNA sequences characteristic of centric heterochromatin is unacetylated, a result consistent with the immuno-fluorescence analysis of metaphase chromosomes (see below).

The inconsistencies noted above may be more apparent than real. There seems little doubt that transcribed or transcribable DNA is packaged into acetylated histones. Whether they are highly acetylated or moderately acetylated and on what histones is a matter for discussion and further experiment. It also seems that histones in non-coding heterochromatin are underacetylated. There is also general agreement that acetylation is not a consequence of, or dependent upon, ongoing transcription, but provides the *potential* for transcription. The question that remains to be resolved is whether genes that are permanently silent in a particular cell type (ie. are never going to be expressed) are *always* packaged in underacetylated nucleosomes. As so often happens, the answer may be, "it depends....".

10.2   YEAST MUTANTS

Mutants of the yeast *Saccharomyces cerevisiae*, with large deletions of the amino-terminal domains of particular core histones, grow in culture almost as well as wild type cells, leading to the conclusion that the amino terminal domains of the core histones are not, individually, essential for the basic cellular functions necessary for growth and division of yeast cells. However, two observations save us from the conclusion that the amino terminal domains of these histones are of no importance. The first is that double mutants (ie. ΔH3/ΔH4 or ΔH2A/ΔH2B) are not viable. In view of this, the viability of the *single* mutants may be just another example of the redundancy that evolution has built into living cells. Secondly, the H4 deletion mutants showed reduced mating efficiency, the effect being particularly dramatic in those in which all acetylatable H4 amino-terminal lysines were absent, i.e. having deletions of residues 4-19. This observation has proved particularly informative.

For yeast to mate successfully, two crucial genes called HMLa, and HMRa must be switched off. The cell can become either mating type a or mating type a by inserting either of these two genes into a new chromosomal locus (the MAT locus), where it is expressed. Cells of different mating types can (if they so choose) mate. In those H4 mutants in which mating efficiency was severely depressed, the normally silent HMLa and HMRa genes were constitutively transcribed (ie. permanently switched on). The inappropriate expression of genes at these normally silent loci is the cause of the severely reduced mating efficiency. Note that this effect was not part of a general disruption of gene regulation; for example, the PHO gene in mutants was regulated just as in wild type cells by changes in phosphate concentration.

These observations have been extended by substituting various amino acids for some or all of the amino terminal lysine residues of H4. Mutants in which all four amino-terminal lysines were substituted by glutamine or glycine (ie. permanently neutral) gave a phenotype very similar to the 4-19 deletion mutant, with severely reduced mating efficiency. Crucially, substitution of any or all of lysines 5, 8 or 12 with neutral residues was without measurable effect, whereas substitution of lysine 16 (or immediately adjacent residues) resulted in de-repression of the HMLa and HMRa genes and severely reduced mating efficiency. These results are consistent with the hypothesis that silencing of the mating type genes requires most of the amino terminal domain of H4 and that lysine 16 should not be acetylated. Acetylation of lysine 16 (mimicked in the mutants by substitution with a neutral amino acid) leads to activation of these genes and a consequent mating defect. A similar experimental approach has been used to prepare and analyse mutants of the amino terminal region of histone H3. Deletions in this region (residues 4-15) or substitution of acetylatable lysines (residues 9, 14 and 18) allowed hyperactivation of genes such as GAL1, that are required for metabolism of galactose and are regulated by the transcription factor GAL4. They can also, in association with H4 mutations, influence expression of the mating type genes.

The genetic approach in yeast has provided strong support for the contention that acetylation of histones H3 and H4 plays a central role in regulating the expression of specific genes. However, it is important to bear in mind that the neutral amino acids glycine and glutamine are structurally quite different to acetyl lysine, so to suppose that they are equivalent when interpreting mutant phenotypes is not without risk. It is important to back up the genetic data with biochemical anayses. In the case of the mating type genes, this has been done by using immunoprecipitation of chromatin fragments with antibodies to acetylated H4 to show that the mating type genes are packaged with underacetylated H4 when silent and acetylated H4 when active. Significantly, the acetylation status of H4 along the HML locus correlates with silencing and not with transcription itself. Thus, H4 acetylation levels along HML increase under circumstances where expression is appropriate, even in mutant strains in which this locus cannot actually be transcribed.

Further genetic experiments have strongly suggested that the mechanism of silencing of the yeast mating type genes involves the interaction of defined non-histone proteins, not only with specific sites on DNA (the so called Silencing Elements) but, directly or indirectly, with the amino-terminal region of H4. This deduction has recently been confirmed by demonstrating binding of the Silent Information Regulator proteins SIR3 and SIR4 to the N-terminal domain of H4 in vitro. In view of the results with H4 mutants, it is reasonable to propose that this interaction can be regulated by the acetylation status of H4, though this hypothesis remains to be proved by direct, biochemical analysis.

10.3   TRANSCRIPTION FACTOR BINDING TO NUCLEOSOMES IN VITRO

Recent experiments have attempted to measure directly the effect of histone acetylation on transcription factor binding. These used, as a model system, the 5S RNA gene from the toad Xenopus. This small gene was assembled into nucleosomes by mixing with core histones in buffers high in salt and urea and then progressively lowering the salt and urea concentrations by dialysis. The 5S RNA gene is transcribed by Pol III and requires the transcription factor TFIIIA for activity. TFIIIA binds to a site near the centre of the gene (not unusual for PolIII transcription factors). Binding of TFIIIA to its cognate sequence was much reduced when the gene was packaged into nucleosomes. However, this inhibition was relieved if the nucleosomes were assembled using either histones from which the amino terminal tails had been removed by proteolysis or histones in which the tails were highly acetylated. It was concluded that acetylation alleviates inhibition by relieving steric hindrance to TFIIIA binding. But this is not surprising. The histone tails are disposed on the nucleosome surface where they will almost inevitably encounter proteins scanning core DNA for a specific base sequence. The reduction in net charge caused by hyperacetylation will certainly reduce the avidity with which they bind core DNA and facilitate their displacement. However, it is important to note that the abolition of transcription factor binding in vitro by positioned nucleosomes is by no means universal and more subtle mechanisms than simple steric hindrance may be involved. Some transcription factors, such as glucocorticoid receptor and the yeast factor GAL4, show (under some conditions at least) only modest reductions in binding to nucleosomal DNA while others, such as NF1 and Heat Shock Factor, show almost complete abolition of binding. Even if histone acetylation is shown consistently to facilitate transcription factor binding *in vitro*, further experiments will be required to establish the functional significance of this, particularly *in vivo*, where adjacent, contiguous DNA, auxiliary transcription factors, structural proteins and enzymes all contribute to the final result.

11. Histone Acetylation And Genomic Marking

Studies on the relationship between histone acetylation and transcription have raised the possibility that one role of acetylation is to define the longer-term transcriptional potential of genomic regions. To put it another way, the results suggest that histone acetylation may serve as a marker by which patterns of gene expression are maintained through the cell cycle or adjusted when changes are needed for differentiation or development. With this possibility in mind, antibodies specific for the acetylated isoforms of core histones have been used to analyse the association of these molecules with different regions of the genome by indirect immunofluorescence microscopy of metaphase or (in insects) polytene chromosomes.

If human metaphase chromosomes are immunolabelled with antibodies specific for the most highly acetylated isoforms of H4, ie. $H4Ac_{2-4}$, the labelling patterns show three consistent properties. Firstly, the blocks of centric heterochromatin in chromosomes 1, 9, 15, 16 and Y are very weakly labelled. Secondly, there is a banded labelling pattern along the chromosome arms, the bright bands having a similar distribution to the R-bands identifiable by conventional banding methods. Thirdly, a single very weakly labelled chromosome is seen in female, but not male, cells. This chromosome has been shown to correspond to the inactive female X, ie. the one X chromosome of the two in females that becomes late replicating and transcriptionally inactive early in development. This inactivation of just one of the two female Xs constitutes the mechanism by which mammals equalize the levels of X-linked gene products in male and female cells, ie. dosage compensation. As will be discussed later, other species achieve the same objective by rather different means. These results show that underacetylation of H4 is a property of both constitutive, centric heterochromatin and facultative heterochromatin (ie. Xi) in mammalian cells, while acetylated H4 is preferentially located in regions of the genome enriched in coding DNA, ie. the R-bands.

The giant, polytene chromosomes found in certain insect larvae provide an alternative means of studying the distribution of acetylated histones through the genome by immunofluorescence microscopy. These chromosomes have been most widely studied in the larvae of the fruit fly *Drosophila melanogaster*. The cells containing these chromosomes have gone through several rounds of DNA replication without intervening cell divisions, resulting in the formation of polytene chromosomes in which a thousand or more strands of DNA are aligned in exact register. Differential packaging of DNA along these chromosomes results in a characteristic and reproducible banding pattern that allows one to navigate along the chromosomes and locate specific genes and genetically defined regions. Immunolabelling of such polytene chromosomes with site-specific antibodies to acetylated H4 has shown that isoforms acetylated at particular residues have characteristic distributions through the genome. In two instances, the distribution of acetylated H4 isoforms has been correlated with particular aspects of genomic function. Firstly, H4 in condensed, genetically inactive (heterochromatic) regions is consistently underacetylated at lysines 5, 8 and 16. However, acetylation at lysine 12 is retained. Secondly, H4 acetylated at lysine 16 is found predominantly (though not exclusively) on the X chromosome in male cells. The significance of this latter finding lies in the fact that in *Drosophila*, dosage compensation (ie the equalisation of X chromosome gene products between XY males and XX females) is achieved by doubling the transcriptional activity of X chromosome genes in male cells. Thus, a single chromosome (the male X) that shows a particular functional characteristic (ie. doubling the rate of transcription of its genes) is marked by a specific acetylated isoform of H4 (H4Ac16). It is interesting to note that, despite the fact that the mechanisms used for dosage compensation are very differention in flies and mammals (discussed above), both use changes in H4 acetylation to mark the functionally altered chromosome.

The existence of such an X chromosome marker is a potentially important functional element in the mechanism of dosage compensation in *Drosophila*. Four genes that are essential for dosage compensation in male *Drosophila* have been identified through their mutants. They are known, collectively, as the Male Specific Lethals (and individually as MSL1, 2 and 3 and MLE). Immunostaining of polytene chromosomes with antisera to the protein products of these genes shows that they are all localised, almost exclusively, on the X chromosome in male cells and co-localize along the chromosome in many discrete bands. Double immunolabelling experiments with rabbit antiserum to H4Ac16 and goat antisera to MSLs have shown that the proteins are, with only rare exceptions, located in exactly the same chromatin regions. This finding supports the proposition that H4Ac16 and the MSL proteins are all components of a multi-subunit complex required to increase the transcriptional activity of genes on the male X chromosome.

12. Histone Acetylation, Nucleosome Surface Markers And Intranuclear Signalling

As is usually the case with a new area of research in biology, much of the scientific data relating to the function of histone acetylation has been descriptive, providing correlations rather than evidence of molecular mechanisms. However, it is now possible to build on the evidence that has been assembled and devise experiments with a more functional and mechanistic aims. In theory, histone acetylation can exert functional effects in either of two ways. Firstly, it may, by itself, alter the structure or physical properties of the nucleosome. This could involve a change in shape or a reduction in the strength of binding of the amino-terminal domains to DNA. The latter is to be expected, given that acetylation causes a loss of positive charge and has been shown to be large. Such a reduction in binding avidity is very likely responsible for the enhanced binding in vitro of some transcription factors to acetylated nucleosomes and for the displacement of histones by protamines during sperm maturation. But it is hard to explain the significance of acetylation of specific histones at specific sites observed

Table 4. An outline of the nucleosome signalling hypothesis

1.	The histone octamer has a signalling function
2.	This function is carried out by the N-terminal domains of the core histones
3.	Variety of signals is provided by post-translational modification
4.	Modifications may be changed in response to extranuclear signalling pathways
5.	Functional responses are mediated by proteins that bind to the tails in a modification-dependent manner

in *Drosophila* and yeast on the basis of facilitated displacement. To accomodate these results, a second, indirect, mode of action has been proposed, namely that the cell will take advantage of the enormous variability provided by the post-translational modification of histones to target specific proteins to defined regions of the genome. These proteins will, in turn, alter the structure of chromatin in a functionally significant way. This mode of action proposes that the nucleosome has a fundamental role in signalling within the nucleus. It is a focal point at which extra-nuclear signalling pathways may impact on chromatin and through which their signals may influence chromatin function. The basic proposals of the nucleosomal signalling hypothesis are outlined in Table 4. Evidence consistent with this hypothesis has been covered in the preceding sections, but definitive proof, particularly regarding the interaction of extranuclear signalling pathways with chromatin and the existence of proteins with the properties specified in part 5 remains elusive. Several yeast proteins have been identified (in addition to SIR3 and SIR4 noted above) that regulate chromatin function by interacting with histones. Of particular interest is the finding that binding of the yeast suppressor protein Tup1 to H3 and H4 is inhibited by high levels of acetylation. These very recent results make one reasonably optimistic that further evidence in support of the hypothesis will not be long in arriving. But in the meantime it provides a useful working model with which to try and rationalise the diverse interactions and nuclear functions in which histone modification seems to be involved.

References

Chromatin and histone modifications
1. Van Holde, K. (1989) *Chromatin*. Springer-Verlag, New York.
2. Wu, R.S., Panusz, H.T., Hatch, C.L. and Bonner, W.M. (1984) Histones and their modifications. In: *CRC Critical Reviews of Biochemistry*, vol.20, pp.201-263.

Reviews of histone acetylation
3. Csordas, A. (1990) On the biological role of histone acetylation. *Biochem. J.* **265**, 23-38.
4. Loidl, P. (1994) Histone acetylation: facts and questions. *Chromosoma* **103**, 441-449.
5. Turner, B.M. (1991) Histone acetylation and control of gene expression. *J.Cell Sci.* **99**, 13-20.
6. Turner, B.M. and O'Neill, L.P. (1995) Histone acetylation in chromatin and chromosomes. *Sem.Cell Biol.* **6**, 229-236.

DNA replication and chromatin assembly
7. Jackson, V., Shires, A., Tanphaichitr, N. and Chalkley, R. (1976) Modifications to histones immediately after synthesis. *J.Mol.Biol.* **104**, 471-483.
8. Kaufman, P.D., Kobayashi, R., Kessler, N. and Stillman, B. (1995) The p150 and p60 subunits of chromatin assembly factor 1: a molecular link between newly synthesized histones and DNA replication. *Cell* **81**, 1105-1114.
9. Laskey, R.A., Mills, A.D., Philpott, A., Leno, G.H., Dilworth, S.M. and Dingwall, C. (1993) The role of nucleoplasmin in chromatin assembly and disassembly. *Phil.Trans.Roy.Soc B.* **339**, 263-269.
10. Sobel, R.E., Cook, R.G., Perry, C.A., Annunziato, A.T., Allis, C.D. (1995) Conservation of deposition-related acetylation sites in newly synthesized histones H3 and H4. *Proc.Natl.Acad.Sci. USA* **92**, 1237-1241.

Cell cycle progression
11. Bradbury, M. (1992) Reversible histone modifications and the chromosome cell cycle. *BioEssays* **14**, 9-16.
12. Megee, P.C., Morgan, B.A. and Smith, M.M. (1995) Histone H4 and the maintenance of genome integrity. *Genes Dev.* **9**, 1716-1727.

Transcription
13. Braunstein M., Rose, A.B., Holmes, S.G., Allis, C.D. and Broach, J.R. (1993) Transcriptional silencing in yeast is associated with reduced nucleosome acetylation. *Genes Dev.* **7**, 592-604.
14. Hebbes, T.R., Clayton, A.L., Thorne, A.W., Crane-Robinson, C. (1994) Core histone hyperacetylation co-maps with generalized DNaseI sensitivity in the chicken β-globin chromosomal domain. *EMBO J.* **13**, 1823-1830
15. Lee, D.Y., Hayes, J.J., Pruss, D. and Wolffe, A.P. (1993) A postive role for histone acetylation in transcription factor access to nucleosomal DNA. *Cell* **72**, 73-84.
16. O'Neill, L.P. and Turner, B.M. (1995) Histone H4 acetylation distinguishes coding regions of the genome from heterochromatin in a differentiation-dependent but transcription-independent manner. *EMBO J.* **14**, 3946-3957.
17. Perry, M. and Chalkley, R. (1982) Histone acetylation increases the solubility of chromatin and occurs sequentially over most of the chromatin. *J.Biol.Chem.* **257**, 7336-7347.
18. Thompson, J.S,, Ling, X. and Grunstein, M. (1994) Histone H3 amino terminus is required for telomeric and silent mating locus repression in yeast. *Nature* **369**, 245-247
19. Vettesse-Dadey, M., Walter, P., Chen, H., Juan, L-J. and Workman, J.L. (1994) Role of the histone amino termini in facilitated binding of a transcription factor, GAL4-AH, to nucleosome cores. *Mol.Cell.Biol.* **14**, 970-981.

Genomic marking
20. Bone, J.R., Lavender, J.S., Richman, R., Palmer, M.J., Turner, B.M. and Kuroda, M.I. (1994) Acetylated histone H4 on the male X chromosome is associated with dosage compensation in *Drosophila*. *Genes Dev.* **8**, 96-104.
21. Jeppesen,P. and Turner, B.M. (1993) The inactive X chromosome in female mammals is distinguished by lack of histone H4 acetylation, a marker for gene expression. *Cell* **74**, 281-289.
22. Turner, B.M., Birley, A.J. and Lavender, J. (1992) Histone H4 isoforms acetylated at specific lysine residues define individual chromosomes and chromatin domains in *Drosophila* polytene nuclei. *Cell* **69**, 375-384

Histone-binding proteins and nucleosome signalling
23. Edmondson, D.G., Smith, M.M. and Roth, S.Y. (1996) Repression domain of the yeast global repressor Tup1 interacts directly with histones H3 and H4. *Genes Dev.* **10**, 1247-1259
24. Hecht, A., Laroche, T., Strahl-Bosinger, S., Gasser, S.M. and Grunstein, M. (1995) Histone H3 and H4 N-termini interact with the Silent Information Regulators Sir3 and Sir4: a molecular model for the formation of heterochromatin in yeast. *Cell* **80**, 583-592.
25. Johnson, L.M., Kayne, P.S., Kahn, E.S. and Grunstein, M. (1990) Genetic evidence for an interaction between SIR3 and histone H4 in the repression of the silent mating loci in Saccharomyces cerevisiae. *Proc.Natl.Acad.Sci. USA* **87**, 6286-6290.
26. Turner, B.M. (1993) Decoding the nucleosome. *Cell* **75**, 5-8.
27. Wolffe, A.P. and Pruss, D. (1996) Chromatin: Hanging on to histones. *Current Biol.* **6**, 234-237.

THE ROLE OF STRUCTURE IN COMPLEXES BETWEEN THE p53 DNA BINDING DOMAIN AND DNA RESPONSE ELEMENTS

A.K. NAGAICH[1], P. BALAGURUMOORTHY[1], W.M. MILLER[1], E. APPELLA[2], V.B. ZHURKIN[3] and R.E. HARRINGTON[1]
[1] Department of Biochemistry/330, School of Medicine, University of Nevada Reno, Reno, NV 89557 USA,
[2] Laboratory of Cell Biology and [3] Laboratory of Mathematical Biology, NCI, National Institutes of Health, Bethesda, MD 20892 USA

Synopsis

A recent cocrystal structure of a minimal core p53 DNA binding domain peptide (p53DBD) complexed with a DNA response element has provided valuable insights into the binding specificity of p53 by identifying specific binding contacts. However, many important questions remained unanswered by that study including the full complex size as well as the organization of p53 tetramers bound to the subcomponents of the DNA recognition site.

To further understand the binding behavior of p53 with its response elements, we discuss an extensive series of biochemical and biophysical studies on complexes between the p53DBD peptide (amino acids 98-309) and a naturally occurring p53 response element, the *p21/WAF1/Cip* binding site. These have included cyclization studies, binding affinity and stoichiometry assays using ultracentrifugation, and high resolution chemical probes footprinting. The studies have demonstrated that the p53DBD peptide binds the *p21/WAF1/Cip1* response element as a tetrapeptide and induces DNA bending. The biochemical methods footprint the peptides on the DNA and identify protein-DNA contacts at single nucleotide resolution. Chemical probes studies utilizing the p53 recognition sequence used in the cocrystal structure verify the protein-DNA contacts observed in the cocrystal, but also suggest additional ones that evidently correspond to a second binding site in this sequence. The studies demonstrate tetrapeptide binding of p53DBD with the *p21/WAF1/Cip1* response element and provide the first direct experimental evidence that the four subunits bind to the major groove on the same face of the response element DNA. They also show that the most important contacts are between each monomeric peptide and the invariant guanosine nucleotides in the highly conserved C(A/T)l(T/A)G part of the consensus half sites.

C. Nicolini (ed.), Genome Structure and Function, 173–207.
© 1997 Kluwer Academic Publishers. Printed in the Netherlands.

Molecular modelling utilizing these findings identify specific interpeptide interactions in the form of steric clashes that require the response element DNA to bend away from the bound peptides, in agreement with cyclization and cyclic permutation studies. This type of binding motif appears to be novel and has not been reported previously in other nucleoprotein complexes. The modelling studies further suggest that p53 response elements possess a common characteristic sequence-directed flexibility that may be of critical importance in the DNA sequence specificity of p53-DNA binding. The results suggest a model for p53-DNA interactions in which highly specific protein-protein and protein-DNA interactions play critical roles in the sequence specificity and affinity of binding and in which both interactions are modulated by sequence-dependent flexibility in the response element DNA.

1. The p53 Protein

Wild type p53 is a nuclear phosphoprotein that occurs in a wide variety of organisms and plays a critical role in tumor suppression. It was first described as a cellular protein that co-precipitated with the large T antigen of simian virus 40 (SV 40) [1] and whose synthesis was enhanced in chemically induced tumors [2]. Its inactivation through mutation or deletions (along with loss of the wild type allele) or through interaction with cellular or viral proteins is highly correlated with human carcinogenesis [3-8].

Virtually all of the presently known biological functions of p53 depend critically upon its DNA binding properties. Much of the evidence for this is based upon its role as a transcription factor or transcriptional enhancer for genes that mediate DNA damage repair and growth arrest through their gene products [9-11]. The latter include *Gadd45*, which is implicated in the stimulation of DNA repair [12, 13] and *WAF1* (*Cip1* or *Sdi1*) which codes for p21, a protein that inhibits several cyclin-dependent protein kinases necessary for cell cycle progression from G1 into S phase [14, 15]. In addition, tumor derived mutations in p53 occur predominantly in the DNA binding domain [16] (see Figure 1) and lead to defective binding [17-20] and transcriptional activation [9, 16]. The many regulatory roles for p53 and its large variety of binding sites suggest that its specificity of binding to individual DNA binding sites may hold a clue to p53 function in its interactions with other regulatory proteins.

A schematic of the p53 protein is given in Figure 1. It consists of four functional domains: (1) an acidic N-terminal region extending from amino acids 1-73 containing a transactivation domain from amino acids 1-43 which mediates the binding of transcriptional factors such as MDM2, E1B and TBP; an alanine-rich segment; (2) a minimal core DNA binding domain from amino acids 98-308 (**p53DBD**) contained within the DNA binding region from amino acids 96-309; a flexible linker region from amino acids ~300-318; (3) a tetramerization domain from amino acids 319-360; and (4) a 33 amino acid, lysine-rich basic domain at the C-terminus whose function has been ascribed to non-specific DNA binding and/or negative regulation of specific DNA

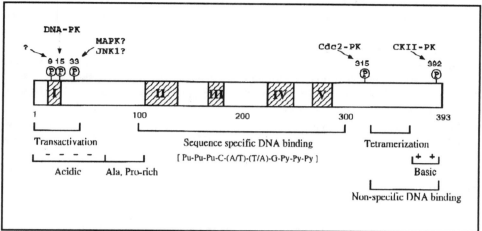

Figure 1. Schematic representation of human p53 (393 amino acids).

binding. The flexible hinge regions on either side of the core DNA binding domain are relatively unstructured [21-24]; reviewed in [25]. At least 5 serine residues in human p53 are phosphorylated and the kinases that phosphorylate three of them have been identified. The role of phosphorylation is not yet clear, although the fact that phosphorylation sites occur in highly conserved regions suggests a functional role [25]. However, no phosphorylation sites occur in the p53DBD (Figure 1).

2. DNA Binding Properties of p53

Wild type p53 may bind to over 100 different naturally occurring DNA binding sites or response elements of which over 50 show functionality, and the human genome may contain between 200 and 300 such sites [26], but so far no consensus functional classes have been identified among all these sites. Functional response elements differ in details of specific base sequence, but all contain 2 tandem decameric elements, each a pentameric inverted repeat. Most decamers follow, or closely follow, a consensus sequence pattern, PuPuPuC(A/T)|(A/T)GPyPyPy [19], where Pu and Py are purines and pyrimidines respectively and the vertical bar indicates the center of symmetry. These decameric elements may be separated by as much as 21 bp without complete loss of p53 binding affinity [19, 27] but *functional* sites, defined as the ability to transcriptionally activate a nearby reporter gene, evidently follow very closely the consensus decamer pattern above with no, or at least very short, intervening spacers [26]. No known functional p53 response elements contain poly-adenine tracts longer than $(A)_3$ [26] or other sequences characteristic of fixed bending [28], but most pentamers contain sites of putative DNA flexibility at one or both ends including CA·TG, TA·TA [29, 30] and GGGC·GCCC [31] sequence elements, suggesting that DNA conformation may contribute indirectly to p53 binding affinity and specificity. This is further substantiated by experiments described later in this document. It is also of interest that the GGGC·GCCC element occurs also in Sp1 GC boxes [30, 32] which are known DNA binding sites. Evidence has also been presented that binding may be allosterically regulated at the protein level [33, 34], which implicates protein conformation. Finally, p53 forms tetramers in solution [35, 36]; tetramerization enhances p53 binding to DNA response elements and is linked with transcriptional activation and enhancer properties through DNA looping [37]. Thus, both direct and indirect readout mechanisms are likely to operate in determining p53 binding specificity. However, few details are yet known about how these various changes in the DNA and protein binding partners affect binding affinity, or how and under what conditions p53 can discriminate among its many DNA response elements.

A cocrystal structure for p53DBD bound to a single consensus DNA pentamer has recently been reported [38]. The response element sequence used in this work was nearly identical (except for 1 base pair) to a binding sequence selected for high p53 affinity by Pab421 antibody binding followed by PCR amplification [39]. Specific contact sites between the p53DBD and bound DNA were identified which provided valuable insight into direct readout binding specificity. However, since the asymmetric unit contained only a single normally bound p53DBD, possible DNA bending induced by protein-protein interactions in a multipeptide complex could not be observed. Nmr structures for the tetramerization domain peptide have also been reported [40, 41] which suggest a cyclically symmetric tetrameric structure ("dimer of dimers") for the tetramerization domain peptide. These structural results are consistent with a tetrameric model for p53 in which each p53DBD binds to a single pentamer and adjacent peptides lie on opposite sides of the DNA helical axis [36] and are fully consistent with the biochemical and molecular modelling studies described here. However, crystallographic and nmr studies reported thus far have been restricted in the size of structural elements that can be investigated; for example, they have not been able to provide information on the interactions of intact p53, or even the p53DBD, with full consensus DNA response elements. Thus, alternative experimental approaches are required to understand the effects of protein-protein interactions and binding induced conformational changes on the specificity and affinity of p53 binding to its specific binding sites.

3. Evidence that Four p53 DNA Binding Domain Peptides Bind Natural p53 Response Elements and Bend the DNA

We have recently shown, using polyacrylamide gel band shift and analytical ultracentrifugation methods, that a peptide consisting of amino acid residues 96-308 of p53, which contains the full p53DBD but none of the tetramerization domain, binds cooperatively to form a stable tetrapeptide complex with either of 2 intact, naturally occurring response elements of quite different sequence [42]. The specific sequences investigated are shown in Figure 2. In this same work, we also used T4 ligase mediated cyclization methods to show that in each case p53DBD binding bends the DNA through an angle of at least 60°, defined here as the supplement to the actual angular deformation or the angular difference between tangents to one arm in the unbent and bent conformations [43].

We have also shown using hydroxyl radical footprinting that each p53DBD binds to a single pentamer in the complex [44]. Molecular modelling utilizing these findings identify specific interpeptide interactions in the form of steric clashes that require the response element DNA to bend away from the bound peptides, in agreement with cyclization and cyclic permutation studies. This type of binding motif appears to be novel and has not been reported previously in other nucleoprotein complexes. The modelling studies further suggest that p53 response elements possess a common characteristic sequence-directed flexibility that may be of critical importance in the DNA sequence specificity of p53-DNA binding. The results demonstrate that the p53DBD possesses a remarkable self-organizing ability since the p53DBD exists as a monomer in free solution [16]. Thus, the results of these investigations suggest a model for p53-DNA interactions in which highly specific protein-protein and protein-DNA interactions play critical roles in the sequence specificity and affinity of binding and in which both interactions are modulated by sequence-dependent flexibility in the response element DNA. It is clear that the DNA bending properties of the p53DBD provide important clues to the complicated protein-protein and protein-DNA interactions at work in the p53DBD nucleoprotein complex.

A comparison of this work to the cocrystal structure of Cho *et al.* discussed above [38] implies that the DNA bending observed in the tetrapeptide complex must be induced by strong *inter*peptide interactions. The molecular modelling studies suggest that these are, at least in part, due to steric clashes in the bound p53DBD tetrapeptide. This has not been reported previously, although peptides that include the N-terminal transactivation domain (amino acids 1 to ~321) have been shown to associate when bound to DNA [24, 37] and evidence has been advanced implicating allosterism as a possible mechanism [33, 34]. However, allosterism associated with DNA binding in the p53DBD alone appears inconsistent with the cocrystal study [38] since this work reported the same structure for both bound and unbound p53DBD peptides. It is also significant that, although we have found that the binding *affinity* of the p53DBD for response element DNA is 10- to 100-fold lower than wild type p53 [42], the "natural" *stoichiometry* of the p53DBD complex, *i.e.*, when the binding domain is not influenced by interactions with the tetramerization domain or with other regions of the p53 molecule, is the same as the wild type p53 complex. Thus, our finding that p53DBD alone can direct specific interactions with response elements is extremely provocative and demonstrates that DNA bending is a vitally important parameter in the binding properties of p53. Thus, careful, quantitative measurements of it are a key prerequisite to understanding the important functional characteristics of the p53 molecule.

```
5'  TCATCAGGAACATGTCCCAACATGTTGAGCTC 3'
    |||||||||||||||||||||||||||||||
3'  AGTCCTTGTACAGGGTTGTACAACTCGAGAGT 5'
    |<--------------->|
         20 WAF1 site
```

```
5'  TCCGATTGCCTTGCCTGGACTTGCCTGGCCTT 3'
    |||||||||||||||||||||||||||||||
3'  CTAACGGAACGGACCTGAACGGACCGGAAAGG 5'
    |<--------------->|
         20 nt RGC site
```

Figure 2. Two p53 response elements located in 32 bp precursor oligonucleotides for T4 ligase mediated cyclization studies: (upper) p21/Waf1/Cip1 site; (lower) Ribosomal Gene Cluster (RGB) site.

3.1 PREPARATION OF THE p53DBD PEPTIDE USED IN THE VARIOUS STUDIES DISCUSSED IN THIS REPORT

A portion of the human p53 cDNA encoding residues 96-308 was amplified by PCR using p53 specific primers

5'-ATATCATATGGTCCCTTCCCAGAAAACCTA-3' and

5'-ATATGGATCCTCACAGTGCTCGCTTAGTGCTC-3'.

The amplified product was cloned in the pET12a expression vector (Novagen) and the core DNA binding domain produced in E.Coli BL21 (DE3). The cells were incubated at 37°C until an O.D.600 of 0.6-1.0, and IPTG (0.25 mM) was added to induce the expression of the recombinant protein. Cells were harvested 2 hr. later by centrifugation, lysed using a French press and sonicated for 2 min in 40 mM MES-Na at pH 6.0, 100 mM NaCl, and 5 mM DTT. The soluble fraction was loaded onto a Resourse S column (Pharmacia) in 40 mM MES-Na at pH 6.0, 5 mM DTT, and was eluted by a 0 to 400 mM NaCl gradient. The pooled fractions were precipitated by 80 % saturated ammonium sulfate and purified further on a Superdex 75 gel-filtration column (Pharmacia) in 50 mM Bis-tris propane-HCl at pH 6.8, 100 mM NaCl, and 1 mM DTT. The purified p53DBD ran as a single band on an SDS polyacrylamide gel.

3.2 DETERMINATION OF BINDING STOICHIOMETRY

Binding stoichio-metries of p53DBD to the response element duplexes (Figure 2) were determined using standard band shift assays [45, 46] and are shown in Figure 3. Varying amounts of p53DBD were incubated with a fixed amount of 5' end-labelled duplex of known concentration in 50 mM Bis-tris-propane-HCl, pH 6.8, containing 10 mM MgCl$_2$ and 10 mM DTT in a total volume of 10 μL on ice for 45 minutes, electrophoresed on a 4% non-denaturing polyacrylamide gel in 0.3 X TBE. Identical free response element DNA controls were run in parallel. After autoradiography, the free DNA and p53DBD DNA complex bands were excised and counted, and the amount of DNA in the complex was calculated as the difference between counts for

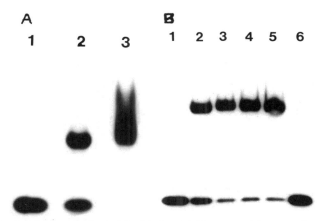

Figure 3. Band shift assays for determining binding stoichiometry and demonstrating cooperativity for p53DBD binding to the Waf1 response element. Similar results were obtained for the RGC site (data not shown). (A) Band shift assay. All lanes contain the same amount of DNA (26.4×10^{-12} mol). Lane 1 duplex DNA alone; Lane 2: DNA plus 55.2×10^{-12} mol p53DBD; lane 3: DNA plus 110×10^{-12} mol p53DBD. (B) Electrophoretic mobility shift assay to demonstrate binding cooperativity. All lanes contain 5 ng DNA. Lanes 1-6 contain the following amounts of p53DBD: 108 ng (1); 216 ng (2); 324 ng (3); 432 ng (4); 540 ng (5); control (no peptide) (6). From [42].

total DNA and that remaining in the free DNA band. The DNA-protein complex bands were then soaked in SDS containing stacking gel buffer for 5 minutes and run on a 12.5% SDS-polyacrylamide gel. This was stained with coomasie blue. The coomasie blue in the stained band was eluted with 25 percent pyridine in water overnight and the absorbance measured at 605 nm at 25°C. Unbound p53DBD did not comigrate with the complex as shown by stained gels with autoradiographs taken from it (data not shown). As controls, the same amounts of protein as used in the binding reactions were run in a SDS-PAGE (12.5%), stained with coomasie blue and the dye absorbance was measured for quantitation. From this, the amount of p53DBD associated with the protein-DNA complex was estimated.

3.3 BINDING AFFINITIES AND COOPERATIVITY USING ANALYTICAL ULTRACENTRIFUGATION

The molar concentration distributions in the ultracentrifugation studies were obtained by using the multi-wavelength scanning technique, which permits transformation of absorbancies as functions of radial position at different wavelengths into the total molar concentrations of each component as a function of radial position [47, 48]. The equilibrium concentration gradients of nucleic acid and protein were fit with appropriate functions for a 4:1 protein to DNA stoichiometry:

$$C_{r,T,N} = C_{b,N} \exp(A_N M_N (r^2 - r_b^2)) + C_{b,N} C'_{b,P} \exp(lnK_{14} + (A_N M_N + 4A_P M_P)(r^2 - r_b^2))$$

$$C_{r,T,P} = C_{b,P} \exp(A_P M_P (r^2 - r_b^2)) + 4C_{b,N} C'_{b,P} \exp(lnK_{14} + (A_N M_N + 4A_P M_P)(r^2 - r_b^2))$$

180

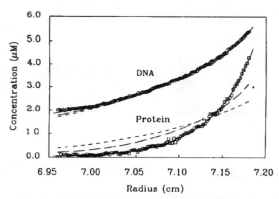

Figure 4. Total molar concentrations of *Waf1* (free and complexed, circles) and p53 DNA binding domain (free and complexed, squares) as a function of radial position at centrifugal and chemical equilibrium at 20°C and 16,000 rev/min in a Beckman XL-A analytical ultracentrifuge. These data are fit with mathematical models for a 4:1 p53DBD to *Waf1* stoichiometry (solid lines), 2:1 stoichiometry (long dashed lines) and 1:1 stoichiometry. Only a model based upon 4:1 stoichiometry is appropriate. From [42].

where K_{14} is the overall equilibrium association constant for the complex, $C_{r,T,N}$ and $C_{r,T,P}$ are the total molar concentrations (uncomplexed plus complexed) of DNA and peptide, respectively; the C_b's and M's are respectively the uncomplexed concentrations of each component at the reference position of the cell bottom (r_b) and the molecular masses (M_N=12384 Da, M_P=23932 Da); and A=$(\partial\rho/\partial c)_\mu \omega^2/2RT$ [$(\partial\rho/\partial c)_\mu$, the density increment at constant chemical potential, is obtained experimentally for each component in the ultracentrifuge, ω is the rotor angular velocity, R the gas constant and T the absolute temperature]. The two data sets were fit simultaneously with these equations, optimizing lnK_{14} as a global parameter, and $C_{b,N}$ and $C_{b,P}$ as local parameters, with a joint fit root-mean-square error of 0.0457 μM.

3.4 THE BINDING STOICHIOMETRY OF p53DBD TO *WAF1* AND *RGC* RESPONSE ELEMENTS IS 4:1 AND THE BINDING IS HIGHLY COOPERATIVE

The binding stoichiometry of the p53DBD to the two response elements was established by the above two methods. Both methods yield a stoichiometry of 4 p53DBD peptides to 1 response element for both *WAF1* and *RGC* response elements. Least squares analysis of the analytical ultracentrifugation data in Figure 4 gives a value of lnK_{14} of 65.17±0.69 at 20°C. This corresponds to an overall dissociation constant of 4.98×10^{-29} M^4 and a binding free energy of -37.9 kcal mol^{-1}.

As shown in Figure 4, attempts to fit the data with models corresponding to a 1:1 or 2:1 peptide to DNA stoichiometry fail to represent the data. Stepwise equilibrium models are also ruled out since the concentrations of species other than free peptide, free DNA and the 4:1 complex are too low to be observable. Thus, the binding of four p53DBD's to each response element occurs in a highly cooperative manner to form a nucleoprotein complex in which the bound p53DBD's are in very close and precise juxtaposition. A similar lack of intermediate species in Figure 3B provides additional evidence for binding cooperativity. Under these circumstances, the cooperativity parameter cannot be determined, and the mean apparent equilibrium dissociation constant per p53DBD monomer (which incorporates the cooperativity parameter) is obtained from the ultracentrifugation results as $(8.3\pm1.4)\times10^{-8}$ M.

3.5 p53DBD BINDING TO RESPONSE ELEMENTS BENDS THE DNA

We have detected and quantitated DNA bending in the 20 bp *p21/WAF1/Cip1* response element [39] and ribosomal gene cluster (*RGC*) [16, 49] response elements (sequences shown in Figure 2) using T4 ligase-mediated cyclization methods similar to those described previously [43, 50-52]. The 20 bp response elements are synthesized in 32 bp precursor oligonucleotides with 3 nt overhanging ends. After end labelling with ^{32}P, the precursors are bound by p53DBD and ligated with T4 ligase. Since the precursors are an integral number of helical repeats in length, DNA bending due to the bound peptide is phased coherently in the linear ligation products. At ligation product lengths such that the two ends are in close proximity due to DNA bending in the precursors, circularization or ring closure occurs with significant probability and competes kinetically with linear ligation, resulting in the formation of both open and covalently closed minicircles whose sizes and size distributions depend upon the induced bending angle.

These species are separated in a 2-dimensional polyacrylamide gel system in which the second dimension contains chloroquine; this intercalator alters the gel mobilities of linear species, open (nicked) circles and covalently closed circles differently and permits their efficient separation [50, 53]. Control experiments in the absence of p53DBD verify that observed DNA bending is induced by the bound peptide.

The response element duplexes (Figure 2) were kinated with γ-^{32}P-ATP and cold ATP. In cyclization experiments, 1.5 µg of the 5' phosphorylated response element duplex in 10 µL of 50 mM NaCl was incubated with 25 µL of p53DBD (1.32 µg/µL) in 50 mM Bis-tris-propane-HCl, pH 6.8, 100 mM NaCl and 1 mM DTT for 45 minutes on ice, and ligations were carried out for 16 hours at 4°C. Control experiments without p53DBD were done in parallel at the same DNA concentration under identical conditions to normalize any DNA concentration effects on microcircle formation. The reaction was arrested with SDS and EDTA and the solution was extracted three times with phenol and twice with a chloroform:amyl alcohol (24:1) mixture. The DNA was ethanol precipitated and analyzed in 5% non-denaturing polyacrylamide gels in the first dimension and run on an 8% non-denaturing polyacrylamide second dimension containing 50 µg/ml of chloroquine. The gel was then autoradiographed. Closed circles (Figures 5) were eluted from the gel and analyzed on denaturing gels as described earlier [43]. These procedures preclude misidentification of possible looped structures as microcircles since loops, resulting from the simultaneous binding of p53DBD to multiple response elements in linearly ligated precursor fragments similar to that described for intact p53 protein [37], would be lost when the bound peptides are removed prior to electrophoresis.

Clear evidence for DNA bending in the tetrapeptide complexes between p53DBD and the *p21/WAF1/Cip1* and *RGC* response elements has been obtained from T4 ligase mediated cyclization studies (Figures 5). The determination of DNA cyclization or ring closure as described above is a powerful and unambiguous method of demonstrating curvature in protein-DNA complexes [43, 50-52, 54]. Recent studies on the binding of the λ phage Cro protein using the present methods [43] have been corroborated by direct imaging using scanning force microscopy [55].

Second dimension gels for p53DBD bound to the precursor oligonucleotides shown in Figure 2 are shown in Figures 5A and 5B for the *p21/WAF1/Cip1* and *RGC* response elements, respectively. Figures 5C and 5D show the corresponding free DNA controls. Topologically relaxed, covalently closed microcircles appear as the second row of spots from the top in each figure. Correspondingly sized nicked microcircles appear just above these and linear species appear on the diagonal. Each of the two response elements bound to the p53DBD give essentially similar patterns (Figure 5A,B) and the smallest microcircles that appear with full intensity correspond to hexamers of the 32 bp precursor oligonucleotides (192 bp). This suggests that each bound response element has a minimum *absolute* bending angle of ~50-60°, defined as the angular difference between tangents to the two arms in the bent conformation [43, 52]. Control experiments in the absence of p53DBD show no microcircles for the *p21/WAF1/Cip1* response element (Figure 5C) and only weak spots for *RGC* (Figure 5D). Thus, essentially all curvature must originate in the peptide bound response elements.

The weak appearance of relatively large microcircles in the free DNA controls for the *RGC* site may be associated with the run of 5 half-helically phased TG·CA sequence elements as these chiefly differentiate the *RGC* from the *p21/WAF1/Cip1* sequence which showed no evidence for cyclization in the control (Figure 5C). Computer modeling [30] showed no tendency for this sequence to cyclize as observed in Figure 4D unless hinge kinking by roll was permitted in both the positive (into the major groove) and negative (into the minor groove) sense in each TG·CA element. In this case, almost perfect end alignments were predicted for the *RGC* site-containing sequence, whereas the *p21/WAF1/Cip1* sequence showed similar circularity but with a small but significant superhelical writhe. These results suggest that DNA flexibility may be an important attribute of p53 response elements. Alternatively, it is possible that differences in intrinsic bending may differentiate these two response elements since the *p21/WAF1/Cip1* element contains helically phased AA·TT dinucleotides.

The estimate of ~50° for DNA bending angles may be a *lower limit* value. The bending angles in the cyclization studies may be limited by steric restrictions imposed by the bound protein since we do not observe circular spots smaller than ~6x32=192 bp for either of the peptide-bound systems (Figures 5A,C). These 192 bp circular spots appear with relatively high spot intensities, and yet the larger circle size spots in Figures 5A and 5C have a distribution of intensities that corresponds to the larger circle size portion of the thermal intensity distribution normally observed with pure DNA (*e.g.*, Figure 5D) and with DNA bound by smaller proteins [43, 52]. This suggests that the intensity distributions are cut off at a ring size of 192 bp, presumably by steric restrictions. Steric restrictions appear to account for discrepancies between bending angles obtained using present methods for the λ phage Cro protein bound to an operator site [43] and those obtained by direct imaging using scanning force microscopy [55].

However, the fact that microcircles are observed for both the *p21/WAF1/Cip1* and *RGC* sites is an unambiguous reflection of curvature in the 32 bp precursor oligomers beyond that due to the DNA alone. Out-of-plane writhe induced in the bound precursors generally leads to no microcircle formation and torsional or angular mismatch of the single strand ends can seriously reduce it [52]. The latter two may affect the *distribution* of spot intensities with circle sizes since they are cumulative with ligation. Strong microcircle formation also suggests that the complexes form with little change in *overall* DNA winding since the precursor sequences, before complexation, have integral helical repeat by design. This is clearly the case for the bound *RGC* site (Figure 5B), which has a typically thermal distribution of circular spot intensities. The distribution for the *p21/WAF1/Cip1* site in Figure 5A suggests that the bound peptides may induce a small writhe or overall winding distortion in the DNA, but any such distortion must be small because of the relatively high intensity of the 6x32=192bp spot.

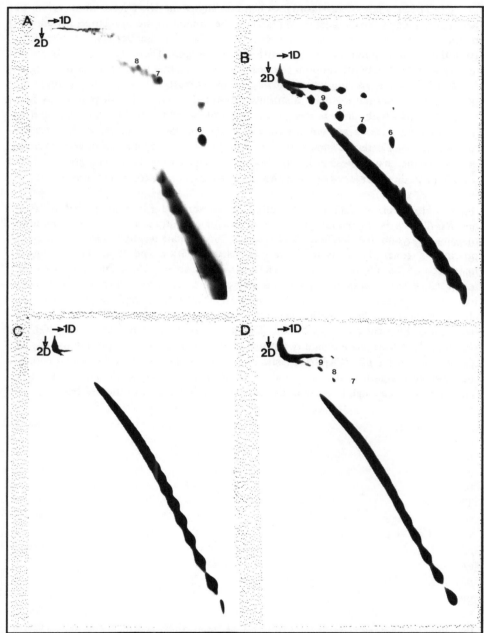

Figure 5. 2-dimensional polyacrylamide gel electrophoresis assays for ligation products of precursor duplexes shown in Figure 2. (A & B) in the presnece and C & D) in the absence of p53DBD. (A & C) Waf1; (B & D) RGB. The upper and central rows of spots correspond to nicked and covalently closed microcircles, respectively, and linear species migrate on the diagonal. Migration directions: 1D, first dimension; 2D, second dimension. From [42].

4. Chemical Probes Footprinting of Specific Protein-DNA Contacts in the Complexes of p53DBD with *WAF1* and the Crystallographic Response Elements

We have carried out chemical footprinting, chemical probes and crosslinking experiments on p53DBD complexed with the *p21/WAF1/Cip1* response element and with the recognition sequence used in a recent cocrystal structure of p53DBD with DNA [38], denoted hereafter as the Cho sequence. These studies complement the crystal structure and provide the first direct experimental evidence that four p53DBD subunits bind to the response elements in a regular staggered array having dyad (or pseudodyad) symmetry. The highly conserved C(A/T)|(T/A)G parts of the consensus half sites make very important contacts with the bound peptides. In particular, the invariant guanosine nucleotides of each consensus pentameric site seem critical for p53 binding. Model building using these contacts shows that when four subunits of p53DBD bind the response element, the DNA has to bend by ~50° to relieve the steric clashes among different subunits. This bending is consistent with the DNA cyclization studies described above and with recent cyclic permutation studies [44]. The p53 response elements exhibit sequence dependent structural heterogeneity in the consensus binding sites, suggesting that this may modulate the specificity and the energetics of complex formation through an indirect readout mechanism.

A number of chemical footprinting/probes methods are highly sensitive to the location-specific nucleoprotein contacts between the p53DBD peptide and the response element DNA, with single base resolution routinely attainable. These include hydroxyl radical footprinting, missing nucleoside analyses, methylation and ethylation interference assays. Glutaraldehyde crosslinking experiments verify binding stoichiometry under the conditions that the probes experiments were conducted. Oligonucleotides investigated include a 65 bp oligonucleotide which contains the 20 bp *p21/WAF1/Cip1* response element; this response element contains both a consensus and a non-consensus half site (Figure 6A, left-hand and right hand boxes, respectively). To facilitate comparison of the present work with the crystallographically determined contact sites [38], we also studied a 67 bp oligonucleotide containing the same 20 bp binding sequence used in the cocrystal structure (Cho sequence) as shown in Figure 6B. These solution results generally agree with the crystallographic contacts for the Cho sequence, although there is evidence for a second binding site in reverse orientation within this sequence that was not observed in the cocrystal. Our results for the *p21/WAF1/Cip1* site show unequivocally that four p53DBD peptides bind this response element in a staggered array and that each consensus pentanucleotide in the *p21/WAF1/Cip1* site makes specific contacts with the p53DBD with the invariant guanosine nucleotide playing a critical role. Hydroxyl radical footprinting demonstrates structural microheterogeneity in the consensus binding sites, suggesting that sequence-dependent structural variability of response elements plays a critical role in the binding of p53 with DNA. Model building of the p53DBD-*p21/WAF1/Cip1* nucleoprotein complex using the results of various chemical probes and foot-printing shows that the four bound p53DBD peptides bend the response element by ~50° to relieve the steric

186

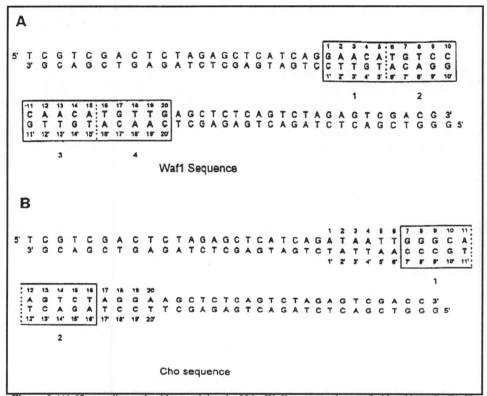

Figure 6. (A) 65-mer oligonucleotide containing the 20 bp Waf1 response element (bold and boxed). (B) 67-mer oligonucleotide containing the 20 bp Cho *et al.* sequences [38]. Each half site is seperated by a solid line and a quarter site by dashed line.

clashes among the bound subunits. This model is in quantitative agreement with the T4-DNA ligase mediated cyclization and cyclic permutation studies discussed above [42, 44].

4.1 HYDROXYL RADICAL FOOTPRINTING OF p53DBD BOUND TO THE *p21/WAF1/Cip1* AND CHO SEQUENCES:

Hydroxyl radical cleavage [56-58] was used to footprint p53DBD complexed with the *p21/WAF1/Cip1* and Cho sequences. Figures 7A and 7B show hydroxyl radical footprinting data for the p53DBD-*p21/WAF1/Cip1* and p53DBD-Cho complexes respectively. Densitometric plots of the various lanes are shown in Figures 8A and 8B. The top strand of the naked *p21/WAF1/Cip1* response element (Figure 6A) shows reduced cleavage at CATG and TGTT base sequences within the 20 bp consensus binding site, with clear minima at TG sequence elements (marked with arrows in Figure 8A, plot a) and a higher cleavage at the central CCCAAC bases (Figure 7A, lane C;

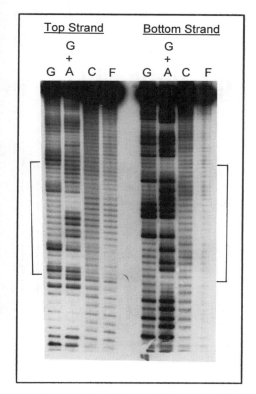
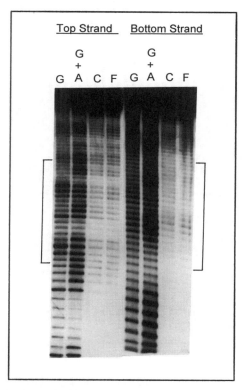

Figures 7A,B. Hydroxyl radical cleavage (foot-printing) studies. Lanes G and G+A are Maxam-Gilbert guanosine and guanosine+adenosine sequencing reactions. Vertical brackets indicate the 20 bp consensus p53 binding sites. (A) 65-mer oligonucleotide containing the p53DBD-Waf1 complex (Figure 1A), and (B) 67-mer oligonucleotide containing the p53DBD-Cho complex (Figure 1B). Lanes C and F are in the absence (control) and presence of p53DBD, respectively.

Figure 8A, plot a). The two TG base elements are separated by 10 bp and hence are spatially in phase along the helix. p53DBD binding shifts the cleavage frequency minima by one base toward the 3' end and further diminishes the cleavage frequency at ATGT and TGTT sequences, while the central CCCAAC bases show a relative increase in the cleavage frequency (Figure 7A, lane F; Figure 8A, plot b). The bottom strand exhibits a significantly reduced cleavage at the GTTGGG and GTTC sequences and an increased cleavage at the ACAT and CAT base sequences with maxima at TA sequence elements (marked as arrows in dashed brackets, Figure 7A, lane C; Figure 8A, plot c). The binding of p53DBD further enhances the cleavage frequency of the CAT sequence while reducing it in the central GTTG and GTT sequences (Figure 7A, lane F; Figure 8A, plot d).

The unusual hydroxyl radical cleavage profiles suggest that the *p21/WAF1/Cip1* response element has narrowed minor grooves at CATG and TGTT sequences and a relatively wider minor groove at the CCCAA sequence. p53DBD

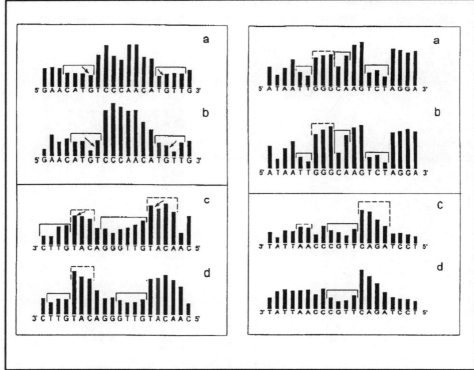

Figure 8A,B. Densitometric plots of gels shown in 7A and 7B, respectively. Plots (a) and (b) are hydroxyl radical cleavage patterns of the top strand in the absence and presence of p53DBD, respectively. Plots (c) and (d) are similar but for the bottom strand. The arrows indicate the minima in the cleavage pattern.

binding further narrows the minor grooves involving ATGT and TGTT bases in the two half sites, compresses the major groove at the GTTGGG bases and shields their sugar-phosphate backbone from the minor groove side, leading to their reduced cleavage. This may, in turn, further expose the sugar-phosphate backbone of the complementary CCCAA sequence from the minor groove side, making it more susceptible to hydroxyl radical cleavage. X-ray crystallography [59] and model building [60] on a variety of DNA sequences indicate that the minor groove width in the ACAITGT sequence in the first half site should be characterized by the distances between sugar moieties of G7 and G4' and the phosphate distances of G7-T3' and G4'-T8 and by the similarly positioned nucleotides in the second half site (Figure 6A,B). Hydroxyl radical data indicate that p53DBD binding drastically reduces the cleavage frequency in the G7 and T8 in the first half site and in G17 and T18 in the second half site (Figure 8A,B), whereas in the bottom strand, the cleavage frequency of G4 and G14' is reduced upon p53DBD binding (Figure 8C,D). Since hydroxyl radical cleavage is a diffusion controlled process, the data clearly point to a relatively narrow minor groove in the CATG regions in the two half sites, which is in accord with the cocrystal structure results [38].

The higher cleavage at the ACAT and CAT bases in the bottom strand as compared to the ATGT and TGTT sequence at the top strand probably indicates asymmetric distortion of the double helix in this region, leading to differential exposure of the sugar-phosphate backbone of the two strands to hydroxyl radicals. The relatively A/T rich regions of the consensus binding site, which have narrow minor grooves, are spaced at integral helical periodicity and occur on the same face of the double helix (Figures 6A and 15A). It is also of interest that GGGC sequences, previously shown to have major groove directed bending [61-63], occur in the central region of many functionally important p53 response elements. The helically phased CA:TG sequence elements in the highly conserved region of the consensus binding site have been shown to be kinked in the CAP nucleoprotein complex [64] and may be similarly kinked in other regulatory complexes [30]. Thus, it is likely that intrinsic flexibility in these sequence elements may promote the formation of a more stable p53-DNA complex by facilitating specific protein-DNA and protein-protein interactions.

The results presented earlier have shown that p53DBD binds the *p21/WAF1/Cip1* response element as a tetrapeptide with high cooperativity and that the DNA is bent by ~50-60° in solution [42]. Thus, it is likely that a region of compressed major groove located between two regions of narrow minor groove, as suggested by the present hydroxyl radical footprinting, may provide the structural basis for such a bending. There is also an inherent asymmetry in the footprinting patterns for the two half sites (Figures 7A, 8A) which otherwise might be expected to be identical. Such an inherent asymmetry in the complex may play a crucial role in terms of bending directionality, as it in the case of the TBP-TATA complex [65, 66].

The hydroxyl radical cleavage patterns for the Cho sequence (Figure 6B) and its complex with p53DBD are shown in Figures 7B and densitometry plots in 8B. The top strand of the unbound Cho sequence shows reduced cleavage at the TCT and the CA sequence elements within the consensus half site, while a higher cleavage frequency occurs at the GGG bases (marked with dashed bracket). The cleavage frequency is also markedly reduced at a TT element outside this half site (Figure 7B, lane C; Figure 8B, plot a). Binding with p53DBD further diminishes the cleavage frequency of the TCT and TT sequences while it enhances the cleavage frequency in the GGG bases (Figure 7B, lane F; Figure 8B, plot b). However, cleavage in the bottom strand is enhanced in the complementary AGAC and AA sequences and reduced in the TTGC bases. Binding with p53DBD further reduces the cleavage pattern in the TTGC sequence. The data clearly point to a narrow minor groove in the TCT sequence and a relatively compressed major groove in the GGGC region. These footprinting data are in general agreement with the crystallographic results [38] in which a narrowed minor groove was observed in the TCTAG sequence due to high propeller twisting. However, the cleavage pattern in the GGGCAA region demonstrates a relatively compressed and shielded major groove and indicates binding of p53DBD in this region, also consistent with the crystallographic results.

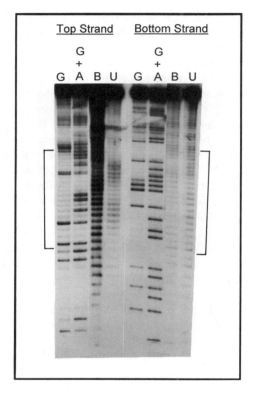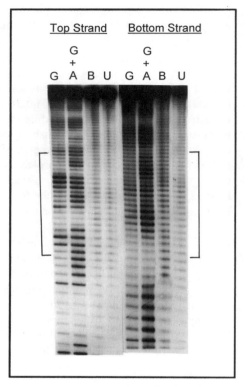

Figure 9A,B. Results of missing nucleoside experiments. Captions and lane identification markers are the same as in Figures 7A,B, respectively, except that lanes B and U refer to p53DBD bound and unbound fractions, respectively.

4.2 MISSING NUCLEOSIDE EXPERIMENTS

Specific base contacts for both complexes were determined using missing nucleoside experiments [67]. Figures 9A and B show the missing nucleoside data for the p53DBD-*p21/WAF1/Cip1* and p53DBD-Cho complexes respectively; corresponding densitometric plots are shown in Figures 10C and D. The bound fraction of DNA in the top strand of the p53DBD-*p21/WAF1/Cip1* complex (Figure 6A) leads to weaker bands for bases in the two ATGT sequence elements in each half site and intense bands for bases in the central CCCAAC sequence (Figure 9A, lane B; Figure 10A plot a). The opposite pattern is observed in the unbound fraction, *i.e.*, bands corresponding to bases in the central CCAA region are weak whereas bands for bases ATGT in both the half sites are more intense (Figure 9A, lane U; Figure 10A, plot b). In the bottom strand, the protein-bound DNA shows weaker bands for bases in ATGT in both half sites whereas, in the unbound fraction, these bases show intense bands (Figure 9A, lane B and lane U; Figure 10A, plots c and d). These data clearly show that ATGT sequence elements in both half sites of the top and bottom strands make important contacts with the bound protein. Modification or absence of either of these base contacts greatly reduces the

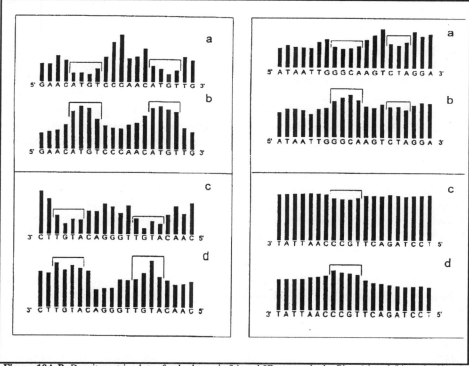

Figure 10A,B. Densitometric plots of gels shown in 9A and 9B, respectively. Plots (a) and (b) are hydroxyl radical cleavage patterns of the top strand in the absence and presence of p53DBD, respectively. Plots (c) and (d) are similar but for the bottom strand. The arrows indicate the minima in the cleavage pattern.

affinity of p53DBD with DNA. It is significant that these bases show reduced cleavage in direct hydroxyl radical footprinting (Figures 7 and 8), suggesting a narrow minor groove in these regions, and further substantiating their direct involvement in the stability of the nucleoprotein complex.

Missing nucleoside data for the p53DBD-Cho complex are shown in Figure 9B with corresponding densitometric plots in Figure 10B. The top strand of the bound fraction shows relatively weak bands in the GGCA and CTA sequence elements (Figure 9B, lane B; Figure 10B, plot a). In the unbound fraction, strong bands are observed in this sequence element, indicating direct contact of these bases with the bound p53DBD (Figure 9B, lane U; Figure 10B, plot b). The bottom strand in the bound fraction shows weak bands in the TGCC element, while in the unbound fraction, these bases exhibit much stronger bands (Figure 9B, lane B & U; Figure 10B, plots c and d). Thus, the data clearly demonstrate that bases in the GGCA element in the top strand and in the TGCC element in the bottom strand make critical contacts with the bound peptide. Most of the contact sites probed by hydroxyl radical footprinting and missing nucleoside probing in the p53DBD-Cho complex are also observed in the crystal structure data [38].

4.3 METHYLATION AND ETHYLATION INTERFERENCE ASSAYS

Methylation interference assays have been widely used to identify contacts between bound proteins and methylated guanosines in the major groove of the DNA [68]. Figure 11A shows the methylation interference results for the p53DBD-*p21/WAF1/Cip1* complex. Corresponding densitometric plots for the different lanes are shown in Figure 12. The bound DNA fraction of the top strand (sedquence in Figure 6A) shows reduced cleavage (greater interference) at the guanosines in the two TG sequence elements (G-7 and G-17 in Figure 12) compared to more intense cleavage at these sites in the unbound fraction (Figure 11A, lane B and lane U; Figure 12, left hand plots a and b). This suggests that these guanosines are in direct contact with p53DBD in the major groove. In the bottom strand, the bound DNA fraction shows reduced cleavage of guanosines at the two TG elements (G-4' and G-14') compared to much stronger bands in the unbound DNA fraction (Figure 11A, lane B and lane U; Figure 12, left hand plots c and d). The residues in the central GGG region of the bottom strand show reduced cleavage in the bound DNA fraction as compared to the control DNA (lane C) but these guanosine signals are missing entirely from the unbound fraction.

These data clearly show that the central GGG guanosines in the bottom strand contact the p53DBD differently from those in the two TG elements. Methylation of the former does not affect p53DBD binding, whereas the guanosines in the TG doublets evidently make structurally important contacts with the protein, probably at the N-7 position, and when these sites are modified by methylation, p53DBD does not bind to DNA. It is important to note that TG guanosines are present in each pentamer quarter-site and constitute the invariant base in most of the high affinity p53 binding sites reported so far. All four guanosines in the TG base doublets (G-7, G-17, G-4' and G-14') are tandemly arrayed in the alternate major groove of the double helix, suggesting that all four subunits of p53DBD bind in the major groove.

The methylation interference data for the p53DBD-Cho complex are shown in Figure 11B with corresponding densitometric plots in Figure 12. In the top strand, the p53DBD-bound fraction shows relatively weaker bands for the GGG (G-7, G-8, G-9) sequence and G-13 compared to the unbound fraction (Figure 11B, lane B and U; Figure 12, right hand plots a and b). In the bottom strand, both the bound and unbound fractions show relatively intense bands for guanosine G10', with the stronger band observed in the unbound DNA fraction. Essentially all guanosine contacts reported in the crystal structure [38] are also observed in the methylation interference assays described here with the following exceptions. (i) contacts between p53DBD and bases in the GGA sequence element at the end of the top strand, *i.e.* G-18, G-19 and A-20, are not observed. (ii) Guanosine G-13 (top strand) either makes direct contact with p53DBD at its N-7 position or is in very close proximity with the peptide in the complex.

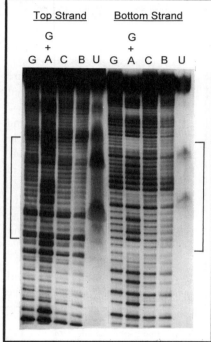

Figure 11A. Methylation interference assays on the *p21/WAF1/Cip1* response element. Captions and lane identification markers are the same as in Figures 2 except that lanes C are controls with dimethyl sulfate treatment of the DNA in the absence of p53DBD.

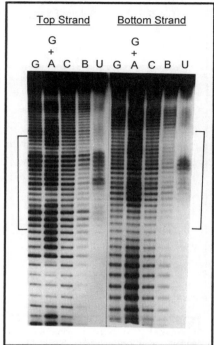

Figure 11B. Methylation interference assays on the Cho sequence. Captions and lane markers are the same as in 11A.

Sugar-phosphate contacts were also probed by ethylation interference [69]. Results for the p53DBD-*p21/WAF1/Cip1* complex are shown in Figure 13A with corresponding densitometric plots of the bound and unbound fractions shown in Figure 14. For the top strand (sequence in Figure 6A), the bound fraction shows weaker bands for bases in the ATGT sequence elements in the two half sites with clear minima at the two TG base doublets (Figure 13A, lane B; Figure 14, left hand plot a) whereas in the unbound fraction, these two guanosine residues show significantly more intense bands (Figure 13A, lane U; Figure 14, left

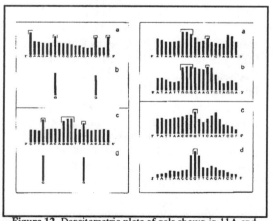

Figure 12. Densitometric plots of gels shown in 11A and 11B, respectively. Plots (a) and (b) are methylation interference patterns of the top strand in the absence and presence of p53DBD, respectively. Plots (c) and (d) are similar but for the bottom strand. The arrows indicate the minima in the cleavage pattern.

194

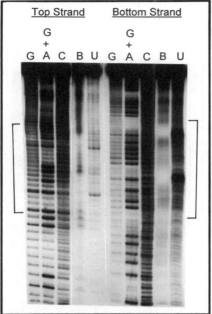

Figure 13A. Ethylation interference assays on the *p21/WAF1/Cip1* response element. Captions and lane identification markers are the same as in Figures 2 except that lanes C are controls with N-ethyl-N-nitroso urea modification of the DNA in the absence of p53DBD.

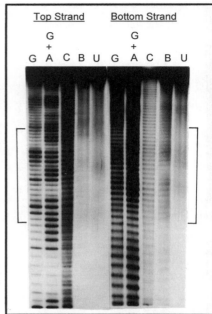

Figure 13B. Ethylation interference assays on the Cho sequence. Captions and lane identification markers are the same as in 13A.

hand plot b). The AA sequence elements at the center of the binding site also appear as bands of medium intensity. For the bottom strand, the bound fraction (Figure 13A, lane B; Figure 14, left hand plot c) shows weaker bands in TGT sequence elements in both half sites, with a clear minimum at the G residue (marked by arrows). On the other hand, bases in the TG elements appear as highly intense bands in the unbound fraction (Figure 13A, lane U; Figure 14, left hand plot d). These data clearly indicate that the DNA backbone at the ATGT sequences in the two half sites is in contact with the hydroxyl radical and missing nucleoside data. In particular, the sugar-phosphate backbone at G-7, G-17, G-4' and G-14' makes very critical contacts with p53DBD and ethylation of phosphates at these residues greatly reduces the binding affinity of the p53DBD-*p21/WAF1/Cip1* complex.

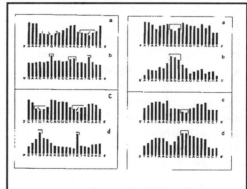

Figure 14. Densitometric plots of gels shown in 13A and 13B, respectively. Plots (a) and (b) are ethylation interference patterns of the top strand in the absence and presence of p53DBD, respectively. Plots (c) and (d) are similar but for the bottom strand. The arrows indicate the minima in the cleavage pattern.

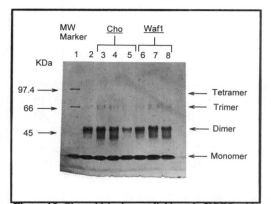

Figure 15. Gluteraldehyde crosslinking of p53DBD with oligonucleotides containing the *p21/WAF1/Cip1* and Cho sequences. 8% SDS polyacrylamide gels were run in the presence of different concentrations of glutaraldehyde followed by silver-staining. Lane 1: protein molecular weight marker. Lane 2: glutaraldehyde crosslinking of p53DBD in the absence of DNA. Lanes 3, 4 and 5: different oligomeric states of p53DBD bound to the Cho oligonucleotide. Lanes 7, 8 and 9: different oligomeric state of p53DBD bound to the Waf1 oligonucleotide.

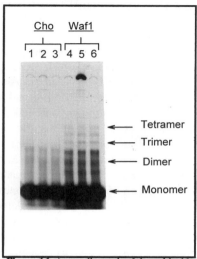

Figure 16. Autoradiograph of the gel in 12 showing crosslinked DNA species. Lanes 1, 2 and 3: labeled DNA associated with lanes 3, 4 and 5 in Figure 15. Lanes 4, 5 and 6: labeled DNA associated with lanes 6, 7 and 8.

Figure 13B shows ethylation interference results for the p53DBD-Cho complex with densitometric plots in Figure 14. For the top strand (Figure 6B), the bound fraction shows relatively weak bands in the GGC element of the consensus binding site, whereas in the unbound fraction, these bands are much stronger (Figure 13B, lane B and lane U; Figure 14, right hand plots a and b). For the bottom strand, the bound fraction shows much weaker bands in the TT region but much stronger bands in the unbound fraction. The ethylation interference signal is not as strong at G-13 in the Cho half site as in the *p21/WAF1/Cip1* duplex where the conserved motif is CATG. This suggests that the CATG element may be an important sequence feature for p53DBD binding. If the T in this element is replaced with an A, as in the Cho sequence, the affinity of the p53DBD-DNA interaction is reduced. Earlier studies have shown that a thymine at the underlined position PuPuPuC(A/T|T/A)GYYY allows more favorable interactions in the minor groove as compared to the adenine [38, 70]. It is clear that the backbone contacts sites inferred from the present results correlate well with those observed in the cocrystal structure [38].

4.4 GLUTERALDEHYDE CROSSLINKING EXPERIMENTS OF *p21/WAF1/Cip1* SEQUENCES WITH p53DBD

Gluteraldehyde crosslinking [71, 72] was used to determine the stoichiometry of p53DBD bound to the *p21/WAF1/Cip1* and Cho response elements. Figure 15 shows an SDS-PAGE analysis of crosslinked species of the p53DBD-*p21/WAF1/Cip1* complex (Figure 15, lanes 6, 7 and 8) and the p53DBD-Cho complex (lanes 3, 4 and 5) in the presence of 0.01%, 0.05% and 0.1% glutaraldehyde concentrations. The p53DBD-*p21/WAF1/Cip1* complex shows four bands of molecular weight 23Kd, 46Kd, 69Kd and 92Kd, representing monomer, dimer, trimer and tetramers respectively, of p53DBD cross-linked with the *p21/WAF1/Cip1* response element. Direct autoradiography confirms that each band is associated with its corresponding DNA fragment (Figure 16, lanes 4, 5 and 6). This demonstrates clearly that the p53DBD peptide associates with the full *p21/WAF1/Cip1* response element as a tetrapeptide, even in the absence of the wild type p53 oligomerization domain, in agreement with the gel band shift and ultracentrifugation results discussed above [42]. With the p53DBD-Cho half-site, cross-linked species include trimers and tetramers in the SDS-PAGE gel (Figure 15, lanes 3, 4 and 5) but only monomers and dimers appear in the autoradiography (Figure 16, lanes 1, 2 and 3). This indicates that p53DBD associates with the Cho sequence primarily as a monomer or dimer. The p53DBD peptide crosslinked in the absence of response elements can also weakly tetramerize, but the primary crosslinking products are monomers and dimers (Figure 15, lane 2). In Figure 15, the DNA and protein bands move as doublets in the gel; this may occur since the faster band in each doublet can arise from a species in which additional intrapeptide crosslinking leads to a more compact structure for the denatured complex and therefore to slightly higher mobility [71].

4.5 A COMPARISON OF THE FOOTPRINTING RESULTS FOR p53DBD COMPLEXED WITH THE *p21/WAF1/Cip1* AND CHO SEQUENCES

The above results are summarized for the p53DBD-*p21/WAF1/Cip1* and p53DBD-Cho complexes in Figures 17A and 17B, respectively. For the p53DBD-*p21/WAF1/Cip1* complex, certain structurally related characteristics become immediately apparent. (i) Most of the contact signals occur in the major groove of the DNA in a staggered array along the two helical turns of the response element in which each of the four bound p53DBD peptides occupies a single pentanucleotide quarter site and faces outward, away from the DNA, but in the same direction. The data indicate that each p53DBD binds to the major groove, in agreement with the cocrystal structural study [38]. (ii) Four guanine bases, G-7, G-17 in the top strand and G-4' and G-14' in the bottom strand which are part of an invariant sequence motif in each pentamer quarter site, are identified as contacts using all four of the footprinting methods. These bases are in phase along the helix with a separation of 10 bp and are located in the major groove of the DNA. (iii) Most of the observed contact points are clustered in the major groove with relatively few contacts in the minor groove also in agreement with the specifically bound p53DBD in the crystal structure [38]. (iv) The GTTG sequence of the bottom strand shows higher protection from hydroxyl radical cleavage than does the complementary CAAC sequence on the opposite strand, suggesting that it is shielded by bound peptide from the minor groove side whereas the complementary CAAC sequence is exposed.

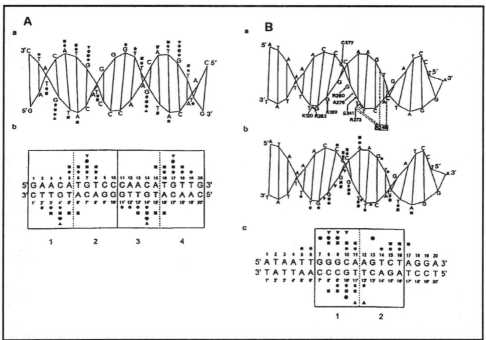

Figure 17. (A) Helical representation (a) and sequence (b) of the 20 bp p21/Waf1/Cip1 response element showing contact sites identified by the chemical footprinting techniques. The helix was generated using wedge angle parameters as described in Bolshoy *et al.* [76] The software [76] simulates the double helix by joining C1' atoms in paired nucleotides. For simplicity of presentation and comparison with the sequence (b), the helical DNA is represented as unkinked. Indicated contacts are as determined by: (●) protection from hydroxyl radical cleavage; (■) missing nucleoside experiments; (▲) ethylation interference signals in the sugar-phosphate backbone; and (△) methylation interference assays. (B) Helical representations (a,b) and sequence (c) of the 20 bp Cho sequence. Generation of the helical models and symbols indicating peptide-DNA contacts are as in (A). The helical models show (i) contact sites observed in the cocrystal structure [38], and (ii) contact sites observed in the present chemical probe studies.

For the p53DBD-*p21/WAF1/Cip1* nucleoprotein complex, the chemical probes/footprinting results are most consistent with a model in which all four p53DBD peptides make their most important contacts between the major groove of the response element DNA with their loop sheet helix motif [38]. The p53DBD peptides are bound to the major groove of each pentameric quarter site in an alternating array so that, apart from DNA bending, the full complex possesses a *quasi*- C_2 symmetry. The data also require that the conserved guanosines of each pentamer half-site play a crucial role in the binding. This model is not generally appropriate for the p53DBD-Cho complex,

198

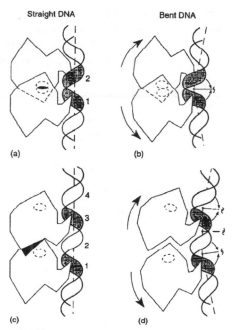

Straight DNA   Bent DNA

(a)   (b)

(c)   (d)

Figure 18. Cartoons showing how molecular modelling determines steric clashes between the four p53DBD peptides bound to the p21/Waf1/Cip1 response element as directed by the chemical probes results, and how these clashes are relieved by bending the DNA by roll at CA:TG sequence elements.

however. In this complex, the hydroxyl radical and the missing nucleoside contacts are symmetrical along the DNA and the methylation and ethylation interference signals and some of the missing nucleoside contacts show a marked asymmetry. The footprinting contacts are arranged very differently from those in the *p21/WAF1/Cip1* complex and suggest that there are two binding sites in the Cho sequence in opposite orientations along the DNA. One is to the GGGCA quarter site element in the upper strand (Figure 6B; Figure 17B) and the second is to the TGCCC element in the lower strand. Contacts in the first of these are consistent with those observed in the cocrystal structure (Figure 17B) [38], and since this is the only specific binding site observed crystallographically, it is likely that it is a site of higher binding affinity. The gluteraldehyde crosslinking results as well as the lower quality of the footprinting data for the p53DBD-Cho complex compared to the *p21/WAF1/Cip1* complex suggest that, in solution, a dynamic equilibrium may exist between complexes involving these two sites. In the *p21/WAF1/Cip1* response element, all the four invariant guanines occur in the TG sequence elements. By contrast, in the Cho sequence, only one guanine occurs in the TG element whereas the other occurs in an AG element. Since intense contact signals are observed for the guanines in the *p21/WAF1/Cip1* TG elements but not in the Cho half-site AG, this suggests that AG elements are very poor contact sites for p53 binding. Again, these results are consistent with earlier observations [38, 70].

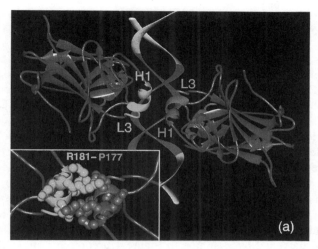

Figure 19. Interactions between the H1 helices and L3 loops in the p53-DNA complexes for unbent DNA. Nomenclature for p53DBD structural elements is from Cho *et al.* [38]. The H1 helix from the left domain interacts specifically with the pentamer GGGCA [38]. The right domain is located symmetrically, so that its H1 helix interacts with the pentamer AGACT. Inset: the H1 helices inter-digitate, producing unacceptable steric clashes (the heavy atoms in the protein are shown as spheres of 50% van der Waals radii). The zinc atoms show as large spheres. The clashes specifically involve Arg 181 and Pro 177. Similar clashes occur using other unbent DNA models including canonical B-form DNA.

5. Stereochemical Model for the p53DBD-DNA Complex

To rationalize the footprinting and crosslinking data, a computer-assisted structural model for the p53DBD-*p21/WAF1/Cip1* complex has been developed. In this model, each of the four bound p53DBD peptides is attached to a single pentameric quarter site as suggested by the probes/footprinting results. The p53DBD coordinates are taken from the cocrystal structure [38]; in that study, Cho *et al.* observed no changes in the p53DBD structure upon DNA binding. Docking of the peptides to the DNA is determined by the specific contacts found here (Figure 17A). The model requires that, in order to avoid severe interpeptide steric clashes, the four p53DBD peptides must induce bending of the response element DNA by ~50° (Figure 18), a result in close agreement with the cyclization results discussed above [42] and with cyclic permutation gel retardation assays [44].

Two sets of steric clashes are observed in the modelled complex between four p53DBD peptides and "straight" DNA. These are illustrated in cartoon form in Figure 18 and are modelled in Figure 19. The first clash occurs between the Zn-binding H1 helices [38] of peptides bound to two adjacent pentamers (Figures 18a and 19); these two p53 domains are bound to opposite sides of the duplex in opposite orientations and "embrace" the DNA. The second clash occurs between the two peptides in similar orientation and bound to the same side of the helix; they are separated by a single helical turn (Figure 18c and 19). These clashes were observed in all modelled complexes involving unbent DNA that were generated as described.

In the description of the modelled complexes presented below and in Figures 18-21, nomenclature for p53DBD structural elements is taken from Cho *et al.* [38]. In Figure 18, the Zn-binding H1 helices are shown as ellipses. The H2 recognition helices are shown as sharply pointed regions buried in the major grooves of the DNA. The four pentamer response element quarter-sites are indexed 1-4. Pentamers 1 and 3, 2 and 4 are in a parallel orientation while pentamers 1 and 2, 3 and 4 are anti-parallel. Conclusions from the models are as follows. (a) With unbent DNA, binding of two p53DBD peptides to pentamers 1 and 2 (ti-parallel orientation) is accompanied by steric hindrance between the H1 helices (overlapped ellipses; Figure 18a). (b) This clash is relieved by bending the DNA toward the major groove along the dyad axis by positive roll at the TG:CA sequence elements (Figure 18b). (c) In unbent DNA, binding of two p53DBD peptides to pentamers 1 and 3 (parallel orientation) is accompanied by steric hindrance. The darkened overlap area in Figure 18 represents steric clash between amino acid residues 99, 167, 170 and 210 from one peptide, and residues 224, 140, 199 and 201 from the other; a steric clash is presumed if the distance between two heavy atoms in a residue pair is less than 2.4A. (d) This clash is relieved by bending the DNA toward the major grooves along the dyad axes at two points separated by 10 bp. This bending is the same as in (b) and is effected by positive role at the TG:CA sequence elements.

It is significant that the steric clashes discussed above and shown in Figures 18 and 19 are relieved if the *p21/WAF1/Cip1* DNA is bent by positive roll through a 15°

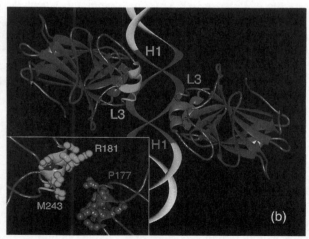

Figure 20. Interactions between the H1 helices and L3 loops in the p53-DNA complexes for bent DNA. Nomenclature is as in Figure 16. Bending the DNA toward the major groove at the junction between the pentamers relieves the H1-H1 clash (the bend is directed away from the viewer). The important spatial complementarity between the two H1-L3 moieties is shown in the inset; in this view, the protruding residues are Arg 181 (in the H1 helix) and Met 243 (in the L3 loop).

angle toward the major grooves in the CA:TG dimers, indicated by centered bullets: GAAC•AlT•GTCCCAAC•AlT•GTTG. Such axial bending does not interfere with the specific interaction of the p53DBD peptides with their cognate pentamers. The two adjacent rolls in the CA:TG dimers induce a bend toward the major groove as indicated in Figures 18b and 18d. It is important to note that the *same* DNA bends relieve both types of clashes: anti-parallel, as in Figures 18a and 19, and parallel, as in Figure 18c. The putative rolls in the CA:TG steps are consistent with the well known flexibility of these dinucleotides [30], as demonstrated in numerous x-ray structures of pure B-DNA and protein-DNA complexes [64, 73, 74] as well as by gel electrophoresis [29]. According to energy calculations, these dinucleotides can also bend anisotropically toward the major groove [60].

The direction of DNA bending observed in the models is entirely consistent with the consensus p53-binding sequence. All four possible tetramers allowed by the consensus, *i.e.*, CAlTG, CTlTG, CAlAG and CTlAG, contain dinucleotide elements that can flex toward the major groove. In addition to CA:TG, the CT:AG element also prefers bending toward the major groove [74-76]. AA:TT or AT:AT dinucleotides are the central elements in three of the tetranucleotides. Both dinucleotides prefer a negative or zero roll angle [74]; thus, the minor groove is expected to be relatively narrow, which is desirable for interactions with the highly conserved Arg 248 [38]. The CTlAG tetranucleotide contains the TA:TA element which can assume roll angles between -6.4° in free DNA [77] to +12° in the *trp* repressor complex [78] and appears, in addition, to possess unusual torsional flexibility [63, 64]. However, the CTlAG element is reported to be extremely rare in functional genomic p53 response elements [26].

The molecular models shown in Figures 19-21 were generated using the MidasPlus program of Ferrin *et al.* [83]. In the bent DNA model shown in Figures 20 and 21, the roll angles for CA:TG are taken as 15°. This leads to an overall bend of ~50°, which is consistent with the experimental results described above [42, 44]. CA:TG roll angles between 15° and 20° are sufficient to relieve all peptide-peptide clashes; 15° seems preferable, however, as this leads to a structure in which the p53DBD peptides are in close proximity to one another and hence can interact to produce the observed binding cooperativity [42].

According to the stereo representation of the bent DNA model shown in Figure 21, the C-terminal residues of p53 are located on the inside of the DNA curve. The C-terminal domains are involved in oligomerization, protein-protein interactions and the modulation of DNA binding [27]. This arrangement may be advantageous for the stabilization of the entire superstructure in the binding of the wild type protein. On the other hand, the N-terminal region, which contains the trans-activational domain, is located on the outside of the loop where it may be more accessible to other proteins. Finally, we note that when DNA is packed tightly in chromatin, it must in general retain access to target sites and response elements. The model presented in Figures 20 and 21 suggests that the intrinsic bendability of p53 response elements favors the same

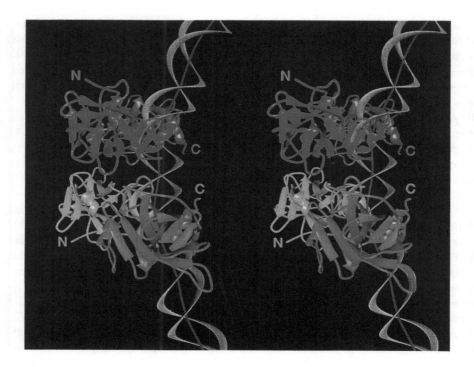

Figure 21. Stereo view of the putative complex between four p53DBD peptides and the Waf1 response element. The DNA axis passes through the centers of the base pairs (the centers of the C6-C8 vectors). In this model, the DNA is bent by ~50° which relieves the steric hindrances between the bound peptides (Figures 15, 17), consistent with biochemical studies (see text). The positioning of the p53DBD and the DNA bending is the same as that shown in Figure 17. In this model, the N-terminal ends of the p53DBD, which attach to the transactivational domain of intact p53, are exposed on the *outside* of the DNA bend and are hence fully accessible. The C-terminal ends, which attach through a flexible region to the oligomerization and C-terminal "tail" domains, are located at the *inside* of the bend, thus facilitating tetramerization in the wild type protein and bringing the C-terminal basic regions into closer proximity with the DNA.

direction of DNA bending as in the p53 nucleoprotein complex. Thus, p53 response elements may be wrapped in chromatin in such a way as to facilitate the approach and binding of the protein. In other words, p53 may prefer to bind response element DNA from the side that is normally exposed in chromatin structures.

6. Summary

The *p21/WAF1/Cip1* site is one of the most important functional sites for p53 binding presently known [14, 15, 79]. The full site consists of a consensus [19] and a non-consensus half site, making it generally representative of a broad class of p53 response elements [26]. The data discussed here provide an overall picture of the spatial arrangement of the peptides on the *p21/WAF1/Cip1* response element DNA along with critical p53DBD-DNA contacts. The picture derived for the p53DBD-*p21/WAF1/Cip1* complex underlines the symbiosis between absolute structural and biochemical methods as represented by the chemical probes-molecular modelling approach described here. Although no high resolution cocrystal structure is yet available for the p53DBD-*p21/WAF1/Cip1* complex, the results from the existing structural study [38] complemented by the biochemical and biophysical results presented here allow a reasonable model for the full tetrapeptide complex to be proposed.

It is known that TG:CA sequence elements are ubiquitous in genomic regulatory sites, and particularly so in those in which DNA bending may occur in the regulatory nucleoprotein complexes [30]. Various studies have shown that these elements are flexible, with potential for abrupt bending or kinking [29, 51, 60, 80-82], and such kinking has been observed structurally in a limited number of nucleoprotein complexes [64, 73]. Hence, it is quite likely that helically phased TG:CA or other flexible sequence elements may play a critical role in the formation of p53 nucleoprotein complexes through the ability of p53 response elements to bend and architecturally accommodate the protein, thus enhancing the stability of the complex through more precise, and perhaps more extensive, protein-DNA contacts. Most functional p53 binding sites contain helically phased TG:CA sequence elements in approximately the same relative positions [19]. This requirement may also provide a structural explanation for the fact that p53 response elements which contain spacers between the palindromic half-sites have been generally observed to be highly unstable and non-functional [26]. The presence of such spacers, unless they are long enough to effectively loop out of the complex [11, 37], will alter the spatial and helical phasing relationships among the TG:CA elements as well as the patterns of helical groove width, which the evidence discussed here suggests may be important to the formation of stable (and possibly functional) p53 nucleoprotein complexes.

In this discussion, arguments has been presented and supported by experiments that sequence dependent structural and dynamical properties of response element DNA may modulate the stability and possibly the structure of p53 nucleoprotein complexes. This is consistent with earlier studies, discussed above, which have demonstrated substantial DNA bending in several p53DBD-DNA complexes [42]. This suggests a possible role for the variation observed in the sequences of known naturally occurring functional p53 response elements [26] and supports the concept that both the sequence specificity and the stability of p53-DNA nucleoprotein complexes are controlled by a complex interplay of specific sequence contacts whose efficacies are modulated by inherent structural features of the response element DNA including its "softness" and ability to flex or kink at specific sequence elements. Such sequence specific deformations of the DNA structure along with the sequence directed structural requirement of the two narrow minor grooves and a compressed major groove as suggested by the studies described above provide a mechanism of exquisite selectivity for the specific binding of p53 to DNA sequences. Thus, the binding of p53 is almost certainly modulated by a (possibly variable) set of specific protein-DNA contacts which

204

may, in turn, be modulated by specific DNA sequences and by a complicated set of specific and nonspecific inter-p53DBD and intra-p53 protein-protein interactions.

Acknowledgements

The authors wish to thank Professors Jeffrey Hayes and Ilga Winicov and Drs. G. Marius Clore, Stewart Durell, Angela M. Gronenborn, and Robert Jernigan for many helpful suggestions. This work was supported by research grants MCB 9117488 from the National Science Foundation, HG00656 and CA70274 from the National Institutes of Health and US Department of Agriculture Hatch Project NEV032D through the Nevada Experiment Station (REH) and from the AIDS Targeted Anti-Viral Program of the Office of the Director of the National Institutes of Health (EA).

References

1.   Lane, D.P. and Crawford, L.V. (1979) *T antigen is bound to a host protein in SV 40 transformed cells*, Nature **278**, 261-263
2.   DeLeo, A.B., Jay, G., Appella, E., DuBois, G.C., Law, L.W. and Old, L.J. (1979) *Detection of a transformation related antigen in chemically induced sarcomas and other transformed cells of the mouse*, Proceedings of the National Academy of Sciences of the United States of America **76**, 2420-2424
3.   Levine, A.J., Momand, J. and Finlay, C.A. (1991) *The p53 Tumour Suppressor Gene*, Nature **351**, 453-456
4.   Hollstein, M., Sidransky, D., Vogelstein, B. and Harris, C.C. (1991) *p53 Mutations in Human Cancers*, Science **253**, 49-53
5.   Lane, D.P. (1992) *p53, gaurdian of the genome*, Nature **358**, 15-16 (News and Views)
6.   Vogelstein, B. and Kinzler, K.W. (1992) *p53 Function and Dysfunction*, Cell **70**, 523-526 (Minireview)
7.   Levine, A.J. (1993) *The tumor supressor genes*, Annual Reviews of Biochemistry **62**, 623-651 (Review)
8.   Meltzer, P.S. (1994) *MDM2 and p53: A question of balance*, Journal of the National Cancer Institute **86**, 1265-1266
9.   Prives, C. and Manfredi, J.J. (1993) *The p53 tumor suppressor protein: meeting review*, Genes & Development **7**, 529-534
10. Milner, J. (1994) *Forms and functions of p53*, Seminars in Cancer Biology **5**, 211-219
11. Prives, C. (1994) *How loops, beta sheets, and alpha helices help us to understand p53*, Cell **78**, 543-546
12. Marx, J. (1994) *New link found between p53 and DNA repair*, Science **266**, 1321-1322
13. Smith, M.L., Chen, I.T., Zhan, Q.M., Bae, I.S., Chen, C.Y., Gilmer, T.M., Kastan, M.B., Oconnor, P.M. and Fornace, A.J. (1994) *Interaction of the p53-regulated protein Gadd45 with proliferating cell nuclear antigen*, Science **266**, 1376-1380
14. Hartwell, L.H. and Kastan, M.B. (1994) *Cell cycle control and cancer*, Science **266**, 1821-1828
15. Pines, J. (1994) *Arresting developments in cell cycle control*, Trends in Biochemical Sciences **19**, 143-145 (review)
16. Pavletich, N.P., Chambers, K.A. and Pabo, C.O. (1993) *The DNA-binding domain of p53 contains the 4 conserved regions and the major mutation hot spots*, Genes & Development **7**, 2556-2564
17. Kern, S.E., Kinzler, K.W., Baker, S.J., Nigro, J.M., Rotter, V., Levine, A.J., Friedman, P., Prives, C. and Vogelstein, B. (1991) *Mutant p53 proteins bind DNA abnormally in vitro*, Oncogene **6**, 131-136
18. Kern, S.E., Kinzler, K.W., Bruskin, A., Jarosz, D., Friedman, P., Prives, C. and Vogelstein, B. (1991) *Identification of p53 as a Sequence-specific DNA-binding Protein*, Science **252**, 1708-1711
19. El-Deiry, W.S., Kern, S.E., Pietenpol, J.A., Kinzler, K.W. and Vogelstein, B. (1992) *Definition of a consensus binding site for p53*, Nature Genetics **1**, 45-49
20. Bargonetti, J., Manfredi, J.J., Chen, X., Marshak, D.R. and Prives, C. (1993) *A Proteolytic Fragment from the Central Region of p53 Has Marked Sequence-specific DNA-binding Activity When Generated from Wild-type But Not From Oncogenic Mutant p53 Protein*, Genes & Development **7**, 2565-2574

21. Unger, T., Nau, M.M., Segal, S. and Minna, J.D. (1992) *p53: A transdominant regulator of transcription whose function is ablated by mutations occurring in human cancer*, EMBO Journal **11**, 1383-1390

22. Lin, J.Y., Chen, J., Elenbaas, B. and Levine, A.J. (1994) *Several hydrophobic amino acids in the p53 amino-terminal domain are required for transcriptional activation, binding to mdm-2 and the adenovirus 5 ELB 55-kd protein*, Genes & Development **8 (10)**, 1235-1246

23. Picksley, S.M., Vojtesek, B., Sparks, A. and Lane, D.P. (1994) *Immunochemical analysis of the interaction of p53 with MDM2; Fine mapping of the MDM2 binding site on p53 using synthetic peptides*, Oncogene **9**, 2523-2529

24. Wang, P., Reed, M., Wang, Y., Mayr, G., Stenger, J.E., Anderson, M.E., Schwedes, J.F. and Tegtmeyer, P. (1994) *p53 domains: structure, oligomerization, and transformation*, Molecular and Cellular Biology **14**, 5182-5191

25. Appella, E. and Anderson, C.W. (1994) *Tumor suppressor protein p53: response to DNA damage, cell cycle control and mechanisms of inactivation in cancer*, Biochimica in Italia **1**, 19-28 (Review)

26. Tokino, T., Thiagalingam, S., El-Deiry, W.S., Waldman, T., Kinzler, K.W. and Vogelstein, B. (1994) *p53 tagged sites from human genomic DNA*, Human Molecular Genetics **3**, 1537-1542

27. Waterman, J.L.F., Shenk, J.L. and Halazonetis, T.D. (1995) *The dihedral symmetry of the p53 tetramerization domain mandates a conformational switch upon DNA binding*, EMBO Journal **14**, 512-519

28. Yoon, C., Prive, G.C., Goodsell, D.S. and Dickerson, R.E. (1988) *Structure of an alternating-B DNA helix and its relationship to A-tract DNA*, Proceedings of the National Academy of Sciences of the United States of America **85**, 6332-6336

29. McNamara, P.T., Bolshoy, A., Trifonov, E.N. and Harrington, R.E. (1990) *Sequence-dependent kinks induced in curved DNA*, Journal of Biomolecular Structure and Dynamics **8**, 529-538

30. Harrington, R.E. and Winicov, I. (1994) *New concepts in protein-DNA recognition: sequence directed DNA bending and flexibility*, Progress in Nucleic Acids Research and Molecular Biology Academic Press, San Diego, pp. 195-270

31. Brukner, I., Susic, S., Dlakic, M., Savic, A. and Pongor, S. (1994) *Physiological concentration of magnesium ions induces a strong macroscopic curvature in GGGCCC-containing DNA*, Journal of Molecular Biology **236**, 26-32

32. Bargonetti, J., Friedman, P.N., Kern, S.E., Vogelstein, B. and Prives, C. (1991) *Wild-type but not mutant p53 immunopurified proteins bind to sequences adjacent to the SV40 origin of replication*, Cell **65**, 1083-1091

33. Halazonetis, T.D., Davis, L.J. and Kandil, A.N. (1993) *Wild-type p53 adopts a "mutant"-like conformation when bound to DNA*, EMBO Journal **12 (3)**, 1021-1028

34. Halazonetis, T.D. and Kandil, A.N. (1993) *Conformational shifts propagate from the oligomerization domain of p53 to its tetrameric DNA binding domain and restore DNA binding to select p53 mutants*, EMBO Journal **12**, 5057-5064

35. Friedman, P.N., Chen, X.B., Bargonetti, J. and Prives, C. (1993) *The p53 protein is an unusually shaped tetramer that binds directly to DNA*, Proceedings of the National Academy of Sciences of the United States of America **90**, 3319-3323

36. Sakamoto, H., Lewis, M.S., Kodama, H., Appella, E. and Sakaguchi, K. (1994) *Specific sequences from the carboxyl terminus of human p53 gene product form anti-parallel tetramers in solution*, Proceedings of the National Academy of Sciences of the United States of America **91**, 8974-8978

37. Stenger, J.E., Tegtmeyer, P., Mayr, G.A., Reed, M., Wang, Y., Wang, P., Hough, P.V.C. and Mastrangelo, I.A. (1994) *p53 oligomerization and DNA looping are linked with transcriptional activation*, EMBO Journal **13**, 6011-6020

38. Cho, Y.J., Gorina, S., Jeffrey, P.D. and Pavletich, N.P. (1994) *Crystal structure of a p53 tumor suppressor DNA complex: Understanding tumorigenic mutations*, Science **265**, 346-355

39. Funk, W.D., Pak, D.T., Karas, R.H., Wright, W.E. and Shay, J.W. (1992) *A transcriptionally active DNA-binding site for human p53 protein complexes*, Molecular and Cellular Biology **12 (6)**, 2866-2871

40. Clore, G.M., Omichinski, J.G., Sakaguchi, K., Zambrano, N., Sakamoto, H., Appella, E. and Gronenborn, A.M. (1994) *High-resolution structure of the oligomerization domain of p53 by multidimensional NMR*, Science **265**, 386-391

41. Lee, W., Harveyo, T.S., Yin, Y., Yau, P., Litchfield, D. and Arrowsmith, C.H. (1994) *Solution structure of the tetrameric minimum transforming domain of p53*, Nature Structural Biology **1**, 877-890

42. Balagurumoorthy, P., Sakamoto, H., Lewis, M.S., Zambrano, N., Clore, G.M., Gronenborn, A.M., Appella, E. and Harrington, R.E. (1995) *Four p53 DNA-binding domain peptides bind natural p53-response elements and bend the DNA*, Proceedings of the National Academy of Sciences of the United States of America **92**, 8591-8595

43. Lyubchenko, Y.L., Shlyakhtenko, L.S., Chernov, B. and Harrington, R.E. (1991) *DNA bending induced by cro protein binding as demonstrated by gel electrophoresis*, Proceedings of the National Academy of Sciences of the United States of America **88**, 5331-5334

44. Nagaich, A.K., Zhurkin, V.B., Sakamoto, H., Gorin, A.A., Clore, G.M., Gronenborn, A.M., Appella, E. and Harrington, R.E. (1996) *Architectural accommodation in the complex of four p53 DNA binding domain peptides with the p21/WAF1/Cip1 DNA response element* (unpublished)

45. Garner, M.M. and Revzin, A. (1990) *Analysis of nucleic acid-protein interactions*, Gel Electrophoresis of Nucleic Acids: A Practical Approach IRL Press, New York, NY, pp. 201-223

46. Hames, B.D. (1990) *Introduction to PAGE*, Gel Electrophoresis of Proteins: A Practical Approach IRL Press, New York, NY, pp. 1-91

47. Kim, S.J., Tsukiyama, T., Lewis, M.S. and Wu, C. (1994) *Interaction of the DNA-binding domain of Drosophila heat shock factor with its cognate DNA site: A thermodynamic analysis using analytical ultracentrifugation*, Protein Science 3, 1040-1051

48. Lewis, M.S., Shrager, R.I. and Kim, S.-J. (1994), pp. 94-118

49. Greenblatt, M.S., Bennett, W.P., Hollstein, M. and Harris, C.C. (1994) *Mutations in the p53 tumor suppressor gene: Clues to cancer etiology and molecular pathogenesis*, Cancer Research 54, 4855-4878

50. Zahn, K. and Blattner, F.R. (1987) *Direct evidence for DNA bending at the lambda replication origin*, Science 236, 416-422

51. Lyubchenko, Y.L., Shlyakhtenko, L.S., Appella, E. and Harrington, R.E. (1993) *CA runs increase DNA flexibility in the complex of ■ cro protein with the OR3 site*, Biochemistry 32, 4121-4127

52. Harrington, R.E. (1993) *Studies of DNA bending and flexibility using gel electrophoresis*, Electrophoresis 14, 732-746

53. Ulanovsky, L., Bodner, M., Trifonov, E.N. and Choder, M. (1986) *Curved DNA: design, synthesis and circularization*, Proceedings of the National Academy of Sciences of the United States of America 83, 862-866

54. Kahn, J.D. and Crothers, D.M. (1992) *Protein-induced bending and DNA cyclization*, Proceedings of the National Academy of Sciences of the United States of America 89, 6343-6347

55. Erie, D.A., Yang, G.L., Schultz, H.C. and Bustamante, C. (1994) *DNA bending by Cro protein in specific and nonspecific complexes: Implications for protein site recognition and specificity*, Science 266, 1562-1566

56. Tullius, T.D., Dombroski, B.A., Churchill, M.E.A. and Kam, L. (1987) *Hydroxyl Radical Footprinting: A High Resolution Method for Mapping Protein-DNA Contacts*, Methods in Enzymology 155, 537-558

57. Burkhoff, A.M. and Tullius, T.D. (1987) *The unusual conformation adopted by the adenine tracts in Kinetoplast DNA*, Cell 48, 935-943

58. Pogozelski, W.K., McNeese, T.J. and Tullius, T.D. (1995) *What species is responsible for strand scission in the reaction of [Fe(II)EDTA](2-) and H2O2 with DNA?*, Journal of the American Chemical Society 117, 6428-6433

59. Fratini, A.V., Kopka, M.L., Drew, H.R. and Dickerson, R.E. (1982) *Reversible Bending and Helix Geometry in a B-DNA Dodecamer: CGCGAATT(Br)CGCG*, Journal of Biological Chemistry 257, 14686-14707

60. Zhurkin, V.B., Ulyanov, N.B., Gorin, A.A. and Jernigan, R.L. (1991) *Static and statistical bending of DNA evaluated by Monte Carlo simulations*, Proceedings of the National Academy of Sciences of the United States of America 88, 7046-7050

61. Goodsell, D.S., Kopka, M.L., Cascio, D. and Dickerson, R.E. (1993) *Crystal structure of CATGGCCATG and its implications for A-tract bending models*, Proceedings of the National Academy of Sciences of the United States of America 90, 2930-2934 ("Cat" sequence shows a smooth 23 deg bend.)

62. Brukner, I., Dlakic, M., Savic, A., Susic, S., Pongor, S. and Suck, D. (1993) *Evidence for opposite groove-directed curvature of GGGCCC and AAAAA sequence elements*, Nucleic Acids Research 21, 1025-1029

63. Dlakic, M. and Harrington, R.E. (1995) *Bending and Torsional Flexibility of G/C-rich Sequences as Determined by Cyclization Assays*, Journal of Biological Chemistry 270, 29945-29952

64. Schultz, S.C., Shields, G.C. and Steitz, T.A. (1991) *Crystal structure of a CAP-DNA complex: The DNA is bent by 90o*, Science 253, 1001-1007

65. Kim, J.L. and Burley, S.K. (1994) *1.9 angstrom resolution refined structure of TBP recognizing the minor groove of TATAAAAG*, Nature Structural Biology 1, 638-653

66. Burley, S.K. (1996) *The TATA box binding protein*, Current Opinion in Structural Biology 6, 69-75

67. Hayes, J.J. and Tullius, T.H.D. (1989) *Missing nucleoside experiment: a new technique to study recognition of DNA by protein*, Biochemistry 28 (24), 9521-9527

68. Wissman, A. and Hillen, W. (1991) *DNA contacts probed by modification protection and interference studies*, Methods in Enzymology Academic Press, New York, pp. 365-378

69. Siebenlist, U. and Gilbert, W. (1980) *Contacts between Escherichia coli RNA polymerase and an early promoter of phage T7*, Proceedings of the National Academy of Sciences of the United States of America 77, 122-126

70. Thukral, S.K., Lu, Y., Blain, G.C., Harvey, T.S. and Jacobsen, V.L. (1995) *Discrimination of DNA binding sites by mutant p53 proteins*, Molecular and Cellular Biology 15, 5196-5202

71. Wang, Y., Schwedes, J.F., Parks, D., Mann, K. and Tegtmeyer, P. (1995) *Interaction of p53 with its consensus DNA-binding site*, Molecular and Cellular Biology 15, 2157-2165

72. Tao, X., Zeng, H.Y. and Murphy, J. (1995) *Transition metol ion activation of DNA binding by the diphtheria tox repressor requires the formation of stable homodimers*, Proceedings of the National Academy of Sciences of the United States of America **92**, 6803-6807

73. Steitz, T.A. (1990) *Structural studies of protein-nucleic acid interaction: the sources of sequence-specific binding*, Quarterly Reviews of Biophysics **23**, 205-280

74. Gorin, A.A., Zhurkin, V.B. and Olson, W.K. (1995) *B-DNA twisting correlates with base-pair morphology*, Journal of Molecular Biology **247**, 34-48

75. Satchwell, S., Drew, H.R. and Travers, A.A. (1986) *Sequence periodicities in chicken nucleosome core DNA*, Journal of Molecular Biology **191**, 659-675

76. Bolshoy, A., McNamara, P.T., Harrington, R.E. and Trifonov, E.N. (1991) *Curved DNA without AA: experimental estimation of all 16 DNA wedge angles*, Proceedings of the National Academy of Sciences of the United States of America **88**, 2312-2316

77. Urpi, L., Tereshko, V., Malinina, L., HuynhDinh, T. and Subirana, J.A. (1996) *Structural comparison between the d(CTAG) sequence in oligonucleotides and trp and met repressor-operator complexes*, Nature Structural Biology **3**, 325-328

78. Otwinowski, Z., Schevitz, R.W., Zhang, R.-G., Lawson, C.L., Joachimiak, A., Marmorstein, R.Q., Luisi, B.F. and Sigler, P.B. (1988) *Crystal structure of Trp repressor/operator complex at atomic resolution*, Nature **335**, 321-329

79. Namba, H., Hara, T., Tukazaki, T., Migita, K., Ishikawa, N., Ito, K., Nagataki, S. and Yamashita, S. (1995) *Radiation-induced G(1) arrest is selectively mediated by the p53-WAF1/Cip1 pathway in human thyroid cells*, Cancer Research **55**, 2075-2080

80. Trifonov, E.N. and Ulanovsky, L.E. (1988) *Inherently Curved DNA and its Structural Elements*, Unusual DNA Structures Springer Verlag, New York, pp. 173-187

81. Barber, A.M. and Zhurkin, V.B. (1990) *CAP binding sites reveal pyrimidine-purine patterns characteristic of DNA bending*, Journal of Biomolecular Structure and Dynamics **8**, 213-232

82. Nagaich, A.K., Bhattacharyya, D., Brahmachari, S.K. and Bansal, M. (1994) *CA\TG sequence at the 5' end of oligo(A)-tracts strongly modulates DNA curvature*, Journal of Biological Chemistry **269**, 7824-7833

83. Ferrin, T.E., Haugn, C.C., Jarvis, L.E. and Langridge, R. (1988) *The MIDAS display system*, J. Mol. Graphics **6**, 13-27

GENETICS OF HUMAN ASTROCYTIC TUMORS

Webster K. Cavenee, Ph.D.
Ludwig Institute for Cancer Research, Department of Medicine and Center for Molecular Genetics, University of California at San Diego, 9500 Gilman Drive, La Jolla, California, 92093-0660, USA

1. Introduction

One model for the malignant progression exhibited by human cancer invokes clonal evolution and selection of population subsets with increasing deviance from the norm [1]. Such a hypothesis predicts that tumor cells of the ultimate stage will carry each of the events, cells of the penultimate state will carry each of the events less that last one and so on. This requires extraordinary frequencies of "mutation" to achieve an end stage tumor or that each of the "mutations" has many effects on the features of the process.

2. Genetic Defects in Brain Tumors

We [2] and others have defined these genetic defects in the progression of astrocytic brain tumors and have systematically tested their significance for aspects of the neoplastic phenotype (Table I). For example, mitotic recombination mapping has pinpointed the distal region of chromosome 17p as a target for alteration in cells of each malignancy stage [3]. This chromosomal region encompassed an excellent candidate as the mutational target: the p53 gene. This gene is now generally accepted to be the most frequently mutated in a variety of human cancers and it also could be shown to be commonly mutated in astrocytic tumors [4]. Such mutations provided an exquisitely sensitive and specific means with which to test the notion of a direct descendancy relationship among cells from tumors undergoing malignant progression in the same person. We obtained several progressive and recurrent cases in which the same tumor was increasing in malignant stage over time. We then identified those high stage tumors with p53 mutations and used this information to synthesize synthetic oligonucleotides corresponding to the mutation as well as to a region of the gene outside that which was mutated. These probes were used as allele-specific discriminators on plaque lifts of PCR-generated p53 DNA libraries from tumors of each stage. These experiments clearly showed that the low-stage disease was composed of a majority population with no or few p53 mutations and a minority population with the mutation which dominated the high stage tumor [5].

C. Nicolini (ed.), Genome Structure and Function, 209–216.
© 1997 *Kluwer Academic Publishers. Printed in the Netherlands.*

Table I. Genetic lesions in astrocytic tumors

WHO Grade	Chromosomal Region	Lesion Type	Target Gene
II	17p13	mutation/LOH	p53
	22q	LOH	?
	4	↑ expression	PDGFR-α
III	13q14	mutation/LOH	RB-1
	19q	LOH	?
	9p	deletion/LOH	IFN,p16,p15
	12q13	amplification	CDK4
	12q13-q14	amplification	MDM2
IV	7p/episome	amplification/ rearrangement	EGFR
	4	amplification	PDGFR-α
	7pter-p21,22q12.3-q13.1	↑ expression	PDGF-A,PDGF-B
	10p,10q	LOH	?

3. Effects of Gene Replacement - Intrinsic

We then tested the effect of replacement of wild-type p53 genes in end-stage gliomas with either absent or mutant endogenous genes. Infection of either type of cell with retroviruses carrying wild-type human p53 under viral promoter control caused dramatic morphologic changes (such as enlargement, multiple nuclei, and syncytia formation) together with cell growth arrest in both the G1 and G2/M phases of the cell cycle [6]. These experiments suggest that one reason for inactivation of the p53 gene is to remove the growth constraints and regulation imposed by its wild-type activity.

4. Effects of Gene Replacement - Extrinsic

The progressive growth of solid tumors is dependent on the process of neovascularization to develop the blood vessels that provide nutrients to and remove waste products from the interior regions of neoplasms; in its absence, tumors generally cease to grow beyond a few cubic millimeters in volume [7]. Similarly, de novo angiogenesis is also required at metastatic sites for continued tumor growth. Tumor cells influence this process by producing both inhibitors and positive effectors of angiogenesis.

Since p53 mutations occurred at each stage of the disease and since they could be demonstrated to be functional growth suppressors in glioma cells, we sought to determine whether there was a relationship between such mutations and neovascularization. We placed the wild type p53 gene under the conditional inducibility afforded by a tetracycline-regulated promoter-activating system in the p53-null human LN-Z308 glioblastoma cell line. Two resultant clones, LNZTA3p53WT4 (WT4) and LNZTA3p53WT11 (WT11) showed undetectable levels of immunoprecipitable p53 protein in the repressed state which was substantially increased in the induced state. We also isolated one clone LNZTA3p53WT1 (WT1) which was tTA inducible but did not express detectable p53 protein or activity; this clone served as a control for potential non-specific aspects of the system.

Table II. Inhibition of endothelial cell migration by p53 induction

Conditional Medium	Tet	bFGF	Cells migrated (% of parental)
LNZ-308	+	-	90±15
	-	-	100 ± 10
WT1	+	-	100 ± 8
	-	-	120 ± 10
	+	+	105 ± 5
	-	+	95 ± 8
WT4	+	-	90 ± 7
	-	-	25 ± 5
	+	+	110 ± 10
	-	+	35 ± 5
WT11	+	-	120 ± 5
	-	-	20 ± 5
	+	+	120 ± 5
	-	+	35 ± 8

We then [8] tested the effect of restoration of wild-type p53 activity on the angiogenic phenotype of LN-Z308 cells by preparing medium conditioned (CM) by the parental and genetically modified cells and analyzing its activity in stimulating capillary endothelial cell migration across a gelatinized membrane in a modified Boyden chamber *in vitro* and on inducing neovascularization in the rat cornea *in vivo* (Table II). The uninduced CM from each cell line induced migration to a similar extent; this activity was not appreciably increased if bFGF was added to the CM. Further, the activities in each case were generally comparable to non-conditioned medium to which bFGF was added. In contrast, when CM was prepared from clones WT4 and WT11 grown in the absence of tetracycline, where wild-type p53 was induced, the level of activity was significantly reduced. These results suggest that the presence of p53 causes either a reduction in the amount of angiogenic factors secreted by these cells or it elicits an inhibitor of the migration stimulatory activity. We tested the latter possibility, by adding bFGF (25 ng ml⁻¹) to the CM; the results indicated that the CM from the induced cells could also inhibit bFGF stimulation of endothelial cell migration. Thus these studies indicate that the presence of wild type p53 in glioma cell precursors serves in the homeostatic regulation of angiogenesis through its modulation (either directly or indirectly) of an inhibitor we have called glioma-derived angiogenesis inhibitory factor (GD-AIF).

We then tested whether this anti-endothelial migratory activity could inhibit the process of angiogenesis *in vivo* (Table III). The same CM were mixed with Hydron casting solution, pellets were implanted into rat corneas and their ability to recruit new blood vessels from the limbus was assessed. Positive responses were obtained with both the parental cell line, LN-Z308, and the WT1 clone which does not induce p53 and similar results were obtained with clones WT4 and WT11 in the absence of p53 induction. However, CM prepared from cells where wild-type p53 was induced, strongly inhibited neovascularization. GD-AIF was also able to prevent the strong angiogenic stimulation of bFGF. Thus, the vascularization of human astrocytic tumors appears to be under both positive and negative influences.

Table III. Inhibition of corneal neovascularization by p53 induction

Conditional Medium	Tet	bFGF	% corneal response
DMEM	-	-	0
	-	+	100
LNZ-308	+	-	100
	-	-	100
WT1	+	-	100
	-	-	100
	+	+	100
	-	+	100
WT4	+	-	100
	-	-	20
	+	+	100
	-	+	20
WT11	+	-	100
	-	-	0
	+	+	100
	-	+	25

5. Angiogenesis

As stated above, tumors play an active role in angiogenesis by producing and secreting a number of angiogenic factors, including basic fibroblast growth factor (bFGF), angiogenin, tumor necrosis factor-a, (TNF-a) and vascular endothelial growth factor (VEGF) [9,10]. Several lines of evidence have suggested that VEGF may be a major factor in the neovascularization and growth of human cancers. First, VEGF and its receptors are expressed at high levels in human tumors of the brain, kidney, bladder, breast and ovary [11-13]. Secondly, forced overexpression of VEGF by tumor cells has, in some cases, enhanced their tumorigenic behavior [14]. Thirdly, the intraperitoneal injection of a specific anti-VEGF mouse monoclonal antibody into nude mice inhibited tumor growth *in vivo*, although it could not be directly applied to the tumor cell inoculation sites, required repeated injections, and resulted in incomplete suppression of tumor growth [15]. Finally, dominant-negative *flk-1* mutants lacking their C-termini and which presumably form heterodimers with wild-type *flk-1* upon VEGF binding but no longer transduce signals, cause tumor growth suppression *in vivo* [16]. However, the role of another high affinity receptor for VEGF, *flt-1* that is also required for angiogenesis, remains unaddressed by these latter studies. Thus, VEGF is one of several angiogenic factors produced by tumor cells and it appears to play a role in neovascularization; whether it is a, or even the, critical angiogenic factor in tumor angiogenesis has yet to be directly tested. Thus, we decided to apply the strategy of exogenous expression of an antisense VEGF construct in a highly tumorigenic glioblastoma cell line to this question.

The human glioblastoma cell line, U87MG, is highly tumorigenic in immunodeficient animals [17], expresses high levels of VEGF mRNA and secretes the protein, and expresses three of the four VEGF splice variants, $VEGF_{121}$, $VEGF_{165}$ and $VEGF_{189}$. Since the largest VEGF isoform in U87MG cells, $VEGF_{189}$, contains all of coding sequence of the smaller forms, a $VEGF_{189}$ antisense construct and, to control for clonal variation, construct containing the bacterial β-galactosidase (LacZ) gene under human cytomegalovirus (CMV) promoter control were assembled and each separately transfected into U87MG cells. Sixteen clones expressing $VEGF_{189}$ antisense message and four clones expressing the LacZ gene were identified; three of each were chosen for further analysis. The antisense VEGF-expressing clones, 10, 13, and 17 each expressed $VEGF_{189}$ antisense message with a concomitant reduction in endogenous VEGF transcripts. In comparison, the controls (either the parental U87MG cells or the three LacZ-expressing clones, 4, 22, and 27) had comparable endogenous VEGF levels with no other mRNA species detected indicating that the reduction of steady-state VEGF mRNA was due to antisense expression and not to intrinsic clonal variation in the parental population. We also used enzyme-linked-immunosorbent-analysis (ELISA) to analyze the amounts of secreted VEGF proteins in conditioned media (CM) collected after cells were cultured for 48 hr. Whereas the parental U87MG and the LacZ-expressing clones 4, 22, 27 cells secreted VEGF at concentrations of 15 to 22 ng ml^{-1} per 10^6 cells, the antisense VEGF-expressing clones 10, 13, and 17 produced VEGF at levels that were 3- to 6-fold less.

We then sought to determine if the expression of antisense VEGF mRNA affected the ability of the U87MG cells to secrete factors into their conditioned media (CM) critical to the induced migration of human microvascular endothelial cells (HMVEC). The data shown in Table III indicate that CM from either the parental U87MG or the LacZ-expressing clones 4, 22, and 27 were highly stimulatory for HMVEC migration while CM from the antisense VEGF-expressing clones 10, 13, and 17 were greatly diminished in this regard. The remaining slight HMVEC migration in the presence of CM from the latter clones may be due to other stimulatory substances, such as bFGF, produced by the tumor cells. The notion that VEGF was the major element in the CM that promoted HMVEC migration was tested by including a VEGF neutralizing monoclonal antibody in the CM during the migration assay. This antibody blocked the stimulation of cell migration elicited even by relatively high concentrations (10 µg ml^{-1}) of purified human recombinant $VEGF_{165}$ and also by CM from the parental U87MG cells and the LacZ-expressing clones. In contrast, it had no effect on the limited promotion of the chemotactic response induced by the CM from the antisense VEGF-expressing clones. The specificity of this blockage was demonstrated by the absence of the effect on the migration by an isotype matched control monoclonal antibody. These data indicate that VEGF was the major component in the CM from the parental U87MG and the LacZ-expressing cells, that stimulated HMVEC migration and that the expression of an antisense VEGF gene construct in the U87MG cells effectively reduced VEGF secretion to levels below those required for one of its primary biological functions.

We next sought to determine the effects of VEGF diminution on *in vivo* behavior. We simultaneously implanted the parental U87MG cells and the LacZ- or antisense VEGF-expressing clones subcutaneously into alternate flanks of immune-deficient nude mice. The data in Table IV show that cells from the parental U87MG and the LacZ-expressing clones 4, 22, and 27 formed tumors with similar kinetics and of similar sizes, while tumor growth was remarkably suppressed when the same numbers of cells from the antisense VEGF-expressing clones 10, 13, and 17 were injected. The latter cells formed small, but palpable, masses in the mice at the same time as the parental U87MG cells but the tumors did not progress in size.

Table IV. Suppression of endothelial cell migration by antisense VEGF (# migrated cells/10 hpf)

Cells	Conditioned Medium	Neutralizing Ab	Isotype Control
PBS	25 ± 5	N/A	N/A
rVEGF	200 ± 10	60 ± 5	205 ± 15
U87	215 ± 5	100 ± 10	220 ± 5
AS-10	55 ± 5	50 ± 5	55 ± 5
AS-13	50 ± 5	45 ± 5	50 ± 5
AS-17	60 ± 5	50 ± 5	65 ± 5
LacZ-4	190 ± 5	50 ± 5	190 ± 10
LacZ-22	150 ± 10	50 ± 5	155 ± 10
LacZ-27	200 ± 10	70 ± 15	190 ± 10
Revertant	223 ± 10	110 ± 2	240 ± 15

We then examined whether the tumor growth suppression observed in heterotopic subcutaneous inoculations was also apparent with ectopic stereotactic implantation to the brain (Table V). Comparable numbers of cells of the following types were implanted: the parental U87MG; an equal admixture of the LacZ-expressing clones 4, 22, and 27; and, an equal admixture of the antisense VEGF-expressing clones 10, 13, and 17. The latter two admixtures were done to assure that any results obtained were not due to variability among the clones and because of the laborious nature of the procedure. One animal in each group that received the parental U87MG cells or the LacZ-expressing clone admixture died after 5 weeks from overgrowth of the tumors, which reached volumes greater than 60 cubic millimeters. The remainder of the mice in these two groups each developed tumors of 45 cubic millimeters or larger. In contrast, 5 of 6 nude mice implanted with the same numbers of the antisense VEGF-expressing clones developed masses of less than 5 cubic millimeters; the remaining mouse only developed a small tumor of less than 15 cubic millimeters at the time of sacrifice. Thus, suppression of tumorigenicity expression was apparent at both heterotopic and ectopic sites, and was elicited by levels of inhibition of VEGF secretion attainable by expression of the antisense constructs.

Finally, we were able to take advantage of the well-described phenomenon whereby exogenous gene expression systems tend to lose efficiency after transfer into cells or animals. One mouse in each group which had been inoculated with the antisense VEGF-expressing clones 10 and 17 developed subcutaneous tumors which appeared to be slow-growing but then grew with kinetics similar to tumors caused by inoculation of the parenteral U87MG cells. One possibility for these observations was that some subset of the cells had genetically or epigenetically reverted so that they were again at wild-type levels with respect to VEGF secretion. We analyzed one of these potential revertants and found that it had regained high levels expression of VEGF, the ability to induce endothelial cell migration *in vitro* and neovascularization *in vivo* and was again highly tumorigenic (Tables III, IV and V).

Table V. Suppression of tumorigenesis by antisense VEGF

Tumor Cells	Subcutaneous Tumor Volume ($mm^3 \pm$ S.E.)	Intracerebral Tumor Volume ($mm^3 \pm$ S.E.)
U87	1957.7 ± 120.5	50.9 ± 6.6
AS-10	50.6 ± 19.9	
AS-13	4.9 ± 2.5	6.7 ± 4.5
AS-17	2.0 ± 0.9	
LacZ-4	2021.0 ± 109.0	
LacZ-22	2453.0 ± 573.0	50.5 ± 18.0
LacZ-27	3674.0 ± 363.0	
Revertant	2500.0 ± 149	84.6 ± 10.9

6. Conclusions

The process of astrocytic tumorigenesis can be reasonably well described within the context of a clonal evolution process [1]. It appears that many of the mutational events which drive the process are ones with pleiotropic effects which alter both cell intrinsic growth regulation and cell extrinsic interaction of tumor cells with normal surrounding cells of the host. These alterations may offer tantalizing targets for therapeutic intervention in the future.

Acknowledgements

The author gratefully thanks the many colleagues and collaborators who have allowed these studies to be accomplished. For the experiments described here particular recognition is due to H.-J. Su Huang, E. van Meir, S. Cheng, R. Nishikawa, M. Nagane, T. Mikkelsen, M. Rosenblum, K. Schwechheimer, P. Polverini, X.-D. Ji, and C. H. M. Huang.

216

References

1. Nowell, P.C. (1986) The clonal evolution of tumor cell populations, *Science* **194**, 23-28.
2. James, C.D., Carlbom, E., Dumanski, J.P. et al. (1988) Clonal genomic alterations in glioma malignancy stages, *Cancer Res.* **48**, 5546-5551.
3. James, C.D., Carlbom, E., Nordenskjöld, M., Collins, V.P., Cavennee, W.K. (1989) Mitotic recombinantion of chromosome 17 in astrocytomas, *Proc. Natl. Acad. Sci. USA* **86**, 2858-2862.
4. Van Meir, E., Kikuckhi, T., Li, H., Diserens, A.-C., Wojcik, B., Huang, H.-J.S., Friedmann, T., de Tribolet, N., Cavenee, W.K. (1994) Analysis of the p53 gene and its expression in human glioblastoma cells, *Cancer Res.* **54**, 649-652.
5. Sidransky, D., Mikkelsen, T., Schwechheimer, K., Rosenblum, M.L., Cavenee, W.K., Vogelstein, B. (1992) Clonal expansion of p53 mutant cells leads to brain tumor progression, *Nature* **355**, 846-847.
6. Van Meir, E., Roemer, K., Rempel, S.A., Kikuchi, T., Diserens, A.-C., de Tribolet, N., Haas, M., Huang, H.-J.S., Friedmann, T., Cavenee, W.K. (1995) Single-cell monitoring of growth arrest and morphological changes induced by transfer of wild type p53 alleles to glioblastoma cells, *Proc. Natl. Acad. Sci. USA* **92**, 1008-1012.
7. Folkman, J. (1985) Tumor angiogenesis, *Adv. Cancer Res.* **43**, 175-203.
8. Van Meir, E., Polverini, P., Chazin, V.R., Huang, H.-J.S., de Tribolet, N., Cavenee, W.K. (1994) Release of an inhibitor of angiogenesis upon induction of wild type p53 expression in glioblstoma cells, *Nature Genet.* **8**, 171-176.
9. Liotta, L.A., Steeg, P.S. and Stetler, S.W. (1991) Cancer metastasis and angiogenesis: an imbalance of positive and negative regulation, *Cell* **64**, 327-336.
10. Blood, C.H. and Zetter, B.R. (1990) Tumor interactions with the vasculature: angiogenesis and tumor metastasis, *Biochim. Biophys. Acta* **1032**, 89-118.
11. Brown, L.F. et al (1995) Expression of vascular permeability factor (vascular endothelial growth factor) and its receptors in breast cancer, *Hum. Pathol.* **26**, 86-91.
12. Plate, K.H., Breier, G., Millauer, B., Ullrich, A. and Risau, W. (1993) Up-regulation of vascular endothelial growth factor and its cognate receptors in a rat glioma model of tumor angiogenesis, *Cancer Res.* **53**, 5822-5827.
13. Wizigmann, V.S., Breier, G., Risau, W., and Plate, K.H. (1995) Up-regulation of vascular endothelial growth factor and its receptors in von Hippel-Lindau disease-associated and sporadic hemangioblastomas, *Cancer Res.* **55**, 1358-1364.
14. Zhang, H.T. et al. (1995) Enhancement of tumor growth and vascular density by transfection of vascular endothelial cell growth factor into MCF-7 human breasts carcinoma cells, *J. nat. Cancer Inst.* **87**, 213-219.
15. Kim, K.J., et al. (1993) Inhibition of vascular endothelial growth factor-induced angiogenesis suppresses tumour growth in vivo, *Nature* **362**, 841-844.
16. Millauer, B., Shawver, L.K., Plate, K.H., Risau, W., and Ullrich, A. (1994) Glioblastoma gorwth inhibited in vivo by a dominant-negative Flk-1 mutant, *Nature* **367**, 841-844.
17. Nishikawa, R. et al. (1994) A mutant epidermal growth factor receptor common in human glioma confers enhanced tumorigenicity, *Proc. Natl. Acad. Sci. USA* **91**, 7727-7731.

TUMOR PROGRESSION AND METASTASIS

Georgii P. GEORGIEV, Sergei L. KISELEV and Evgenii M.
LUKANIDIN
*Department of Molecular Genetics of Cancer, Institute of Gene Biology,
Russian Academy of Sciences, 34/5 Vavilov St., 117334Moscow, Russia;
and Danish Cancer Society*

In this chapter, the genes involved in the control of tumor progression and tumor metastasis are described. The present state of the problem is briefly overviewed in the first part. As a great number of papers appeared in the field, we do not provide the chapter with a detailed list of references, but send a reader to very extensive recent reviews containing practically all important references [1].

In the second part, we describe the strategy for search and characterization of novel genes involved in the control of tumor progression and tumor metastasis using as examples two genes extensively studied in our laboratories. This gives a reader an idea about the experimental approaches to solve the problem.

1. "Metastatic Genes" And "Metastatic Proteins"

1.1. TUMOROGENESIS

In previous lectures [Cavenee, this volume], the main mechanisms of tumorigenesis was discussed. The role of oncogenes and tumor suppressor genes is demonstrated. Huge number of genes that normally encode growth factors, growth factor receptors, proteins involved in signal transduction, transcription factors, and many others being mutated or over-expressed may become oncogenes. The genes encoding inhibitors of replication, proteins involved in control of cell cycle, regulators of apoptosis, etc. play a role of tumor suppressors and their inactivation strongly enhance the risk of neoplastic cell transformation.

Because of the existence of numerous oncogenes and tumor suppressor genes, each tumor has its own genetic portrait. Still some features are more common than the others. For example, human tumors very frequently have the mutations in the *p53* tumor suppressor gene [2]. Sometimes, the *p53* gene is just inactivated, but more frequently it is mutated in such a way that the p53 protein cannot properly function and at the same

C. Nicolini (ed.), Genome Structure and Function, 217–237.
© 1997 *Kluwer Academic Publishers. Printed in the Netherlands.*

time acquires the ability for longer survival than wild type protein. In this case, the normal homologue can not compete with mutated one and p53 function is lost [this volume].

The members of *ras* gene family are also the usual subjects for mutation and conversion to oncogene as well as different receptors containing protein kinase domains [3, 4]. These events are associated with many tumors. However, it is necessary to have more than one mutation for transformation (in human, approximate estimates give the average figure of five mutations), and this immediately creates an extremely high genetic variability among tumors.

1.2. TUMOR PROGRESSION

On the other hand, the tumor is not a static formation. It is developing and changing all the time. This depends on genetic instability of the tumor. As a result of transformation, tumor cells acquire the partial independence from regulatory signals coming from neighboring cells and grow more or less independently of these signals. Another important feature is the loss of elimination of the cells with damaged DNA. Normally such cells can not overcome the cell cycle checkpoints and go through the apoptotic process to their death. The p53 protein plays an important role in the direction of damaged cells along the apoptotic way. Many mutations in the *p53* gene lead to the loss of ability to induce apoptosis and to downregulate cell proliferation [2]. As a result, cells become able to survive DNA damage, and this results in the increase of mutation rate in such cell population. Interestingly, another protein involved in the control of apoptosis, product of the *bcl-2* gene that downregulates apoptosis is also a potential oncogene [reviewed in 5].

This and possibly some other processes result in accumulation of different types of mutations in tumor cells: translocations, loss of heterozygosity, point mutations and the enhanced transposition rate. In its turn, this leads to accumulation of heterogeneity in tumor cell population. If the mutated cell acquires some advantage for rapid growth or other properties useful for the cell itself, it has a good chance for survival and multiplication. Such changed cells may replace with time the original population of tumor cells. This phenomenon is designated as tumor progression, that usually leads to appearance of more malignant phenotype. As a rule, the same tumor contains cells with genotype differences and several clones obtained from the same tumor are different in the level of malignancy.

It should be pointed out that in some cases, the malignant phenotype can appear just at the first stage of tumor development, as for example, in tumors induced by activation of *neu* oncogene [6]. However, in many cases, the malignancy is being developed in the course of tumor progression. One of the major features of malignant tumor cells is its ability to give metastases, i.e., the new foci of tumor growth in distantly located regions of the organism.

1.3. MAIN STEPS IN TUMOR PROGRESSION AND METASTASIS

Tumor metastasis is the major cause of mortality from cancer. This process is very complex. It includes several independent steps: 1) vascularization of a primary tumor node; 2) detachment of tumor cell from the primary focus; 3) invasion into the blood vessel; 4) transfer to the new site and arrest there; 5) adhesion to the endothelial cell; 6) extravasation; 7) invasive growth at a new place; 8) vascularization of a novel focus, etc. Each step depends on new special properties of tumor cell, ability to induce angiogenesis, detachment-attachment of cell from or to cell aggregate, cell invasiveness and cell motility.

Each step is a complex one and is controlled by a number of genes and proteins. Therefore, one can expect the involvement of a number of genes or proteins in the control of tumor metastasis. Several activated oncogenes themselves can create metastatic phenotype [7]. However, in many other cases, the processes of oncogenesis and metastasis are uncoupled, and special genes are responsible for the appearance of metastatic phenotype.

It should also be pointed out that the genes of tumor progression and metastasis can be separated into two groups: effector genes and upstream regulatory genes. The protein products of effector genes directly determine the invasiveness and other features of metastatic tumor, while proteins encoded by the genes of the second group act in an indirect way. They either control the expression of different effector genes or control some general cell functions indirectly determining the features' characteristic for malignant tumor cell.

1.4. GENES CONTROLLING VASCULARIZATION

Until vascularization, or angiogenesis, occurs, the tumor can not grow up to more than ca. 2 mm in diameter due to the shortage of nutrition. If vascularization has occurred, tumor cells can grow to much larger dimensions. Another consequence is that tumor cells can penetrate into the blood vessels that becomes the first step for metastasis development. The angiogenesis is also extremely important for the successful development of metastases (see below).

The invasion of endothelial cells into the tumor node and formation of capillary sprout is influenced by many different cytokines produced either by tumor cells or by normal inflammatory cells if inflammation associates tumor development. There are positive and negative cytokines. Among positive cytokines inducing tumor angiogenesis are fibroblast growth factors, especially bFGF, vascularization endothelial growth factor (VEGF), interleukin-8 (IL-8) and many others [8]. Such cytokines as interferons (IFN-a and IFN-b) angiostatic cartilage-derived inhibitors and several others inhibit angiogenesis in tumor foci [8-10].

The production of corresponding cytokines by tumor cells depends on the complex interactions between tumor and surrounding host cells. Human renal cell carcinoma (HRCC) cells implanted into kidney produces high incidence of lung metastasis, whereas those implanted subcutaneously do not. In the former case, HRCC possess ten- to twenty-fold higher amount of bFGF mRNA and induce active angiogenesis [11]. IFN-a and IFN-b were found to inhibit the bFGF synthesis at the transcription level in several different tumor cell lines [12].

Human melanoma cells actively express VEGF, bFGF and IL-8. After subcutaneous transplantation, tumor cells actively synthesize IL-8, while after intraliver transplantation, the production of IL-8 by melanoma cells is very low. The level of IL-8 correlates with angiogenesis and tumor progression. Moreover, cocultivation of melanoma cells with keratinocytes enhances while cocultivation with hepatocytes suppresses IL-8 synthesis [13]. In certain cases, contacts with tumor cells activate production of VEGF in endothelial cells [14]. Thus, angiogenesis is a result of a complex interplay between different factors produced both by the tumor and host cells.

As was mentioned above, angiogenesis may become a target for the treatment of tumors. The important point is that the function of angiogenesis in the adult organism is restricted by wound healing and processes in the uterus. Therefore, angiogenesis may temporarily be blocked without damaging the organism as a whole. The interesting example for this is the experiments with angiostatin, peptide derived from plasminogen after proteinase cleavage [15]. Its application to nude mice with different human cancers leads to treatment not only the primary tumor but also metastases. This indicates the extremely important role of angiogenesis in tumor development.

Sometimes, angiostatic cytokines are produced by a primary tumor itself and this prevents overgrowth of metastases. In such cases, surgical removal of the primary focus may lead to induction of metastasis.

1.5. DETACHMENT OF TUMOR CELLS FROM THE ORIGINAL CELL CLUSTER

Next important step in metastasis is the detachment of tumor cell from aggregate to further enter the blood vessel. Cell aggregation depends on several systems, the most important being represented by cadherins, immunoglobulins, integrins and selectins.

Aggregation of the cells plays an especially important role in the progression of the cancers, or epithelial tumors, as epithelial cells originally form firm aggregates and have a very little motility. They bind to each other and to the basal membrane. Adherence junction between normal and tumor epithelial cells depends on cadherin family of proteins. The best characterized member of the family, E-cadherin, is a ca. 120 kD protein with a large N-terminal extracellular part containing four calcium-binding domains, transmembrane domain and C-terminal intracellular domain. The extracellular domains of E-cadherins located on different cells interact in calcium-dependent way [16]. The C-terminal domain interacts with the C-terminal region of β-catenin (plakoglobin), whose N-terminal part binds to α-catenin in its turn interacting with cytoskeleton [17]. Such chain of protein-protein interactions firmly fixes epithelial cells at their position. Other members of the family are N- (neural), P- (placental), M- (muscle), OB- (osteoblast), LI- (liver and intestinal) cadherins [16, 18-21].

There is a strong correlation between cell behavior and cadherin content. In transfection experiments, stable transfection and expression of the *E-cadherin* gene strongly suppress the metastatic phenotype of different tumor cell lines [22, 23]. On the other hand, antibodies to E-cadherin make the tumor cells more aggressive in some cases [24]. Mutations in the *E-cadherin* and β-*catenin* genes leading to their inactivation have been observed in some tumors and found to be associated with enhancement of metastatic phenotype [25, 26]. Downregulation of *E-cadherin* gene expression leads to a similar effect [27-29].

Attachment to a basal membrane is typically realized through integrins α6β4 and α6β1 interaction with membrane laminin and α1β1 with collagen IV. In addition, the cells are attached to extracellular matrix through integrin α5β1-fibronectin interaction. In fact, the number of proteins involved in these interactions is higher, and we mention only some characteristic ones. The role of corresponding genes in tumor progression seems to be more controversial, although quite convincing data have been obtain for integrin α5β1 that possesses metastasis-suppressive functions [30, 31].

Detachment from the cell aggregate leads to acquirement by epithelial cells of several properties typical for mesenchimal cells: mesenchimal transformation of epithelial cells takes place. This phenomenon well correlates with the increase of invasiveness and appearance of metastatic phenotype in tumor cells [32, see also below].

1.6. INVASIVE GROWTH

Different groups of genes/proteins are involved in determination of invasive tumor growth. As examples, one can mention tyrosine protein kinase-type receptors, autocrine motility factor (AMF) and its receptor and different types of degradative enzymes.

Among the genes encoding tyrosine kinase receptors, *c-met, c-neu, c-ret* and *c-ros* were shown to be associated with the invasiveness of tumor cells. In more detail, the pair of ligand, scattering factor (SF)/hepatocyte growth factor (HGF), and of receptor, c-Met, was studied. Their interaction leads to the induction of liver morphogenesis [33] and at the same time to the increase of motility and invasiveness of tumor cells into collagen matrix [34]. The synthesis of both cMet and SF/HGF is increased in different malignant tumors, in particular, in many cases of metastatic human breast cancer [35]. In contrast to effector proteins, the system SF/HGF-cMet belongs to the control proteins determining signal transduction that is probably realized in change of expression of several different genes.

AMF is a 64 kD protein that interacts with the receptor, 78 kD glycoprotein, gp78, and activates cell motility [36-38]. The synthesis of the components of this ligand-receptor system is activated in parallel with progression of some tumors, for example, bladder carcinoma [39].

A lot of data have been obtained on the role of different degradative enzymes in tumor progression and factors controlling their activity, for example, different metalloproteinases and serine proteases, including catepsins, collagenases, stromelysins; plasminogen activator and its inhibitor, as well as gyaluronidase and several other enzymes destroying extracellular matrix. The genes encoding these degradative enzymes are frequently activated in metastatic and, in general, in invasive cancer cells, although correlation is not absolute. This group of enzymes may obviously play a role in degradation of extrcellular matrix helping overgrowth and invasion of cancer cells, in particular their invasion into blood vessels.

1.7. ATTACHMENT TO ENDOTHELIAL CELLS (ARREST IN CAPILLARY BED) AND EXTRAVASATION

The process of attachment of tumor cells to endothelial cells is induced by cytokines produced in tumor or inflammatory cells. These are IL-1, TNF, lypopolysacharides (LPS), etc. Weak attachment is mediated by selectins. Synthesis of E-selectin is induced in endothelial cells [40, 41]. Its extracellular part interacts with tumor cell or leukocyte carbohydrates through N terminal lectin-like domain [42].

Strong attachment is mediated by interaction of integrins of tumor cells with the members of immunoglobulin superfamily located on the surface of endothelial cell. For example, $\alpha 4\beta 1$ integrin usually present in melanomas and sarcomas binds to VCAM-1 (vascular cell attachment molecule) [43, 44], while $\alpha 6\beta 1$ (present in colon carcinomas) and $\alpha 6\beta 4$ (present in lung carcinomas) bind to ICAM-1 (intercellular cell attachment molecule) [45, 46]. These interactions are responsible for a firm attachment.

Several reports have appeared indicating the special role in tumor metastasis of the CD-44 transmembrane hyaluronate receptor. Both metastatic and non-metastatic tumors contained the major variant of this protein [47], but only metastatic ones contained some minor variants of protein characterized by the presence of additional domains, that were found responsible for intercellular interactions [48]. The appearance of such variants was a result of alternative splicing that led to inclusion into mRNA of additional small exon(s). Stable transfection of non-metastatic cells with the construction expressing CD-44 variant in some cases led to the enhancement of metastatic potential [48, 49], although some opposite results were also obtained. Since both the attachment and detachment of cells play a role in metastasis just at different stages of the process, these controversial results may not be too surprising. Further experiments are needed for final conclusions.

1.8. PUTATIVE METASTASIS SUPPRESSOR GENE *nm23*

There are several other genes whose function in tumor metastasis, although exists, is less understandable. For example, the *nm23* gene has been claimed to serve as metastasis suppressor gene [50]. It encodes an enzyme, nucleoside diphosphate kinase, whose relation to metastatic process is unclear [reviewed in 51]. In some metastatic tumors, the expression of *nm23* is inhibited [52]. Stable transfection with the construction expressing *nm23* in some cases reduces metastatic potential [52]. Still the results remain to be controversial.

It was found recently that nm23 protein might be involved in the control of the *c-myc* gene [53]. Upstream region of the latter contains at position -101 - -160 three sequences CCCACCC/GGGTGGG that bind PuF protein activating factor. The latter was found to be encoded by the *nm23* gene. It is not yet clear why activation of *c-myc* expression should interfere with metastatic process.

Now it is important to find those genes that play a key role in creation of metastatic phenotype. One of markers for such gene is its over- or under- expression in many metastatic tumors comparing to non-metastatic ones. Another important marker is enhancement or suppression of metastatic phenotype after stable transfection of several different tumor cells with constructions actively expressing the gene of interest in sense or antisense orientation.

1.9. STRATEGY FOR THE SEARCH OF NEW GENES WITH UNKNOWN FUNCTION INVOLVED IN THE CONTROL OF TUMOR METASTASIS

Each of above mentioned genes play a certain role in the acquirement by tumor cells of an invasive and metastatic phenotype. Probably, in different cases, different genes are involved making the situation to remind that with tumorigenesis in general. As many genes are involved, one could expect that several genes controlling metastasis are still unknown. In particular, this may be "upstream genes" that are not directly involved in cell functioning but control other gene or protein activity. An important direction of further studies is to clone such genes and to understand the role of their protein products in tumor progression and metastasis. In this respect, the most interesting are the genes that either play a key role in creation of metastatic phenotype in many different tumors or which can potentially be used for metastasis diagnostics and/or treatment. Below, we describe such a strategy on example of two genes discovered in the laboratory. One of them may play such a key role in some metastatic tumors. This is the *mts1* gene, encoding the protein designated as Metastasin 1, or Mts1.

2. The *mts1* Gene

2.1. ISOLATION OF THE *mts 1* GENE AND ITS PROPERTIES

The general approach for the search of new genes that may play a role in the control of tumor progression and appearance of metastases is the isolation of genes up or down regulated in metastatic tumor cells comparing to non-metastatic tumor cells of the same origin.

The *mts1* gene has been cloned in experiments on cDNA libraries' subtraction using two mouse tumor cell lines, CSML-100 and CSML-0. The CSML-0 cells have originated from spontaneous adenocarcinoma of mammary gland [54]. They were maintained in cell culture and by subcutaneous transplantation in isogenic mice. CSML-0 cells were non-metastatic, but after several transplantations, they acquired the ability to give metastases to the lungs and lymph nodes. Thus, the CSML-100 cells were obtained. They differed from CSML-0 not only in respect of metastatic potential, but also by morphology - more extended fibroblast-like shape.

The cDNA library subtraction protocol used in this work allowed one to isolate only abundant genes, and in several experiments one and the same gene was fished out. It gave rise to 0.55 kb mRNA, which was abundant in CSML-100 cells and absent from CSML-0 cells. This mRNA was detected in many metastatic tumor cell lines of different origin, but not in non-metastatic tumors, although several exceptions could be observed. Therefore, the gene was called as *mts1*, a gene encoding Metastasin 1 protein (Mts1) [55].

Next step in such studies is the sequencing of cDNA and deducing of protein structure. Mts1 was found to be a protein 101 aminoacids (aa) long with two typical calcium-binding domains. It belongs to the S-100 sub-family of calcium-binding proteins. The *mts1* gene was described at about the same time in several other groups under different names, but without any relation to tumor metastasis [56-60].

The *mts1* gene is expressed in several normal tissues: embryonic fibroblasts, trophoblasts, lymphoid cells, in particular, in T-lymphocytes and activated macrophages [61]. At least some of these cells possess invasive properties. The level of *mts1* expression can be readily modulated by different lymphokines or calcium ionofores.

The *mts1* has also been cloned from human genome. Human Mts1 protein differs from its mouse counterpart just by 7 aa substitutions.

2.2. THE ROLE IN TUMOR METASTASIS

The next main question is whether the over-expression of the *mts1* gene in tumor cell can or can not change the phenotype from non-metastatic to metastatic one and vice versa, i.e. whether the presence of Mts 1 protein is common for metastatic behavior of the tumor cell or their coincidence is just occasional. Certainly, the over-expression of any cellular gene could not be expected as the only factor responsible for metastasis (see above), but some genes on certain background of expression of other genes may become indispensable for that. These "key genes" can be detected in transfection experiments that allow to switch on or off the gene functioning.

Several cell systems were used in such experiments. First, CSML-100 cells were transfected with the construction containing *mts1* cDNA in antisense orientation under the control of Moloney sarcoma virus promoter/enhancer element present in its LTR (long terminal repeat). The cell lines actively expressing antisense RNA were selected and used for subcutaneous transplantation to isogenic mice. They had a dramatically decreased metastatic potential comparing to highly metastatic CSML-100 cells and mock-transfected cells. Instead of hundreds of metastatic foci in the lungs of mice subcutaneously injected with original CSML-100 cells, one could see either no or single metastases after transplantation of the same cells but expressing *mts1* antisense RNA [62]. Thus, *mts1* expression was necessary for maintaining a metastatic phenotype of these cells.

Another technology to switch off the gene expression is to use construction synthesizing ribozymes, i.e. RNAse that specifically cleave a particular RNA. Ribozyme construct specifically cleaving human *mts1* RNA in the second exon was transfected into human osteosarcoma (OHS) cells. The control OHS cells gave metastases to bone marrow of nude rats after intracardiac injection. The stable transfectants strongly suppressed metastatic phenotype. The Mts1 protein content in such cells was reduced [63].

The reversed experiment with CSML-0 cells stably transfected with construction expressing sense *mts1* mRNA gave negative results [62]. However, it was found that, in spite of active *mts1* transcription, these cells did not contain Mts1 protein. Thus, in addition to the control of *mts1* expression at the transcription level, the control at translational level does also exist and some cells cannot translate *mts1* mRNA. Therefore, these experiments are non interpretable until the translation suppression is overcome.

Still sense constructions were successfully tested in three other cell lines. One of them is Line 1 cells that are mouse small cell lung carcinoma cells highly metastatic to lungs upon the intravenous transplantation. However, after dimethylsulfoxide (DMSO) treatment, they lose an ability to metastasize. DMSO treatment was also shown to inhibit strongly *mts1* expression. Sense constructions were transfected to these cells and the transfectants actively expressing exogenous *mts1* gene acquired the ability to give metastases even after DMSO treatment. The latter did not interfere with *mts1* expression governed by MSV-LTR control elements [62].

A strong increase in metastatic potential was found in rat mammary epithelial Rama37 cells after stable transfection with the *mts1*-expressing constructions. The original cells are benign and do not metastasize, while transfectants gave metastases to lungs and lymph nodes [64].

Finally, the experiments with well-characterized human mammary adenocarcinoma MCF-7 cells were performed [65]. MCF-7 cells are rather benign. They can grow after transplantation to nude mice only in the case of support with estrogen and if transplantation into mammary fat pad is performed. The growth is non-invasive and no metastases could be observed. MCF-7 cells do not contain any significant amount of Mts1 protein. Only in stromal cells, a low level of *mts1* expression could be observed. The expression of the exogenous *mts1* gene induced by stable transfection with the construction containing the *mts1* gene under the control of HMG promoter strongly changed the properties of the MCF-7 cell growth. First, their growth in nude mice became hormone-independent. Second, they could grow after just subcutaneous transplantation. Third, the invasive growth at the primary focus could be detected. Fourth, the metastases to the regional lymph nodes were observed. The tumor cells contained varying amounts of Mts1 protein. Thus, in all mentioned cases, the appearance or disappearance of Mts1 protein led to a significant modulation of metastatic phenotype in the expected direction.

Still, a weak point of transfection experiments is the heterogeneity of cell population used for transfection. For example, CSML-0 cells consist of three morphologically different cell types that are reproduced after cloning from individual cells. It can not easily be excluded that only cells with pre-existing differences in metastatic potential have been selected during transfection experiments. Some other approaches should also be used.

The most clear evidence for the casual role of the *mts1* gene expression for creation of metastatic phenotype was obtained in experiments with transgenic animals [66]. Transgenic mice were obtained with the construction containing the *mts1* gene under the control of MMTV-LTR promoter/enhancer element. Transgenic mice expressed exogenous *mts1* in several tissues. The highest level of expression was found in lactating mammary glands where the MMTV promoter is very active. The endogenous *mts1* gene is not expressed in lactating mammary gland. Interestingly, the phenotype of transgenic mice was not changed comparing to normal mice. Even the presence of high amount of Mts1 protein in lactating mammary glands did not interfere with their functioning and no mammary gland tumors could be observed. Obviously, the *mts1* gene is not an oncogene.

Thereafter, the transgenic mice were crossed with mice from GCR/A strain characterized by a high incidence of mammary gland tumors appearing after several cycles of pregnancy and lactation. These tumors are non-metastatic. As transgenic mice were heterozygous, only half of the offspring carried the transgene, while another half represented the control group. The tumors appeared with the same high frequency in both groups and they were morphologically indistinguishable. The tumor growth rate did not depend on the presence of the transgene also.

However, a dramatic difference in the metastatic potential of tumors belonging to two groups was found. As was mentioned above, non-transgenic tumors never metastasize. Just in only one case (of 30 tested), non-transgenic tumor gave metastases to lungs, but this probably depended on certain additional genetic change. On the other hand, 40% of transgenic tumors were metastatic.

Then, the tumors were subcutaneously transplanted to athymic mice to determine their metastatic phenotype. Both transgenic tumors that had metastasized before and transgenic non-metastasizing tumors gave rise to lung metastases in about half of cases. Thus, 40-50% incidence of metastasis is an intrinsic feature of spontaneous mammary carcinomas expressing the *mts1* gene. Non-transgenic tumors never metastasized after transplantation with one above mentioned exception where metastases developed in all animals.

The distribution of Mts1 protein was detected with the aid of immunostaining. It was found in transgenic tumor cells at the primary focus as well as in the metastatic foci. The concentration varied in a wide range among different cells even in the same tumor. Non-transgenic tumor cells in neither case contained Mts1 protein. The stromal cells in both types of tumors contained the same small amounts of Mts1 expressed from endogenous gene.

All mentioned experiments clearly demonstrate that in some tumors *mts1* expression is sufficient for acquirement of metastatic phenotype. It is quite clear that *mts1* can not be responsible for metastatic phenotype in all cases, as several metastatic tumors, in particular one appeared among non-transgenic tumors do not express it. However, the described results show that the *mts1* gene is one of the key metastatic genes. The question arises, what is a possible mechanism of the action of Mts1 protein.

2.3. A POSSIBLE ROLE OF THE MTS 1 PROTEIN

To answer the question, the intensive studies of the protein had to be performed. For this, one needs high amounts of protein. It was obtained in bacterial system with added olygohistidine tail that allowed an easy purification of the protein on the nickel columns. The purified protein was used for preparing polyclonal and monoclonal antibodies.

Western blotting and immunostaining of cells with these antibodies showed the Mts1 protein to be localized in the cytoplasmic fraction, like other calcium-binding proteins. Experiments on fractionation of cell extracts suggest the presence of a significant fraction of Mts1 protein in cytoskeleton.

To understand Mts1 function, the attempts to determine the targets of Mts1 protein were performed [67]. The *in vivo* labeled proteins were immunoprecipitated with antibodies against Mts1 and the proteins specifically precipitated by these antibodies were analyzed. Such approach resulted in isolation of different components of myosin complex. On the other hand, antibodies to myosin co-precipitated Mts1. Ultracentrifugation of cell extract in sucrose gradient also demonstrated cosedimentation of a rather significant fraction of Mts1 with a much heavier myosin complex suggesting their reversible association. After double immunostaining with antibodies to Mts1 and myosin, one can see the exact coincidence in fluorescence distribution for both antibodies.

In particular, Mts1 interacts only with heavy chain of non-muscle myosin as followed from different overlay experiments with antibodies specific for different types and different chains of myosin. Mts1 does not interact with light chains and with heavy chain of smooth muscle myosin.

To further analyze Mts1-myosin interaction, different deletions in the gene encoding heavy chain of non-muscle myosin were obtained and their products were tested for interaction with Mts1 in protein overlay experiments (paper in preparation). The only peptide responsible for this interaction is located on the carboxy-terminal non-helical region (VPRRMAR aminoacids) of myosin.

Then the effect of such binding on protein kinase mediated phoshorylation of heavy chain of non-muscle myosin was tested. Mts1 specifically inhibited phoshorylation of serine residue No 1917 by protein kinase C without any effect on other phosphorylation sites and without interference with casein protein kinase action. The target serine residue is located just near the binding site for myosin. One can suggest that at least one of Mts 1-induced effects is an inhibition of this phosphorylation reaction. The latter was claimed to play a role in non-muscle myosin functioning putatively leading to changes in cell motility. This may be a possible way for changing the metastatic phenotype of tumor cells. However, at the moment this is just a current hypothesis. It should be pointed out that myosin may not be the only target for Mts 1. In particular, another 35 kD protein was eluted from Mts 1 affinity column.

It should be pointed out that another important function of Mts1 protein may be a mesenchimal transformation of epithelial cells. Strutz *et al.* found that expression of the *mts1* gene in epithelial tissues induces there the mesenchimal transformation [68]. We found that appearance of Mts1 proteins in the cells of transgenic mice is accompanied by the loss of E-cadherin [unpublished]. The reverse correlation between Mts1 and E-cadherin content may be especially important as E-cadherin is one of most clear-cut "metastatic proteins" with well-understood function. Mechanism of Mts1 influence on E-cadherin remains unclear.

One can expect interesting r esults from knockout experiments with the *mts1* gene that have not yet been done. In any case, the functional studies of Mts1 protein are now in progress and can lead to understanding its role in tumor metastasis.

2.4. CONTROL OF THE *mts 1* GENE EXPRESSION

It has been found that no sequence rearrangement usually takes place during the change of tumor phenotype for metastatic one, as followed from Southern blot hybridization experiments. The only exception was observed in the myelomonocytic leukaemia WEHI-3B cell line where IAP retrovirus-like mobile element was found to be inserted into the first intron of the *mts1* gene [69]. As a result, the transcription was started within IAP sequence and chimeric mRNA was synthesized. However, protein remained unchanged. In other cases no rearrangements were found.

Therefore, activation of *mts1* mRNA synthesis should result from changes in concentration of certain trans-regulatory factors. Finding of some trans-regulatory factors responsible for *mts1* activation may lead to finding of a new gene involved in creation of metastatic phenotype, upstream in respect to the *mts1* gene.

The *mts1* gene consists of three exons. The first exon is small and does not contain translated sequences. In human, the alternative splicing may give an additional exon, which also lacks coding sequences [70]. The gene is located in the gene cluster containing several other members encoding proteins belonging to S-100 family [71, 72]. The distances between genes in the cluster are rather small. Examination of *mts1* upstream region up to the 3'-end of the neighbor gene of the cluster has led to finding of no cis-regulatory elements but TATA-box containing promoter. All cis-regulatory elements have been found within the first intron.

The first element is the enhancer of a moderate strength located in a position from +296 to +312 from cap site [73]. Its sequence, TGTTGCTATAGTGTATA, does not possess homology with known enhancers. Protein binding to this sequence was visualized by bandshift assay and by DNA footprinting. The removal of the sequence reduced transient transcription about fivefold. However, the protein binding to this enhancer was detected in the nuclear extracts from both metastatic CSML-100 and non-metastatic CSML-0 cells. Also, the same level of activation by the enhancer was found in transient transfection experiments with both cell types. Only *in vivo* footprinting showed the difference between them. The enhancer was protected against micrococcal nuclease action in CSML-100 but not in CSML-0 cells. Thus, the enhancer selectively works in metastatic cells in spite on the presence of activating protein in both.

One could suggest the involvement of structural changes in chromatin. Really, the test for DNA methylation showed the absence of *mts1* methylation in CSML-100, while in CSML-0 cells the gene was heavily methylated [74]. The latter might interfere with binding of activating protein to the enhancer. The effect should be indirect as the enhancer core sequence does not contain CpG dinucleotides.

Another cis-regulatory element is represented by the sequence TGACTCG located in the first intron at position +351 which also binds nuclear proteins [75]. The sequence is similar to the minus chain of the consensus AP-1 binding sequence (TGAGTCA) differing from the latter by one base substitution. As a result of substitution, CpG dinucleotide appears that is the subject for deoxycytidine methylation. Both methylated and non-methylated sequences interact with nuclear proteins as followed from band shift experiments. However, the protein bound to methylated sequence is different from that bound to non-methylated one and is much more abundant in nuclear extracts.

Consensus AP-1 binding sequence competes with methylated but does not compete with non-methylated element. This suggests that only methylated sequence binds AP-1 factor. The conclusion was proved in supershift experiment where antibodies to Jun and Fos proteins slowed down the mobility of complexes between nuclear factor and methylated sequence. Thus, methylation of CpG creates a novel site for AP-1 binding. The mechanism of specificity change by methylation depends on the similarity of meCpG sequence to TpG in complementary strand of the AP-1 binding site: meC mimics T.

In the *mts1* gene, methylated AP-1 binding element plays a role of transcription silencer of a moderate strength. This inhibition seems to play a role *in vivo*, as a particular CpG sequence in CSML-0 cells is not methylated, while in CSML-0 cells, it is completely methylated.

The third cis-regulatory element is located further downstream (+ 780,+794) and represented by a sequence reminding NFkB consensus binding site: GGGGTTTTTCC [76]. The sequence does really bind NFkB factor, but this binding does not change *mts1* transcription at least in experiments with transient expression. On the other hand, the same sequence binds another factor of higher molecular weight. As was shown in experiments with different constructions, the latter factor bound the longer sequence and was responsible for the activation of transcription in transient transfection assay. Moreover, the protein factor is present in nuclear extracts from CSML-100 but not from CSML-0 cells indicating the possible role of the factor in the *in vivo* activation of *mts1* transcription.

The sequence containing microsatellite motif CAG is located closely to previous element (+800-+836). It is protected both in *in vivo* and *in vitro* footprinting assays from nucleases. Presumably, it binds the protein interacting with microsatellite. The element plays a role of a weak silencer-like element [76]. Finally, in the 3'-part of the first intron, the last positive cis-regulatory element was detected, which has not been well characterized yet.

The importance of these experiments is in attempts to detect the factors which are present only in metastatic cells and play a key role in *mts1* activation. The activation of the gene(s) encoding such factor(s) may play a role in creation of a metastatic phenotype at a more upstream level. One candidate is the 200 kb protein binding to the NFkB-like binding site. It is being cloned and studied at the moment. Many questions concerning the functioning of the Mts1 protein remain to be answered.

3. A Novel Lymphokine, Tag7, Putatively Playing A Role In Tumor Progression

3.1. DISCOVERY AND GENERAL PROPERTIES

Another interesting example of genes discovered in the course of comparing the gene expression in metastatic and non-metastatic cell lines is a gene that probably is not a typical "metastatic gene", but may be used for some practical purposes. The experiments on this gene are presented below.

An extensive search of genes over- or under- expressed in metastatic tumors comparing to non-metastatic ones was performed on another system where the metastatic phenotype was certainly not connected with the *mts1* gene activation. These are again cell lines originated from mouse mammary adenocarcinoma. The original non-metastatic line was VMR-0 cell line. In the course of transplantations to isogenic mice, the tumors and cell lines were obtained that gave metastases with preferential location either in liver (VMR-L) or in ovaries (VMR-Ov) [54]. "mRNA display" approach was applied to these cell lines to analyze the differences in gene expression pattern.

The first gene found to be over-expressed in metastatic VMR-L and to a lesser extent in VMR-Ov cells was designated as *tag7* [77]. It encodes the protein of 151 aminoacids without any obvious homology to other proteins in the data bases searched. Further studies showed the homology of the 5'-upstream region of the *tag7* gene and that of the murine LT-β (lymphotoxin-beta) gene. The same cis-regulatory elements binding sites for transcription factors, such as Ets 1, NF-kB, Sp-1, Myo-D are located in both and have the same order [78]. Moreover, the expression of both in normal tissues also has the same patterns. These observations led to a more detailed comparison of aminoacids sequences of Tag7 protein and the members of TNF-Lymphotoxin family. The regions of homology were found in the conservative domains of the latter. Thus, one can suggest that Tag7 protein belongs to TNF-Lymphotoxin family of cytokines.

3.2. EXPRESSION OF THE *tag 7* GENE IN NORMAL TISSUES AND TUMORS

Expression of the *tag7* gene takes place in several tissues of normal organism, first of all, in hematopoietic and lymphoid tissues. The level of *tag7* mRNA synthesis and Tag7 protein accumulation depend on activation of lymphocytes. In contrast to TNF and LT-α, *tag 7* mRNA and protein are constitutively synthesized in lymphocytes thus reminding LT-β expression pattern [79]. LPS stimulation of lymphocytes leads to the decrease of both mRNA and protein expression during the first hours of activation and enhancement of *tag7* expression at the later stage of activation (18 hours). The most interesting observation is the changes in protein localization while activation. In freshly isolated lymphocytes Tag 7 protein is cell- associated meanwhile in the later stages of activation it becomes secreted. Besides lymphocytes, *tag7* is also expressed in macrophages and thymocytes. Interestingly, its active transcription was observed in some specialized cortical areas of the brain, hyppocampus and Purkinje cells of the cerebellum. Its expression was found in decidual tissues at 6.5-8.5 days of gestation [80].

Studies on tumors did not show any correlation between *tag7* expression and tumor progression. For example, CSML-0 tumors *in vivo* express it at a significant level, while CSML-100 tumors do not [77]. However, *in situ* hybridization showed some groups of CSML-100 cells to express *tag7* mRNA. Interestingly, even VMR-L cells express the *tag7* gene at a low level while in culture, but have a very high expression *in vivo* after transplantation. It seems that there is no simple correlation between *tag7* expression and malignancy, but this is realized through a complex interaction between tumor and host cells (see below).

3.3. PROPERTIES OF THE TAG 7 PROTEIN

The antibodies to Tag7 protein were used for detection of Tag7 protein produced by tumor cell lines VMR-L and CSML-0. Surprisingly, in spite of high level of *tag 7* transcription in both cultures no protein was detected until their activation by LPS. After 18 hours of LPS stimulation the major fraction (about 90% of total protein present in cell culture) was found in culture media and only less than 10% in cellular fraction [77]. Thus, Tag7 is secreted not only in lymphocytes but is produced like TNF and LT-α [81-83] by tumor cell lines. Tag7 protein possesses a rather strong cytotoxic activity in respect to certain cell lines, for example, L929.

3.4. ACTION OF *tag 7* EXPRESSION ON TUMOR GROWTH

VMR-0 cells growing in culture and non-expressing the *tag7* gene were transfected with constructions over-expressing *tag7*. Cell lines with different level of *tag7* expression were obtained and transplanted to the isogenic mice. The tumor growth rate was dramatically decreased in proportion to the level of *tag7* expression. This was due to a host response to the Tag 7 released by the tumor, since growth resumes normally if the recipients receive anti-Tag 7 antibody (unpublished). Unlike TNF administration [84] no weight loss was observed. Still after a long period of time (2 and more months), the tumors reached the same size as the control ones. At this stage another difference between control and stably transfected cells was noticed. The transfected cells were free from any necrotic foci, while the control tumors were heavily necrotized at even earlier stages. The significance of this phenomenon is not completely clear.

Our understanding of the role for which Tag 7 evolved is too shallow to allow a precise assessment of conditions under which administration of the cytikine would prove to be benefit but our observation lend support to the approach of treating patients, since intratumor Tag 7 release may be helpful whatever the sensitivity of the tumor to Tag 7.

3.5. OTHER GENES DIFFERENTIALLY EXPRESSED IN METASTATIC VMR TUMORS

Besides *tag7*, several other genes either over- or under- expressed in metastatic VMR-L and VMR-Ov cells were detected using the variant of "mRNA display" specially developed to facilitate fishing out the genes encoding proteins with known properties. For this, the sequences corresponding to the most conservative parts of the genes of interest were taken as the second primer. Genes for protein kinases and phosphatases were selected because their products are frequently involved in the control of different cell functions.

Only few genes found in this way were really genes encoding above mentioned enzymes. Others were fished out just due to some occasional homology. The most prominent of them are mouse leukocyte-common antigen related protein tyrosine phosphates (LAR PTP) and a novel serin-threonine kinese.

4. Conclusion

Certainly, understanding of the control of tumor metastasis is important by itself as a special biological phenomenon. It also adds to understanding of the biology of lymphoid cell and to mechanisms of development. Besides that, they have some obvious significance for the development of new approaches for diagnostics and therapy of tumor metastasis.

The biochemical or histochemical analysis can be applied to operationally removed tumor to detect its chances to give metastases and to define the optimal program of postoperational treatment. The genes whose products are most frequently overexpressed in metastatic tumors are candidates for such analysis. This may be, for example, the *mts1, E-cadherin, nm23* and several other genes.

Another approach is just the immunological analysis of the patient serum. In this case the genes encoding the secreted proteins are of special interest. In particular, the genes for some degradative enzymes are candidates for such approach.

234

Some proteins inhibiting the development of tumor metastasis may be used in the therapy of the latter. In this respect Tag7, may be of interest. We already mentioned the use of anti-angiogenetic agents as a very potential tool for metastasis treatment.

Finally the proteins specifically expressed on the surface of metastatic cells may be used for targeting metastatic cells with toxic reagents.

5. References

1.   Gunthert, U., Birchmeier, W., (Eds), (1996) Attempts to understand metastasis formation II, regulatory factors, in *Current topics in microbiology and immunology*, Springer-Verlag Berlin Heidelberg, Vol. 213/II.
2.   Hollstein, M., Sidransky, D., Vogelstein, B., Harris, C.C., (1991) p53 mutations in human cancers, *Science*, **253**, 49-53
3.   Barbacid, M., (1987) Ras genes, *Annu. Rev. Biochem.*, **56**, 779-827
4.   Vogelstein, B., Kinzler, K.W., (1993) The multistep nature of cancer *Trends Genet.*, **9**, 138-141
5.   White, E., (1996) Life, death, and pursuit of apoptosis, *Genes Dev.*, **10**, 1-15
6.   Guy, C.T., Cardiff, R.D., Muller, W.J., (1992) Induction of mamary tumors by expression of polyomavirus middle T onc ogene: a transgenic mouse model for metastatic disease, *Mol. Cell Biol.*, **12**, 954-961
7.   Anderson, M.W., Reynolds, S.H., You, M., Maronpot, R.M., (1992) Role of proto-oncogene activation in carcinogenesis, *Environ. Health Perspect.* **98**, 13-24
8.   Folkman, J., (1992) The role of angiogenesis in tumor growth, *Semin Cancer. Biol.* **3**, 65-71
9.   Folkman, J., (1995) Angiogenesis in cancer, vascular, rheumatoid and other disease, *Nature. Med.*, **1**, 27-31
10.   Fidler, I.J., Ellis, L.M., (1994) The implication of angiogenesis for the biology and therapy of cancer metastasis, *Cell*, **79**, 185-188
11.   Singh, R.K., Bucana, C.D., Gutman, M., Fan, D., Wilson, M.R., Fidler, I.J., (1994) Organ site-dependant expression of basic fibroblast growth factor in human renal cell carcinoma cells, *Am. J. Pathol*, **145**, 365-374
12.   Singh, R.K., Gutman, M., Bucana, C.D., Sanchez, R., Llansa, N., Fidler, I.J., (1995), Interferons alpha and beta downregulate expression of basic fibroblast growth factor in human carcinomas, *Proc. Natl. Acad. Sci. USA*. **92**, 4562-4566
13.   Gutman, M., Singh, R.K., Xie, K., Bucana, C.D., Fidler, I.J., (1995) Regulation of IL-8 expression in human melanoma cells by the organ inviroment, *Cancer. Res.*, **55**, 2470-2475
14.   Warren, R.S., Yuan, H., Matli, M.R., Gillett, N.A., Ferrara, N., (1995) Regulation by vascular endothelial growth factor of human cancer tumorigenesis in a mouse model of experimental liver metastasis, *J. Clin. Invest.* **95**, 1789-1797
15.   O'Reilly, M.S., Holmgren, L., Shing, Y., Chen, C., Rosenthal, R.A., Moses, M., Lane, W.S., Cao, Y., Sage, E.H., Folkman, J., (1994) Angiostatin: a novel angiogenesis inhibitor that mediates the suppression of metastases by a Lewis lung carcinoma, *Cell*, **79**, 315-328
16.   Takeichi, M., (1991) Cadherin cell adhesio n receptors as a morphogenetic regulator, *Science*, **251**, 1451-1455
17.   Hulsken, J., Birchmeier, W., Behrens, J., (1994) E-cadherin a nd APC compete for the interaction with beta-catenin and the cytoskeleton, *J. Cell Biol.*, **127**, 2061-2069
18.   Donalies, M., Cramer, M., Ringwald, M., Starzinski Powitz, A., (1991) Expression of M-cadherin, a member of the cadherin multigene family, correlates with differentiation of skeletal muscle cells, *Proc. Natl. Acad. Sci. USA*, **88**, 8024-8028
19.   Sano, K., Tanihara, H., Heimark, R.L., Obata, S., Davidson, M., St John, T., Taketani, S., Suzuki, S., (1993) Protocadherins: a large family of cadherin-related molecules in centranervous system, *EMBO J*, **12**, 2249-2256
20.   Okazaki, M., Takeshita, S., Kawai S., Kikuno N., Tsujimura A., Kudo, A., Amann, E., (1994) Molecular cloning and characterization of OB-cadherin, a new member of cadherin family expressed in osteoblasts, *J. Biol. Chem.*, **269**, 12092-12098

21. Berndorf, D., Gessner, R., Kreft, B., Schnoy, N., Lajous Peter, A.M., Loch, M., Reutter, W., Hortsch, M., Tauber, R., (1994) Liver-intestine cadherin: molecular cloning and characterization of a novel Ca(2+)-dependent cell adhesion molecule expressed in liver and intestine, *J. Cell Biol.,* **125,** 1353-1369

22. Frixen, U.H., Behrens, J., Saches, M., Eberle, G., Voss, B., Warda, A., Lochner, D., Birchmeier, W., (1991) Cadherin-mediated cell-cell adhesion prevents invasivness of human carcinoma cells, *J. Cell Biol.,* **113,** 173-185

23. Vlemincks K., Vakaek, L., Mareel, M., Fiers, W., Van Roy, F., (1991) Genetic manipulation of E-cadherin expression reveals an invasion suppressor role, *Cell,* **66,** 107-119

24. Behrens, J., Mareel, M.M., Van Roy, F.M., Birchmeier, W., (1989) Dissecting tumor cell invasion: epithelial cells acquire invasive properties after the loss of uvomorulin-mediated cell-cell adhesion, *J.Cell Biol.,* **108,** 2435-2447

25. Becker, K.F., Atkinson, M.J., Reich, U., Becker, I., Sievert, J.R., Hofler, H., (1994) E-cadherin gene mutations provide clues to diffuse type gastric carcinomas, *Cancer. Res.,* **54,** 3845-3852

26. Kanai, Y., Oda, T., Tsuda, H., Ochiai, A., Hirohashi, S.,(1994) Point mutation of the E-cadherin gene in invasive lobular carcinoma of the breast, *Jpn. J. Cancer Res.,* **85,** 1035-1039

27. Moll, R., Mitze, M., Frixen, U.H., Birchmeier, W., (1993) Differential loss of E-cadherin expression in infiltrating ductal and lobular breast carcinomas, *Am. J. Pathol.,* **143,** 1731-1742

28. Kinsella, A.R., Green, B., Lepts, G.C., Hill, C.L., Bowie, G., Taylor, B.A., The role of the cell-cell adhesion molecule E-cadherin in large bowel tumor cell invasion and metastasis, *Br. J. Cancer,* **67,** 904-909

29. Umbas, R., Isaacs, W.B., Bringuier, P.P., Scshaafsma, H.E., Karthaus, H.F., Oosterhof, G.O., Debruyne, F.M., Schalken, J.A., (1994) Decreased E-cadherin expression is associated with poor prognosis in patients with prostate cancer, *Cancer Res.,* **54,** 3929-3933

30. Screiber, C., Fisher, M., Hussein, S., Juliano, R.L., (1991) Increased tumorgenicity of fibronectin receptor deficient Chinese hamster ovary cell variants, *Cancer Res.,* **51,** 1738-1740

31. Stallmach, A., Von Lampe, B.V., Matthes, H., Bornhoft, G., Riecken, E.O., (1992) Diminished expression of integrin adhesion molecules on human colonic epithelial cells during the benign to malignant tumor transformation, *Gut.,* **33,** 342-346

32. Birchmeier, W., Behrens, J., (1994) Cadherin expression in carcinomas: role in the formation of cell junctions and the prevention of ivasiveness, *Biochem. Biophis. Acta,* **1198,** 11-26

33. Schmidt, C., Bladt, F., Goedecke, S., Brinkman, V., Zschiesche, W., Sharpe, M., Gherardi, E., Birchmeier, C., (1995) Scatter factor/hepatocyte growth factor is essential for liver development, *Nature,* **373,** 699-702

34. Weinder, K.M., Sachs, M., Reithmaches, D., Birchmeier, W., (1995) Mutation of juxta membrane tyrosine residue 1001 suppresses loss-of-function mutations of the met receptor in epithelial cells, *Proc. Natl. Acad Sci. NY,* **92,** 2597-2601

35. Yamashita, J., Ogawa, M., Yamashita, S., Nomura, K., Kuramoto, M., Saishoji, T., Shin, S., (1994) Immunoreactive hepatocyte growth factor is a strong and independent predictor of recurrence and survival in human breast cancer, *Cancer Res.,* **54,** 1630-1633

36. Liotta, L.A., Mandler, R., Murano, G., Katz, D.A., Gordon, R.K., Chiang, P.K., Schiffmann, E., (1986) Tumor cell autocrine motility factor, *Proc. Natl. Acad. USA.,* **83,** 3302-3306

37. Watanabe, H., Carmi, P., Hogan, V., Raz, T., Siletti, S., Nabi, I.R., Raz, A., (1991) Purification of human tumor cell autocrine motility factor and molecular cloning of its receptor, *J. Biol. Chem.,* **266,** 13442-13448

38. Siletti, S., Timar, J., Honn, K.V., Raz, A., (1994) Autocrine motility factor induces differential 12-lipoxygenase expression and activity in high and low-metastatic K-1735 melanoma cell variants, *Cancer Res.,* **54,** 5752-5756

39. Siletti, S., Yao, J., Sanford, J., Mohammed, A.N., Otto, T., Wolman, S.R., Raz, A., (1993) Autocrine motility factror receptor in human bladder carcinoma: gene expression, loss of cell-contact regulation and chromosomal mapping, *Int. J. Oncology,* 3, 801-807

40. Lawrence, M.B., Springer, T.A., (1991) Leukocytes roll on a selectin at physiologic flow rates: distinction from and prerequisite for adhesion through integrins, *Cell,* **65,** 859-873

41. Butcher, E.C., (1991) Leukocyte-endothelial cell recognition: three or more steps to specificity and divercity, *Cell,* **67,** 1033-1036

42. Tozeren, A., Kleinman, H.K., Grant D.S., Morales, D., Mercurio, A.M., Byers, S.W., (1995) E-selectin-mediated dynamic interactions of breast- and colon-cancer with endothelial cell monolayers, *Int. J. Cancer,* **60,** 426-431

236

43. Garofalo, A., Chirivi, R.G.S., Foglieni, C., Pigott, R., Mortarini, R., Martin-Padura, I., Anichini, A., Gearing, A.J., Sanchez-Madrid, F., Dejana, E., Giavazzi, R., (1995) Involvment of the very late antigen 4 integrin on melanoma in interlukin-1-augmented experimental metastasis, *Cancer Res.*, **55**, 414-419

44. Mattila, P., Majuri, M.L., Renkonen, R., (1992) VLA-4 integrin on sarcoma cell lines recognizes endothelial VCAM-1. Differential regulation of the VLA-4 avidity on various sarcoma cell lines, *Int. J. Cancer*, **52**, 918-923

45. Perrotti, D., Cimino, L., Falcioni, R., Tibursi, G., Gentileschi, M.P., Sacchi, A., (1990) Metastatic phenotype: growth factor dependence and integrin expression, *Anticancer Res.*, **10**, 587-598

46. Schreiber, C., Bauer, J., Margolis, M., Juliano, R.L., (1991) Expression and role of integrins in adhesion of human colonic carcinoma cells to extracellular matrix components, *Clin. Exp. Metastasis*, **2**, 163-178

47. Haynes, B.F., Liao, H.X., Patton, K.L., (1991) The transmembrane hyaloronate receptor (CD44): multiple function, multiple forms, *Cancer Cells*, **3**, 347-350

48. Gunthert, U., Hofmann, M., Rudy, W., Reber, S., Zoller, M., Hausmann, I., Matzkus, A., Wenzel, A., Ponta, H., Herrlich, P., (1991) A new variant of glycoprotein CD44 confers metastatic potential to rat carcinoma cells, *Cell*, **65**, 13-24

49. Hofmann, M., Rudy, W., Zoller, M., Tolg, C., Ponta, H., Herrlich, P., Gunthert, U., (1991) CD44 splice variants confer metstatic behavior in rats: homologus sequences are expressed in human tumor cell lines, *Cancer Res.*, **51**, 5292-5297

50. Steeg, P.S., Bevilacqua, G., Kooper, L., Thorgeirsson, U.P., Talmadge, J.E., Liotta, L.A., Sobel, M.E., (1988) Evidence for a novel gene associated with low tumor metastatic potential, *J. Natl. Cancer Inst.*, **80**, 200-204

51. De La Rosa, A., Williams, R.L., Steeg, P.S., (1995) Nm23/nucleoside diphosphate kinases: toward a structural and biochemical understanding of its biological functions, *Bioessays*, **17**, 53-62

52. Gunthert, U., Birchmeier, W., (Eds) (1996) Attempts to understand metastasis formation II, regulatory factors, in *Current topics in microbiology and immunology*, Springer-Verlag Berlin Heidelberg, Vol.213/II, p217-219

53. Postel, E.H., Berberich, S.J., Flint, S.J., Ferrone, C.A., (1993) Human *c-myc* transcription factor Puf identified as Nm23-H2 nucleoside diphosphate kinase, a candidate suppressor of tumor metastasis, *Science*, **261**, 478-480

54. Senin, V.M., Ivanov, A.M., Afanaseva, A.V., Buntsevich, A.M., (1984) New organotropic-metastatic transplanted tumors in mice and their use for studing laser effect on dissemination, *Vestnik USSR Acad. Med. Sci.* **5**, 85-91

55. Ebralidze, A., Tulchinsky, E., Grigorian, M., Afanaseva, A., Senin, V., Revazova, E., Lukanidin, E., (1989) Isolation and characterization of a gene spesifically expressed in different metastatic cells and whose deduced gene product has a high degree of homology to a Ca(2+)-binding protein family, *Genes Dev.*, **3**, 1086-1093

56. Barraclough, R., Kimbell, R., Rudland, P.S., (1984) Increased abundance of a normal mRNA sequence accompanies the conversion of rat mammary cuboidal epithelial cells to elongated myoepithelial-like cells in culture, *Nuc. Acids Res.*, **12**, 8097-8114

57. Jackson-Grusby, L.L., Swiergiel, J., Linzer, D.I.H., (1987) A growth related mRNA in cultured mouse cells encodes a placental binding protein, *Nuc. Acids Res.*, **15**, 6677-6690

58. Masiakowski, P., Shooter, E.M., (1988) Nerve growth factor induces the genes for two proteins related to afamily of calcium-binding proteins in PC12 cells, *Proc. Natl. Acad. Sci. USA*, **85**, 1277-1281

59. Goto, K., Endo, H., Fujioshi, T., (1988) Cloning of the sequences expressed abundantly in established cell lines: identification of a cDNA clone highly homologus to S100, a calcium binding protein, *J. Biochem.*, **103**, 48-53

60. Engelkamp, D., Schafer, B.W., Erne, P., Heizmann, C.W., (1992) S100α CAPL, and CACY: molecular cloning and expression analysis of three calcium-binding proteins from human heart, *Biochemistry*, **31**, 10258-10264

61. ' Grigorian, M., Tulchinsky, E., Burrone, O., Tarabykina, S., Georgiev, G., Lukanidin, E., (1994) Modulation of mts 1 expression in mouse and human normal and tumor cells, *Electrophiresis*, **15**, 463-468

62. Grigoriám, M., Tulchinsky, E., Zain, S., Ebralidze, A., Kramerov, D., Kriajevska, M., Georgiev, G., Lukanidin, E., (1993) The mts 1 and control tumor metastasis, *Gene*, **135**, 229-238

63. Maelandsmo, G.M., Hovig, E., Skrede, M., Kashani-Sabet, M., Engebraten, O., Florenes, V.A., Myklebost, O., Lukanidin, E., Grigorian, M., Skanlon, K.J., Fodstad, O., (1995) Reversal of the in

vivo metastatic phenotype of human osteosarcoma cells by an anti-CAPL (mts 1) ribozyme, *Proc. Natl. Acad. Sci. USA,* in press

64. Davies, B.R., Davies, M.P.A., Gibbs, F.E.M., Barrachlough, R., Pilip, S., Rudland, P.S., (1993) Induction of metastatic phenotype by transfection of a benign rat mammary epithelial cell line with the gene for p9Ka, a rat calcium-binding protein, but not with the oncogene EJ-ras-1, *Oncogene,* **8,** 999-1008

65. Grigorian, M., Ambartsumian, N., Lykkesfeldt, A.E., Bastholm, L., Elling, F., Georgiev, G., Lukanidin, E., (1996) Effect of *mts 1* expression on the progression of human breast cancer cells, in preparation

66. Ambartsumian, N.S., Grigorian, M.S., Larsen, I.F., Karlstrom, O., Sidenius, N., Rygaard, J., Georgiev, G., Lukanidin, E., (1996) Metastasis of mammary carcinomas in GRS/A hybrid mice transgenic for the *mts 1* gene, *Oncogene,* in press

67. Kriajevska, M.V., Cardenas, M.N., Grigorian, M.S., Ambartsumian, N.S., Georgiev, G.P., Lukanidin, E.M., (1994) Non-muscle myosin heavy chain as a possible target for protein encoded by metastasis-related *mts 1* gene, *J. Biol. Chem.,* **269,** 19679-19682

68. Strutz, F., Okada, H., Lo, C.W., Danoff, T., Carone, R.L., Tomaszewski, J.E., Neilson, G., (1995) Identification and characterization of a fibroblast marker: FSPI *L. Cell Biol.,* **130,** 393-405

69. Tarabykina, S., Ambartsumian, N., Grigorian, M., Georgiev, G., Lukanidin, E., (1996) Activation of *mts 1* transcription by insertion of a retrovirus-like IAP element, *Gene,* **168,** 151-155

70. Ambartsumian, N., Tarabykina, S., Grigorian, M., Tulchnsky, E., Hulgaard, E., Georgiev, G., Lukanidin, E., (1995) Characterisation of two splice variants of metastasis-associated human *mts 1* gene, *Gene,* **159,** 125-130

71. Engelkamp, D., Schafer, B.W., Mattei, M.G., Erne, P., Heizmann C.W., (1993) Six S100 genes are clustered on human chromosome 1q21: identification of two genes coding for the two previously unreported calcium-binding proteins S100D and S100E, *Proc. Natl. Acad. Sci. USA* **90,** 6547-6551

72. Dorin, J.R., Emslie, E., van Heeningen, V., (1990) Related calcium-binding proteins map to the same subregion of chromosome 1q and to extended region of synteny on mouse chromosome 3, *Genomics,* **8,** 420-426

73. Tulchinsky, E., Kramerov, D., Ford, H.L., Reshetnyak, E., Lukanidin, E., Zain, S., (1993) Characterization of a positive regulatory element in the *mts 1* gene, *Genes Dev.,* **8,** 79-86

74. Tulchinsky, E., Ford, H.L., Kramerov, D., Reshetnyak, E., Grigorian, M., Zain, S., Lukanidin, E., (1992) Transkriptional analysis of the mts 1 gene with specific reference to 5' flanking sequences, *Proc. Natl. Acad. Sci. USA* **89,** 9146-9150

75. Tulchinsky, E.M., Georgiev, G.P., Lukanidin, E.M., (1996) Novel AP-1 binding site created by DNA-methylation, *Oncogene,* **12,** 1737-1745

76. Tulchinsky, E., Prokhorchouk, E., Georgiev, G., Lukanidin, E., (1996) A kB-related binding site is a integral part of the mts 1 gene composite enhancer element located in the first intron of the gene, *J. Biol. Chem.,* in press

77. Kiselev, S., Kustikova, O., Korobko, E., Prokhorchouk, E., Kabishev, A., Lukanidin, E., Georgiev, G., (1996) Tumor growth inhibition and cytotoxic activity of a novel soluble cytokine with structural similarity to TNF-family members, in prepation

78. Pokholok, D., Maroulakou, I., Kuprash, D., Alimzhanov, M., Koslov, S., Novobratseva, T., Turetskaya, R., Green, J., Nedospasov, S., (1995) Cloning and expression analysis of the murine lymphotoxin β gene, *Proc. Natl. Acad. Sci. USA,* **92,** 674-678

79. Browning, J.L., Ngam-ek, A., Lawton, P., DeMarinis, J., Tizard, R., Chow, E.P., Hession, C., O'Brine-Greco, B., Foley, S.F., Ware, C.F., Lymphotoxin β, a novel member of the TNF family that forms heteromericcomplex with lymphotoxin on the cell surface.

80. Kawamoto, S., Okubo, K., Yoshii, J., Katsuki, M., Matsubara, K., (1995) Analysis of gene expression in mouse embriogenesis by 3'-directed cDNA sequencing, EMBL database

81. Rubin, B.Y., Anderson, S.L., Sullivan, S.A., Williamson, B.D., Carswell, E.A., Old, L.J., (1986) Nonhematopoietic cells selected for resistence to tumor necrosis factor produce tumor necrosis factor, *J. Exp. Med.* **164,** 1350-1355

82. Spriggs, D.R., Imamura, K., Rodriguez, C., Sariban, E., Kufe, D.W., (1988) Tumor necrosis factor expression in human epithelial cell lines, *J. Clin. Invest.* **81,** 455-460

83. Garrett, I.R., Durie, B.G.M., Nedwin, G.E., Gillespie, A., Bringman, T., Sabatini, M., Bertolini, D.R., Mundy, G.R., (1987) Production of lymphotoxin, a bone-resorbing cytokine, by cultured human myeloma cells, *N. Engl. J. Med.* **317,** 526-532

84. Oliff, A., Defeo-Jones, D., Boyer, M., Martinez, D., Kiefer, D., Vuocolo, G., Wolfe, A., Socher, S.H., (1987) Tumors secreting human TNF/cachectin induce cachexia in mice, *Cell,* **50,** 555-563

THE FUTURE OF DNA SEQUENCING: AFTER THE HUMAN GENOME PROJECT

CHARLES R. CANTOR[1], CASSANDRA L. SMITH[1], DONG JING FU[2], NATALIA E. BROUDE[1], RON YAAR[1], MARYANNE MALONEY[2], KAI TANG[2], JOEL GRABER[1], DANIEL P. LITTLE[2], HUBERT KOESTER[3], ROBERT J.COTTER[4]
[1]*Center for Advanced Biotechnology and Departments of Biomedical Engineering, Biology, and Pharmacology and Experimental Therapeutics, 36 Cummington St., Boston Univ., Boston MA 02215;*
[2]*Sequenom, Inc., 101 Arch St., Boston MA, 02210;*
[3]*Department of Biochemistry and Molecular Biology, University of Hamburg, Martin-Luther-King-Platz 6, 20146 Hamburg, Germany; and*
[4]*Middle Atlantic Mass Spectrometry Laboratory, Johns Hopkins University School of Medicine, 725 N. Wolfe St., Baltimore MD 21205.*

1. Results of the Human Genome Project

The notion of determining the DNA sequence of the entire human genome was conceived in the mid 1980's. The initial impulse came from a consideration of how improved DNA sequencing methodology might enhance, dramatically, our ability to find mutations and be able to quantitate the effect of radiation and other environmental insults on human inherited mutations. However, a stronger and more constant driving force has been the usefulness of DNA sequence information in finding genes and studying their activity.

The initial plans for the Human Genome Project in the United States included a heavy focus on obtaining low resolution structural information on the 22 autosomes and two human sex chromosomes prior to the onset of large scale genomic DNA sequencing. Such low resolution data is called mapping, and it results in DNA samples that ultimately form the substrate used for subcloning and actual sequence reading. As plans for the Project matured in the late 1980's, a fifteen-year time span was proposed. A key feature of the plans for the Human Genome Project was technology development. It was hoped that a ten-fold improvement in the technology for DNA mapping and sequencing would be achieved during the first third of the project and a similar improvement in sequencing technology would occur again in the middle third of the project [11]. Currently, in 1996, the project is deemed by most observers and participants to be on budget and ahead of schedule. The desired improvements in technology have been handily achieved. It is now likely that the Project itself will be completed successfully within the original time frame using methods which by and large already exist today as commercial instruments or protocols.

239

C. Nicolini (ed.), Genome Structure and Function, 239–260.
© 1997 *Kluwer Academic Publishers. Printed in the Netherlands.*

The Human Genome Project as originally designed include a number of model organisms. These were generally gene rich species where classical and molecular genetics was already well developed including the bacterium *Escherichia coli*, the yeast *Saccharomyces cerevisiae*, the fruit fly *Drosophila melanogaster*, the nematode *Caenorhabditis elegans*, and the mouse *Mus musculus*. As interest in genomics expanded, and the methods improved and became more generally applicable a number of other organisms have joined the priority list for genome sequencing. The haploid sizes of these genomes are listed in Table 1 along with an estimate of the current extent of sequencing progress, where known.

TABLE 1. Sizes of Genomes that are Current Targets for DNA sequencing

Species	Genome Size(bp)	Sequence available
Mycoplasma genitalium	5.8×10^5	completed [14]
Haemophilis Influenza	1.8×10^6	completed [13]
Escherichia coli	4.0×10^6	50%
Saccharomyces cerevisiae	1.3×10^7	completed [12]
Drosophila melanogaster	1.7×10^7	
Arabidopsis thaliana	1.2×10^8	2.4%
Caenorhabditis elegans	1.0×10^8	43%
Mus musculus	3.0×10^9	
Homo sapiens	3.0×10^9	

The Genome Project has already produced a number of surprises. The original plan was to make maps and perform DNA sequencing as a systematic approach to finding all of the genes and making them available for subsequent biological studies. However, very effective alternative strategies for finding most if not all of the genes have emerged and have been implemented on a large scale. There are estimated to be about 10^5 human genes. In a given cell type a few percent of these are expressed at levels high enough to be easily detected as mRNA and immortalized by converting the mRNA into a cloned cDNA copy. By choosing a sufficient set of different cell types in different physiological states, it seems possible to sample almost all of the human genes. Several focused efforts at mass cDNA sequencing have emerged, and, as a result, it is inevitable that most if not all human genes will be found in this way within the next year or so. It is important to note, however, that the cDNA approach provides genes without any direct knowledge about their function or their location. The challenge is now to speed up the processes for mapping known sequences and for obtaining preliminary working hypotheses about their function.

2. Needs for Future DNA Sequencing

When the Human Genome Project started there were less than 10 megabases (Mb) of DNA sequence available in public data bases. Today this number is more than 600 Mb. While the Human Genome Project seems large compared with the currently available information, the needs for DNA sequencing in the future totally dwarf these numbers. One estimate for future DNA sequence demands comes from the pool of human genetic diversity. Between any two homologous chromosomes DNA sequence differences occur at the level of at least 0.1%. Thus to catalog DNA sequence differences alone for the human population would require scoring 3×10^6 sites in 2 chromosomes each in 5×10^9 individuals. The actual database required to contain the resulting 3×10^{16} bases of sequence information is inconceivable by current computer standards but, projecting the current very steady rate technology development in computer storage and processing capabilities, this database will be the equivalent of only a few CD-ROM's fifteen years from now. While this is the ultimate human sequence database, it is not just a fantasy. We will describe below how data on human DNA sequence variation is both necessary and useful in understanding gene function and dysfunction.

2.1 GENE FINDING

Most genes responsible for inherited human diseases are found by a combination of classical human genetics and DNA sequencing. Linkage analysis is use to monitor the pattern of inheritance of the phenotype we know as a human disease and to correlate this pattern with the inheritance of anonymous variable DNA sequences (genetic markers) whose location in the genome is known approximately. From such studies, where successful, eventually emerge two or more such genetic markers that show a pattern of inheritance almost identical to that of the disease and flank its position in the genome. This narrows the location of the disease-causing gene to within a few million base pairs, at the current level of resolution of genetic mapping methods. Within this zone can lie just a few human genes, or scores of them. To decide which gene is responsible for the disease itself requires a great deal of fortune and persistence. The ultimate test is showing DNA sequence differences within the gene that correlate specifically with the disease phenotype. Various strategies are used to focus on particular subsets of the region as targets for these extensive and expensive horizontal sequence comparisons. When diseases caused by more than one gene are at issue, the sequencing problem increases in difficulty markedly because the original linkage analysis is less precise, or may be totally ineffective. Inevitably, however, we will have to sequence the human genome many times over indeed to identify precisely the set of genes that is responsible for more than 4000 known inherited diseases.

2.2. DIAGNOSTIC DNA SEQUENCING

DNA sequencing is poised to serve as an effective diagnostic tool, and it is now recognized that this tool in many cases is a necessity, not a luxury. The earliest inherited diseases known, like sickle cell anemia, and many other hemoglobinopathies, were misleading, in retrospect, because their sequence etiology was very simple. Only a single, constant base change is known to produce the very specific disease phenotype we call sickle cell anemia. There is no need to sequence the hemoglobin gene to test for this disease-causing allele. A specific test can be set up to examine just the one base in question. Such tests can be allele-specific hybridization or allele-specific polymerase chain reaction (PCR) amplification.

In contrast to sickle cell anemia, most genetic diseases that have now been unraveled at the molecular level show a staggering complexity of different DNA changes. Cystic fibrosis is caused by more than 400 [5] known mutations scattered throughout a large gene, and the number of mutations keeps growing. Worse still are human cancers where a spectrum of mutations of almost limitless complexity is responsible for inherited forms of these diseases and a different spectrum, of equal complexity is seen in spontaneous (i.e. environmentally induced) forms of cancers [10].

In all such cases the entire DNA sequence of the gene must be examined in order to identify a mutant disease-causing allele or rule out its presence. The potential need for such DNA sequencing is very large indeed; it approaches the full complexity of human DNA diversity described earlier. The key technical issue is whether efficient strategies can be developed that will allow one to look just for DNA differences against a norm, in such diagnostics, or whether the entire gene must be resequenced over and over again in each individual to be tested.

2.3. EVOLUTIONARY COMPARISONS

DNA sequence data is not only the ultimate biological information, it is also the easiest to utilize on a quantitative comparative basis. Thus, it is inevitable that, as DNA sequencing costs drop, the methodology will be extended on a large scale to more and more organisms. There are already nascent genome projects on a number of plants and animals of commercial importance. Projects are also planned on experimentally important laboratory animal and plant systems like dogs, rats, zebra fish, primates, and so on. Large numbers of microorganisms of environmental or commercial importance are already being sequenced. Where will the trend stop? It will probably not stop until every known organism is sequenced, even fossil organisms, if good enough samples can be found to allow this. Several aims will be achieved by this massive endeavor. From DNA sequence data it should be possible by eventual future technology to recreate any organism that accidentally becomes extinguished, should this ever be desirable or necessary. In the massive set of comparative DNA sequence data will lie the most complete picture about the evolution of species on the planet that we will ever be able to obtain.

2.4. EXPRESSION SCREENING

DNA sequence data is a prelude to studies of biological function. One aspect of this is determining the structure of the proteins or RNA products coded for by the genes. Great strides are needed to speed up this process by both experimental and computational approaches. Even once structures are known, however, functional clues are usually much more readily obtained by studying the action of the gene in one or more biological milieus. One needs to know in which cells is the gene expressed, at what times and in what quantities. Such studies on a single gene are practical today by widely used methods. However a complete knowledge of the pattern of gene expression is desirable to be able to view the interplay of all of the characters in any complex biological scenario. Where this has been done so far, the results have been extremely revealing.

One way to monitor gene expression is to sample individual mRNAs at random through cDNA copies and identify these. This has been done to date principally by direct DNA sequencing of these clones. The results when coupled to specific physiological events have proven to be very useful in establishing working hypotheses about the functions of newly discovered genes. It is, however, too expensive to keep identifying the presence of genes products by DNA sequencing. Two less costly methods are being used with increasing frequency. In differential display, PCR is used to make cDNA copies of the most abundant transcripts in two or more systems, and these are compared by direct side by side high resolution DNA electrophoresis [8]. Hundreds of specific species can be studied at once in such comparisons, and gene products in common among diverse systems, or gene product differences that are characteristic of a particular state of a system are readily identified as bands of DNA with altered mobility. It is not always easy, however, to go from such observed differences to identification of the particular DNA species responsible for them.

Subtraction methods are the other approach to comparison [6, 7]. Here expressed mRNAs in two (or more) different states or systems are compared all at once by cross-hybridization schemes that allow species present in only one of the two samples to be purified. By subtraction one isolates the novel DNA sequences in a form that readily allows them to be studied, but most quantitative information about the differences is lost and must be recovered in subsequent additional experiments.

Expression profiling of all of the genes will reveal a stunningly detailed picture of the state of a cell. It will be a diagnostic tool of such informativeness and sensitivity that the term imaging has already been applied to this approach even thought the image is a virtual one and not a true geometric image in the usual sense. Expression imaging is also likely to become a research tool of major importance since it will provide, in many ways, the ultimately coupling between physiological studies at many levels, and the action of genes. For such approaches to become reality, methods must be developed that will greatly decrease the current costs of expression profiling and increase their informativeness, so that even relatively rare transcripts can be studied. How these goals may be accomplished in the decade ahead will be described later in this article.

2.5. ENVIRONMENTAL MONITORING

Gene expression is a convolution of the intrinsic propensies of genes as modulated by the effect of the environment. Thus, the pattern of gene expression can tell us whether an organism is healthy or sick, well-fed or starving, saddled with mutations or intact, and so on. The genes of most species are distinguishable, and so by examining what genes are present in a sample of the biosphere we can reconstruct what organisms are present and in what relative amounts. By combining both sorts of information we ought to be able to quantitatively assess the state of an ecological system with an accuracy far exceeding anything that is currently feasible. Large scale DNA sequence screening will be needed, and methods for doing this are only in their infancy.

Most of the organisms on the planet are microorganisms and most of these cannot be cultured by any methods known today [1]. Thus, we know of the existence of most of earth's creatures only through their DNA. This is unlikely to change anytime soon since the difficulty of formulating effective culture media for a multitude of single microorganisms staggers the imagination. As a result, we sample an ecosystem by PCR using conserved sequences to promote the amplification of variable, species-specific regions between them. This is the way we currently identify the inhabitants and estimate their characteristics. In the future such monitoring should occur on a much grander scale using methods that are much more powerful.

2.6. DNA AS AN ADDITIVE

The number of possible DNA sequences of length n is 4^n. Thus a particular 20 base sequence will occur by chance only once in the entire human genome if at all. A 60 base sequence is all but unique in the universe. This encourages the possibility that DNA molecules might be custom designed to serve as universal additives that will uniquely identify any sample, e.g. a solution, a bottle. DNA is chemically quite stable; it has been recovered from fossils such as amber, bone and from mummies. DNA can be amplified from single molecules up to levels that allow DNA sequencing to provide an unequivocal identification. If a 60 base identifier sequence were flanked between two 20 base primers, known only to the manufacturer of the sample, the additive could be detected only by those who knew the primer sequences. Thus, DNA sequences may well serve in the future as universal bar codes, allowing the diffusion of substances and objects to be monitored in a silent, invisible, harmless, but extremely informative way. This can have a down side but there are also many potential benefits. For example, every individual person already has unique sequence characteristics (even so-called identical twins.) These differences serve intrinsically as sample identifiers, and tests that included their detection would guarantee that two samples are never interchanged. Such sample mix up is well known to be a major cause of errors in many aspects of medical diagnostics.

3. Techniques and Strategies for DNA Sequencing

The first DNA sequence data was obtained in 1971 by Ray Wu of Cornell University [4]. In several years of heroic effort he managed to work out the 12 single - stranded complementary bases that allow the two sticky ends of bacteriophage lambda DNA to come together to form a circular structure. Since then, virtually all DNA sequence data has been obtained by variations on a basic theme: the production of sequence ladders. Two general approaches to ladder sequencing were pioneered. Maxam and Gilbert started with an intact DNA labeled at one end and fragmented it in a base-specific way [3]. In principle, four independent reactions would be carried out, one for each base. The lengths of the resulting labeled fragments were measured by high resolution polyacrylamide gel electrophoresis, and these lengths indicated the locations of each specific base. The challenge was to develop fragmentation methods that were extremely even, so that all base locations revealed themselves at detectable levels. Maxam-Gilbert sequencing today is rarely used except for specialized applications.

The second general approach to ladder sequencing was developed by Sanger [2]. He formed a complex between a short DNA primer and a complementary DNA template. The enzyme, DNA polymerase, was used to extend the primer in the presence of all four normal deoxynucleoside triphosphates (pppdN's) and a single terminator such as a di-deoxy-pppN. This results in a spectrum of fragments with all lengths that end in a specific base. Four independent termination reactions are used to cover all of the four bases. The analysis is by the same electrophoresis described for Maxam-Gilbert sequencing. A label is introduced either in the primer, or in each of the terminators. Sanger sequencing proved easier to use, adapt and automate and today it accounts for more than 99% of all known DNA sequences.

3.1. CONTEMPORARY DNA SEQUENCING TECHNOLOGY

Initially, radioactive ^{32}P was used as the label for DNA sequencing gel electrophoresis. During the past few years several automated fluorescent gel readers suitable for DNA sequencing have become commercially available and widely adopted. The overwhelming majority of new DNA sequence information is currently obtained with such instruments. In some a single color fluorophore is used, and the four base-specific reactions are run in side by side lanes. In the most commonly used equipment, four different colored fluorescent labels are used simultaneously so that all four ladders can be analyzed in a single gel-electrophoretic lane. Samples for DNA sequencing are made by cloning into plasmid or bacteriophage vectors and sequenced by using the known vector sequenced to position a primer close to the start of the usually unknown sequence of the cloned insert. Alternatively, samples are made by PCR amplification. The reactions required to generate a set of ladder reactions are carried out by x-y robots customized for this purpose. Typically 24 samples times 4 reactions each would be prepared at once. Anywhere from 10 to 40 different samples can be analyzed simultaneously by gel electrophoresis, and the fluorescent bands are detected as they move past a laser beam. The resulting spatio-temporal pattern of fluorescence is decoded into DNA sequence information automatically by existing software. The entire sequence reading process takes a few hours, and anywhere from a few hundred to almost a thousand bases of DNA sequence are obtained for each sample. The resulting data is called raw DNA sequence. It contains various sorts of errors and needs to be processed further both by manual inspection and further computation until finished, highly accurate and continuous sequence data is obtained.

The amount of raw DNA sequence data that must be accumulated to produce one base of finished data depends quite a bit on the types of samples studied and the strategy used to subdivide them into individual targets for DNA sequencing reactions. In *de novo* DNA sequencing, anywhere from two to ten bases of raw DNA sequence are required to produce one base of finished sequence. Usually data is obtained separately from both strands of the original DNA target; regions are sampled more than one time in different phases, and the resulting data is assembled. The additional redundancy used in such strategies increases the accuracy of the overall finished sequence significantly, but it also increases the cost of performing the overall sequencing process in direct proportion to the total number of raw bases that are scanned. Working with existing commercial equipment, a single individual can produce between 10^4 and 10^5 bases of raw DNA sequence per day. The amount depends on skill level as it effects the quality and length of each sequence read and the number of electrophoretic runs per day that are performed.

In comparative or diagnostic DNA sequencing, the data is already available for a reference sequence. The goal is to compare it with the target to see if any differences emerge. The nature of these differences will inform not only that a mutation is present but frequently will provide hints about whether that mutation is likely to have a serious phenotypic consequence. For example a frameshift that leads to premature termination will usually be much more serious than a single base change that results in a single conservative amino acid substitution. In most cases, comparative DNA sequencing corresponds to raw DNA sequencing. The primer can be chosen from the known sequence to avoid or minimize the occurrence of common sequencing artifacts. A major potential difficulty occurs however because humans are diploid. With two copies of each gene, if a single mutant copy is present alongside a normal copy, two different bases will show up at that location in the DNA sequence ladders. Existing software is sometimes incapable of coping with such events. If a complex spectrum of mutations exists, one will not know which specific sequences are present together on one chromosomal copy and which on the other. The analysis of such complex haplotypes can only be done after subcloning the two separate genomic copies. This is rarely possible the time scale required to satisfy the needs of clinical diagnostics.

3.2. IMPROVED *DE NOVO* METHODS

De novo DNA sequencing almost always refers to the production of finished, assembled sequence. The goal is the complete DNA sequence of the target of interest. Current limitations in *de novo* DNA sequencing arise both in the strategy used to divide the target into samples ready for sequencing and in the actual method used to read the sequence. Two basic sorts of strategies are used. In shotgun sequencing one picks sequenceable clones at random and reads the sequence from one or both ends, depending on the vector system used. In directed sequencing there is a preplanned scheme for reading sequence. For example a long clone could be sequenced systematically by primer walking in which, first, the sequence is read from an end as far in as possible. Then a second primer is synthesized that is complementary to residues near the end of the known sequence; this is used in a second round to extend that sequence. The process can be repeated until the entire sequence of the clone is known. Short oligonucleotides can be stacked together or ligated on the template to avoid the need for *de novo* synthesis of the primers needed [15-17].

Improvements in shot gun sequencing must focus heavily on the properties of the libraries of clones used as input. The more even the coverage, the more efficient the statistical sampling of the clones will be. The lengths of the clones are not terribly critical since one sequences just the ends anyway in shotgun approaches. Indeed, some strategies for large scale sequencing propose to start with the end sequences of clones more than 100 kb in average size [18]. Shot gun sequencing is very efficient in the beginning since it can be automated very effectively and many samples can be studied in parallel. However, the assembly of short regions into a long completed sequence can be a formidable task in a large shot gun project. This is especially true where no *a priori* clone ordering information exists and where there are large amounts of interspersed repeated sequences that can confuse the assembly process. Finishing the last bits of DNA sequence of a target can also be very arduous and may in some projects account for a third or more of the total effort. Segments missing from the library must be found by using clones from other libraries or attempting to amplify uncloned regions of DNA by priming from adjacent clones. However, these approaches are not easy to automate, and they are also not guaranteed to succeed.

Optimum random or directed sequencing strategies have not been established. In fact, in almost all large scale sequencing to date, a shotgun process is used for most of the task, and then a directed strategy becomes increasingly important near the end of the project.

For directed or shotgun sequencing both, longer DNA sequence reads are very helpful. Faster sequencing throughput is also an advantage. The new methodologies under exploration aim squarely at these issues. Another factor is the accuracy of the raw sequence in each read. Improved software has greatly increased the accuracy in currented automated fluorescent gel reading to far better than earlier estimates of 99%. This seems adequate for most sequence assembly tasks. How accurate finished sequence should be is debatable. In humans, naturally occurring DNA polymorphisms at the 0.2% level make it questionable whether first pass DNA sequencing need be more accurate than this. Since comparative DNA sequence data is really the goal in the majority of DNA sequencing studies, errors and polymorphisms will be sorted out anyway as more copies of the same gene or region are examined.

3.2.1. Electrophoresis in Thin Gels or Capillaries

The current limitations in DNA sequencing speed arise mostly from heat generated during the process of gel electrophoresis. DNAs will travel faster as the electrical field is increased but so will the heat generated in the sample. To compensate for this, better methods of heat dissipation are required. In practice what this boils down to is thinner samples to allow for better heat conduction to a cooling plate or the equivalent. Both thin slabs and capillaries are being tested in many different laboratories. Other key limiting steps in high throughput DNA sequencing efforts are time spent cleaning the gel plates used in conventional or thin slab electrophoresis and the time consumed in pouring and polymerizing new gels. Electrolytic damage prevents most gel matrices from being reused many times. Thus, there is considerable activity both in exploring reusable gels and pourable liquid gels that can be pumped in and out of capillaries or preassembled thin plates. Various schemes for large arrays of capillaries or their equivalents are being tested including arrays of capillaries etched into glass plates by microfabrication methods [19]. Such systems have shown impressively fast sequencing speeds with overall runs in minutes rather than the hours required for conventional DNA sequencing, but the read lengths of such instruments do not yet match those seen in macroscoping DNA electrophoresis. In high throughput DNA sequencing the cost of making the ladder samples is a considerable fraction of the total cost, and most of this in highly automated systems is the cost of the chemicals, enzymes, and disposables involved. The cost of the former two items scales with the amount of sample needed. Hence, a real advantage of capillary or thin gel slab methods is that the amount of sample required is considerably less than in ordinary DNA sequencing.

Capillary DNA sequencing instruments are already available commercially. Capillaries or thin slabs are likely to dominate the next few years of DNA sequencing, because they represent a gentle perturbation on methods already in wide-spread use. It seems likely that instruments with many capillaries or large thin slabs will be capable of measuring 10^6 bases of raw DNA sequence per day. To go very much above this number may require the introduction of totally different technology, and some of the possible ways in which this may be accomplished will be described next.

3.2.2. Single Molecule Sequencing

The notion of using an exonuclease to remove one base at a time from a DNA molecule and determine sequence by detecting the identity of that base is not new [20]. If one starts with a population of DNA molecules the amount of sequence readable in this way even with perfectly behaved enzymes is quite modest because stochastic fluctuations rapidly dephase the signals arising from different molecules in the sample. In principle, however, if a single DNA molecule could serve as the target, the sequence might be readable along its entire length [41]. This is the great appeal of single molecule sequencing methods. The challenge is the creation of a detection system with the sensitivity of single bases and a minimum number of false negatives and false positives which are devastating in sequencing since they result in frame shifts. Strong single fluorophores can be reliably detected by contemporary methods, but for single molecule sequencing all four bases have to have usable fluorescent yields. One approach to this is to synthesize DNA with four different fluorescent base analogs at once, but thus far it has not been possible to carry this out successfully. An alternate approach is to find a way to make use of the fluorescence of the four natural bases or enzymatically tolerable close analogs of these. This may offer a better chance, but it places very severe demands on the sensitivity of the fluorescence detection scheme.

An alternative approach to single molecule sequencing is microscopy. Early attempts to sequence DNA by electronmicroscopy were unsuccessful because with typical disordered samples, the beam currents required to image the sample damage it beyond the point where the sequence can be recognized reliably. More recently a number of groups have attempted to obtain high resolution images of DNA by scanning tip microscopy. Here, damage is less but it still can be significant. While it is probably technically possible to scan short bits of DNA sequence by one version or another of scanning tip microscopy, it does not, today, seem likely that the speeds that will be allowable in such processes will compete with other high throughput DNA sequencing methods. If double stranded DNA is used, a nice straight contour will be seen on a flat surface because of the stiffness of the double helix. However to read the bases one needs to see them somehow and in the double strand a few bases every turn are blocked from view by the backbones. A single strand may be a better target because one might be able to find a way to hold it on a surface by the backbone so that the bases all stick up and are more easily measured by the scanning microscope tip. However, ordinary preparations of single-stranded DNA show complex contours which, if they cannot be avoided by some sort of special surface chemistry, can complicate the sequence read or even lead to ambiguities if the strand crosses itself.

3.3. IMPROVED COMPARATIVE OR DIAGNOSTIC SEQUENCING METHODS

For many applications of DNA sequencing a reference is already known, and the question of interest is whether a DNA target of interest is identical to this reference or different. Depending on the application one may wish to know where and what all the differences are or just that somewhere there is a difference. The perfect method would reveal just a catalog of the differences. Such a method does not exist, and indeed none has thus far ever been convincingly proposed. However, some methods under development come close to this ideal goal, as we will illustrate below. For all differential or comparative DNA sequencing methods, the overwhelming criterion is throughput. The read length is relatively unimportant, because no assembly is required. The target can be broken into as many short reads as desired as long as the speed of each read is sufficiently high and the cost of each read is sufficiently low. Accuracy is also not a major consideration. In most anticipated uses of differential DNA sequencing, the answer in the great majority of cases will be no differences found. Where differences are seen they will almost certainly be checked by a repeat of the analysis or by re-examination by a different method. Thus it is important to have relatively few false negatives, i.e. one does not want to miss a real difference, but false positives, i.e. typical DNA sequencing errors such as a miscalled base, will not be particularly problematic.

3.3.1. Sequencing by synthesis

In conventional automated DNA synthesis, a base corresponding to the 3' end of the desired compound is attached to a solid support, and the remaining bases of the sequence are coupled to it one at a time by chemical methods. This process is quite efficient and well automated, and thousands of compounds a day can be prepared by a moderately-sized laboratory. Several groups have been exploring using such a process for DNA sequencing instead of DNA synthesis. Here enzyme-catalyzed rather than chemical synthesis is invoked. The sequence to be synthesized is dictated not by a computer program set by the operator - instead it is directed by the DNA template strand. A primer is extended one base at a time along this strand using tricks such as deblockable terminators. All four possible terminators are tried at each step, and which base has been incorporated is determined by which reaction has succeeded. This can be monitored by using fluorescent or chemiluminescently labeled bases, or by detecting the pyrophosphate released when a pppdN is incorporated. The appeal of the approach is that it can be extensively parallelized, is easy to automate, and may be reduced to a very small scale thus greatly reducing the cost of sample preparation. The disadvantage of implementations to date is that the sequence reads have been rather short.

A version of sequencing by synthesis that may have great utility in DNA diagnostics is the addition of just a single fluorescent terminator. Here by the use of four different colors a complete analysis of the bases adjacent to a set of primers can be determined in a single step. This is essentially a variant on some of the hybridization based DNA sequencing schemes to be described in the next section.

3.3.2. Sequencing by Hybridization

Four groups appear to have independently proposed and explored the notion that one might read DNA sequence by words rather than by letters [21-24]. In the most commonly conceived form of sequencing by hybridization (SBH), an ordered array is formed of all possible probes of length n. This array is called a DNA chip. A labeled target sequence is brought to this array and allowed to hybridize. Wherever in the array the target is detected one knows that the complement of the corresponding sequence is present. In principle one should be able to assemble the overlapping pattern of target words and reconstruct the sequence of the target. There are 4^n words of length n in a four letter alphabet. The thermodynamics of DNA hybridization favor the use of the shortest possible probes that still give useable stability since the effects of mismatches become more and more difficult to detect as n grows. In practice probes of length 8 or 9 are favored. The former require an array of more than 65,000 probes to be constructed. This is possible both with chemical methods as well as with the types of photolithographic techniques used in semiconductor manufacturing. An alternative version of SBH uses an array of samples, which can be any size, rather than an array of probes. To the sample array one or more probes are hybridized at a time. This allows a short bit of DNA sequence to be read from, in principle, an arbitrarily large number of target samples.

SBH is very attractive in principle because it is intrinsically a highly parallel method. However several caveats must be noted. First the method is not rigorous for *de novo* DNA sequencing. Where interspersed repeats of length *n-1* occur in the target, ambiguities will occur in reconstructing the order of single-copy DNA sequences flanking these repeats. Where tandem repeats occur in a target longer than n, the length of these repeats will not be rigorously known. Wherever a word of length n is repeated in the target, information about the number of occurrences may be distorted or lost altogether since quantitative hybridization data is not always very reliable. For all of these reasons, although there are relatively simple ways to sidestep most of these problems, most observers of SBH have viewed it as a method more suitable for comparative or diagnostic DNA sequencing than for *de novo* sequencing. This immediately simplifies the design of the probe array since instead of all 4^n words one can use a customized array that corresponds just to the target in question and the spectrum of possible mutations to be tested for.

A second set of difficulties with SBH arises from the impact of mismatches. A mismatch at the ends of a short DNA duplex is not very destabilizing, because the ends of a helix do not have the same structural constraints to form a regular, ordered stack that central residues do. In fact certain end mismatches including G-T and G-A are as stable as normal Watson-Crick base pairs. For this reason, SBH cannot easily distinguish end mismatches from perfect matches. A related problem is the variation of duplex stability on base composition and even base sequence. No one set of conditions will allow G-C rich and A-T rich sequences to be examined at a stringency where mismatches can be discriminated against effectively. For this reason usually the temperature is varied, and an actual melting curve for each target-probe complex is measured. This increases the complexity of the analysis considerably. Alternative some rather complex software can be used to compensate for known variations in behavior.

The ideal target-chip combination would produce useable data from a large percentage of array elements. However, to accomplish this the target needs to have a DNA complexity (number of base pairs in this case) comparable to the number of base pairs represented by the sum of all the chip elements. While it is possible to do high stringency hybridizations with such DNA complexities [42], one must be very careful to avoid background problems and unwanted crosshybridization artifacts. Thus a number of tricks have been introduced to try to make these massively parallel hybridizations as clean as possible.

SBH has been shown in a few pilot studies to work, in practice. Its use as a large scale DNA diagnostic tool is just beginning. There is room for much further development of SBH technology. For example, one way to reduce the problems caused by end mismatches is to use probes that extend beyond the work length of interest. An example of such a probe is $N_2ATGACCAGN_2$, where the N refers to a mixture of all four bases. This set of probes ensures that any mismatches in the hybridized species will be internal. In DNA diagnostics, if a complete set of all n letter words complementary to the normal target is used, any mutation in the target will lead to mismatches with at least several probes. A deletion will lead to the total absence of hybridization to a subset of probes. An insertion except for tandem duplications will also result in some hybridization mismatches. Thus virtually all mutated targets should be distinguishable, and most will be uniquely analyzable.

3.3.3. Mismatch Scanning

For some applications what is desired is a test that reveals that a mutation has occurred and possibly provides some information about the character of that mutation, but the test is really a high throughput screening method. Its utility is to exclude most targets, or most parts of a complex target from the expensive and laborious process of mutation detection by DNA sequencing. Most current approaches to general mutation detection employ a procedure called mismatch scanning. The goal is to see if any bases in a target are different from those in a standard reference. To accomplish this, samples of the target and reference are mixed at equal concentrations, melted and allowed to reassemble. Statistically, half of the reassembled DNA double strands will be heteroduplexes - they will consist of one strand each of the reference sample and the target sample. If these have any differences in DNA sequences the result will be mismatches at the corresponding places in the DNA duplexes. Strictly speaking, a mismatch is an internal loop formed by two single strands of equal lengths. If insertions or deletions occur in the target the result will be bulge loops in which one strand can form a continuous duplex while the other contains one or more bases looped out from this duplex. We will use the term mismatch here to refer to both sorts of structures.

Mismatches can be detected in a number of different ways. They alter the electrophoretic mobility of DNA [26]. They alter the thermal stability of the duplex [25]. They render the DNA susceptible to various chemical reagents, to proteins whose natural role it is to bind to and eventually repair mismatches, and to enzymes that cleave mismatched strands at or near the point of the mismatch [27]. The ideal detection system for mismatches would be cheap, fast, and detect all possible mismatches. Such a system has not yet been found, but it is clear that mismatch detection is potentially a very attractive strategy. One clear advantage of mismatch scanning is that the result for each DNA target is a yes or no answer. Thus, many targets (or fragments of a complex target) can be analyzed at once, in the same gel electrophoretic lane or in the same tube. If mismatches are very rare, and the tests are definitive enough, sample pooling strategies might be introduced to further increase the throughput.

It should be possible to combine mismatch scanning with some of the other schemes described for mutation detection. As one example of this, we have recently demonstrated that bulge loops can be selectively labeled by first nicking the structures with single-strand specific nucleases and then using one of the nicks as a primer to direct DNA synthesis by nick translation or strand displacement in order to incorporate a label [33]. This approach can be used to score the length of a tandem repeat by hybridization, an analysis not possible by conventional SBH. An array of probes with different possible repeat lengths is constructed, all flanked by the same single-copy DNA sequences. The target is allowed to hybridize to the array. A bulge loop will be formed in every case except where the repeat length of the target matches that of the probe in the array. Labels will be introduced at every point except where the exact matches occur. More elaborate protocols can turn this somewhat undesirable negative scoring assay into a positive scoring assay.

3.3.4. Mass Spectrometry

Mass spectrometry (MS) has had a revolutionary impact on protein sequencing during the past half decade. It is not surprising then that MS has attracted considerable attention as a tool for DNA sequencing or other nucleic acid analyses. We will argue here that MS in the near future is likely to become the paradigm for all high throughput DNA analysis. But the path towards this goal has been long and tortuous. In MS the sample must be brought to a high vacuum and given a net charge. What is then measured in a number of different experimental geometries is the ratio of the mass of the molecules to that charge. Traditional MS instruments used in the analysis of smaller organic molecules use magnetic fields to produce bent orbits that depend on the mass to charge ratio. For high molecular weight species different geometries are more effective. In time of flight (TOF) MS the molecular ions are created at a fixed point, accelerated briefly by a high electric field and then allowed to drift in a straight path until they hit the detector. What is measured is the time elapsed before the detector is reached. In ordinary TOF instruments the entire flight path is straight. Variants of TOF use a reflectron which is an electrical field that stops the moving molecules and forces them to reverse their path either at an angle or colinearly. This can be used to compensate for differences in the original kinetic energy of the molecules before they were electrically accelerated, and the result is a sharpening of the mass resolution. The alternative for high molecular weight molecules is fourier transform ion cyclotron resonance (FT-ICR). Here the molecules travel in a circular path confined by a perpendicular magnetic field. In a close analogy to pulsed nmr methods, the molecules are forced to change orbits by an electromagnetic pulse sequence, and the kinetics of their response to this sequence is measured. The fourier transform of the response kinetics yields the mass to charge ratios of all of the species in orbit. Because TOF is a single shot, destructive experiment, one must usually repeat it numerous times to produce high quality spectra. In contrast, FT-ICR is a non-destructive method, and the amount of time used to observe a single orbit change can be varied to alter the quality of the spectrum more or less at will.

MS is not a commonly used DNA analytical tool today because DNA molecules do not like to fly into high vacuum. DNA is a highly charged species that clings tenaciously to most surfaces. When forced into a vacuum by most methods, DNA is left with so much kinetic energy that it usually fragments into a complex set of species that defies further meaningful analysis. The charges on DNA are largely equalized by bound counterions, and in typical biological solutions a mixture of counter ion types is present. Thus the mass of this starting species is really a complex mixture of masses, as are all the products that derive from it by fragmentation. Given these awesome obstacles, early schemes for using MS in DNA sequencing avoided the problem of macromolecular MS entirely and instead resorted to the considerable power and sensitivity of MS as a multiplex, i.e. a multimass, detector. Thus, schemes were developed to label DNA with four different sulfur stable isotopes instead of four different fluorophores. Conventional or capillary gel electrophoresis would be run, and the eluted samples burnt and sulfur detected by MS as SO_2. A variant on this scheme used metal chelate labels with different stable masses. In principle more than 50 different stable metal isotope masses could be worked with simultaneously, allowing, in principle, more than a dozen DNA samples to be sequenced in the same gel lane [28]. Such approaches may provide an incremental improvement in DNA analysis but they still rely on the rate limiting step of gel electrophoresis.

New methods of introducing macromolecules into vacuum have greatly enhanced the power of MS to analyze nucleic acid samples. In electrospray a thin stream of fluid droplets containing sample solution is introduced directly into high vacuum in the presence of an electrical field which leaves a charge on each droplet as it forms. The solvent evaporates, and the macromolecules are left with whatever charge the droplet had. The method is very gentle, but it produces molecules with a broad range of charges, and thus the resulting mass spectra are very complex. In matrix assisted laser desorption ionization (MALDI) the analyte molecules are mixed with a partially aqueous solution of a photoactivatable organic acid-like hydroxypicolinic acid. The solutes crystallize as the solvent evaporates, and it is some unknown features of the cocrystals that are required for subsequent efficient charging. The dried sample is placed into the high vacuum chamber of a TOF instrument or the equivalent, and a laser pulse is applied to the sample. This vaporizes the organic acid which carries aloft the macromolecule in a plume, and collisions with matrix components act somehow to transfer electrons to the DNA to produce, usually, singly-charged species. While the physics of MALDI remains relatively obscure, its utility as a sample injection method for DNA and RNA in both TOF and FT-ICR is truly impressive. At present relatively high resolution TOF spectra (up to 1 part in a thousand) with mass accuracies approaching one Dalton can be obtained routinely with oligonucleotides up to around 80 bases in length, providing that extensive sample purification and washing methods are used to produce a homogeneous composition of counterions. The largest DNA's seen by MALDI-TOF MS are around 500 bases, but these do not have high enough resolution to allow DNA sequencing [29]. With electrospray, molecules up to 80 bases have been analyzed at high resolution, and much larger species have been seen and masses measured accurately [30]. In general, with either technique, the process of vaporization leads to duplex denaturation so, usually, only single-stranded species are seen. In FT-ICR, resolutions of DNA and RNA samples of between a part in 10^5 and a part in 10^6 have been observed. At such resolutions a complicating factor is naturally occurring stable isotopes like ^{13}C, so that each molecular species actually displays itself as a complex pattern of isotopic isomers differing by 1 Dalton in mass and representing different numbers of ^{13}C present.

The analytical power of MS is awesome, with resolutions far exceeding other forms of spectroscopy typically applied to macromolecules. Furthermore, the speed of MS is impressive. In favorable cases high quality TOF spectra can be obtained in milleseconds while FT-ICR measurements may take just a few seconds. The key challenge is to optimize the implementation of these methods in DNA analysis. Several different approaches are being explored. The potentially most powerful approach for DNA sequencing is post source decay. Here collisions with gas molecules or laser irradiation are allowed to fragment the molecular ions after they have been vaporized. In close analogy to traditional MS, the resulting spectrum of fragmentation products is analyzed in an attempt to reconstruct the original sequence. Indeed this can be done, and several successful examples have been reported [31], but the difficulty of the spectral analysis is so great that this approach may not ever achieve the desired high throughput rates. A less ambitious, but at least, today, more practical approach, is to fragment the DNA before vaporization. This can be done by standard Sanger ladder approaches. Then what is analyzed in the MS is all of the same fragments that would normally be analyzed by electrophoresis. In such a scheme the goal is to avoid, as much as possible, any fragmentation except that generated by the original Sanger ladder. Such schemes work well today with DNA samples in the range of 40 to 80 bases. It is not actually necessary to analyze the four different termination reactions separately. The mass differences among each of the four nucleotides are distinctive. The smallest difference is an A versus a T which is 9 Daltons. Thus the four ladder products can be mixed together and analyzed all at once - the difference in mass between each adjacent fragment gives the identity of the 3' terminal base of the larger fragment.

Because of the resolving power of MS, it is not necessary to use labels and not necessary to use traditional Sanger ladder chemistries. The mass itself is a distinctive label. Thus, exonucleases or endonucleases can be used to generate the fragments needed for MS analysis. In essence, this mode of application of MS is essentially electrophoresis in the vapor phase. What one gains is speed, because there is no friction, no heat generation and, thus, no cooling, so high fields can be used; there are also no artifacts due to single strand secondary structure, because this does not effect the mass. The resolution is also much higher. Ordinary gel electrophoresis has a resolution of about one base. TOF, with a 6000 Da twenty base target, has a resolution of around 6 to 10 Da, far exceeding the 300 Da average mass difference corresponding to a single nucleotide.

The high resolution of MS also encourages its use in other forms of DNA analysis. For example, when used as a hybridization detector MS offers the advantage of working with many probes simultaneously. The number of different base compositions of probes of length n are $(n+3)!/n!3!$. For n in the range of 8 to 15, most of these will have distinguishable masses, at the highest resolutions currently achievable. Thus, by varying both the base composition and probe length a very large number of probes with different masses can be created, all of which can be used simultaneously and distinguished uniquely by MS. Thus MS seems ideally suited to serve as an SBH detector. A particularly attractive feature of MS as a hybridization detector is that in the MALDI process the target to be detected is automatically removed from the probe that captured it.

MS also should serve very well as a detector of DNA lengths in other forms of analysis including the ligase chain reaction and short PCR reactions. Here, the mass is such a distinctive signature that artifacts are easily distinguished from the desired true amplification bands. Perhaps the ultimate potential use of MS is in comparative DNA analysis by restriction digestion, Here, one can cleave the target into a large set of very small restriction fragments by combining the activities of several restriction enzymes with frequently occurring cleavage sites. One sees separately the masses of both strands of each resulting fragment. With fragments in the 20 to 60 base range, any change in mass in any of the fragments is potentially detectable. All mutations will lead to mass changes except for extraordinarily unlikely events like a double mutation within one fragment which changes an A to a C at once position and a C to an A at another. Thus

MS will allow a large target to be scanned all at once for the presence of mutations. If any are found one will know the approximate location from the total mass of the fragment, and the nature of the mutation (by the mass change), but one will not know where on the target fragment the mutation has occurred. MS of DNA is still in its infancy, but it promises to become a major if not the dominant force in the way nucleic acids are handled in the future.

3.3.5. Combined approaches and indexing

We have already shown one example of how different DNA analysis approaches can be combined for increased power. Here we present other examples. In each case we view MS as the ultimate analytical tool for these analyses, but they can also be carried out with conventional fluorescent, chemiluminescent or radioisotopic detection. First we note that MS and mismatch scanning are complementary methods for mutation detection. The former tells what mutation has occurred but not where. The latter tells where the mutation is but not what it is. When used together all of the necessary information about the sample will be obtained.

The general theme we explore in this section is enzyme-enhanced DNA detection. A number of the schemes for doing this are illustrated in Figure 1. There are several advantages to using enzymes to supplement DNA hybridization. First, enzymes have evolved to work with a range of DNA base compositions and sequences at relatively constant physiological conditions. Thus, they obviate the need, described earlier, to vary conditions to achieve desirable hybridization stringencies across a wide range of DNA samples. Second, enzymes have, in general, more discrimination against mismatches, particularly end mismatches, than can be obtained in pure physical hybridization. Third, enzymes can be used to reduce background from unwanted adsorption in solid phase assays like hybridization or SBH. If a labeled macromolecular target or probe is used, its will generally be physically quite sticky. With enzymatic steps, frequently the label can be introduced as a single pppdN with much less stickiness. Fourth, the use of an enzyme can provide additional DNA sequence information beyond that already given by the hybridization reaction itself.

Solid phase hybridization is shown in Figure 1a, just as a reference. A simple form of enzyme-assisted hybridization is shown in Figure 1b. This approach, described earlier, uses DNA polymerase to add a single, labeled base by extending the immobilized probe (primer) on the hybridized target (template). With four different colored terminators one can read an additional base of sequence. More complex schemes for lengthening each individual sequence read can be imagined. In Figure 1c, an alternative approach is shown using DNA ligase. Here a duplex probe with a single-stranded overhang is used to capture a target by its end. The fidelity of the capture is ensured by ligation [31]. More complex schemes have been developed, and they look promising. For example, ligation and polymerase extension can be combined as shown in Figure 1d in a detection scheme we have named positional sequencing by hybridization. This format can be used as a high throughput solid state method for generating Sanger sequencing ladders. Targets can be captured from a mixture, extended by DNA polymerase in the normal Sanger way, and then subjected to DNA sequence analysis either by conventional fluorescent gel readers or MALDI-TOF MS [35, 36]. It turns out, with such schemes, that a ligation step is unnecessary. Because only a five or four base overhang is needed, only 1024 or 256 different primer sequences, respectively, are needed to deal with all possible targets. In a variant on this scheme, using a method developed independently for a totally different purpose and called indexing [34], double-stranded DNA fragments are generated from a target by a type IIS restriction nuclease which cuts outside of its recognition sequence. This gives fragments with a mixture of many different types of ends. Most of these can be uniquely captured by ligation and used directly without purification (Figure 1e) to generate DNA sequencing data [35, 36]. The advantage here is that the sequencing can be carried out directly on the double-stranded captured target by nick translation or strand displacement. This avoids the often time consuming step of single-stranded DNA preparation, and it bypasses the extremely undesirable stickiness seen with single-stranded DNA under most conditions.

More complex schemes are being tested. For example, Figure 1f shows a scheme in which, after a target is captured by hybridization with a probe array, it is subjected to a further hybridization and ligation reaction to proofread the original detection and provide additional sequence information. In such schemes and others what is really being done is to use to power of the probe array as a high resolution sequence dependent DNA purification step, whose power far exceeds the gel electrophoretic length fractionations that so dominate current DNA analytical methods.

4. High Throughput DNA Screening Methods

The methods described above for high throughput DNA sequencing can also be used intact or with variations for high throughput DNA screening. As described earlier in this article, such screening methods are needed both for monitoring gene expression and for environmental or ecological surveillance. Certain approaches must be modified if quantitative abundance information is required.

An array of samples or probes serves as a multi-element detector to display, simultaneously the components of a complex mixture of DNA targets. To obtain quantitative abundance information one can carry out a competitive hybridization experiment. Such approaches were first developed using as a detector a metaphase chromosome spread. In comparative genome hybridization (CGH) a chromosome spread from one cell type is used in simultaneous hybridization with a mixture of two different colored target samples [38]. One color represents labeled genomic DNA from the same cell type. The other is genomic DNA from the sample to be compared. The ratio of the two colors seen gives the relative abundance of each chromosomal region in the two targets. The technique is very powerful for spotting small chromosomal differences such as regional amplifications seen in many cancer cells. Note that, because a differential measurement is involved, one can drive the hybridization to completion and, thus, achieve high detection efficiencies without distorting the quantitative information about the samples.

An analog of CGH allows differential gene expression to be measured. Here an array of cDNAs (or probes) is used that presents all genes to be tested. Differentially labeled cDNAs are made from the target and a reference sample, mixed, and allowed to hybridize to the array competitively [39]. Again, as in CGH, the differential nature of the hybridization allows the reaction to be pushed to completion. If the probe is in excess it will dominate the kinetics of the hybridization so that all species, rare, and abundant will hybridize at the same rate and produce a labeled array with a relatively narrow dynamic range. However the quantitative comparison between the two samples will still be maintained as the color ratios. A potential limitation in differential hybridization analysis of cDNA abundance is that the PCR steps typically used to make the samples may lead to some quantitative distortion, particularly if there are things like differentially spliced transcripts with different secondary structure potentials or lengths. Never the less this approach looks very powerful indeed.

Perhaps the ultimate approach to differential analysis suggested to date is serially analyzed gene expression (SAGE) or variants on this original proposal [40].

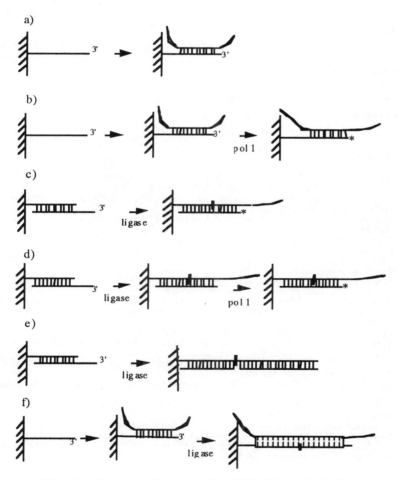

Figure 1. Schemes for solid-state detection of DNA (See text for details)

The basic idea behind SAGE is to reduce a complex mixture of DNA samples, i.e. a whole cell cDNA preparation, to a set of DNA fragments of identical length. Each contains typically 9 to 13 bases of DNA sequence characteristic of one component of the original mixture. In the original SAGE proposal the fragments are ligated together in a complex protocol to make random concatemers which are then amplified and sequenced. In this way one should be able to preserve most of the quantitative features of the original sample mixture without the usual PCR distortions. The resulting concatemers are then subjected to conventional DNA sequencing. In most cases 10 to 13 bases will suffice to uniquely identify the particular gene from which the cDNA derived. Hence, far less sequencing needs to be done to analyze a representative sample of the original mixture of gene products. In thinking about the advantages and implications of SAGE, it is obvious that the general approach is especially well adapted to MS analysis, since the desired target fragments are in a range where MS today is extraordinarily fast and accurate. A number of strategies look appealing, and it should not be hard to come up with one that combines some of the approaches described earlier and allows an entire SAGE analysis to be carried out with a single array of captured short DNA fragments.

5. Acknowledgements

This work was supported in part by grant number DE-FG02-93ER61609 from the U.S. Department of Energy.

6. References

1. Hugenholtz, P., and Pace, N.R. (1996) Identifying microbial diversity in the natural environment: a molecular phylogenetic approach, *TIBTECH* **14**, 190-197.
2. Sanger, F., Nicklen, S., and Coulson, A.R. (1977) DNA sequencing with chain-terminating inhibitors, *Proc. Nat. Acad. Sci. USA* **74**, 5463-5467.
3. Maxam, A.M., and Gilbert, W. (1977) A new method for sequencing DNA, *Proc. Nat. Acad. Sci. USA* **74**, 560-564.
4. Wu, R., and Taylor, E. (1971) Nucleotide sequence analysis of DNA. II. Complete nucleotide sequence of the cohesive ends of bacteriophage lambda DNA, *J. Mol. Bio.* **57**, 491-511.
5. Cystic Fibrosis Genetic Analysis Consortium: Cystic fibrosis mutation database: http://www.genet.sickkids.on.ca/cftr
6. Sverdlov, E.D., and Ermolaeva, O.D. (1993) Subtractive hybridization: Theoretical Analysis and the principle of trapping, *Bioorganic Chem.* **19**, 1081-1088.
7. Ermolaeva, O. D, and Sverdlov, E.D. (1996) Subtractive Hybridization: A Technique for Extraction of DNA Sequences Distinguishing Two Closely Related Genomes, *Genet. Anal. (Biomol. Eng.)* [in press].
8. Liang, P., Bauer, D., Averboukh, L., Warthoe, P., Rohrwild, M., Muller, H., Strauss, M., and Pardee, A. B. (1995) Analysis of altered gene expression by differential display, *Methods in Enzymology* **254**, 5763-5764.
9. Zhao, S., Ooi, S.L., and Pardee, A.B. (1995) New primer strategy improves precision of differential display, *Biotechniques* **18**, 842-846.
10. Beroud C; Verdier F; Soussi T. (1996) p53 gene mutation: software and database, *Nuc. Acids Res.* **24**, 147-150.
11. Collins F; Galas D. (1993) A new five-year plan for the U.S. Human Genome Project, *Science* **262**, 43-6.
12. Yeast genome database: ftp://genome-ftp.Stanford.edu/yeast/goneme_seq/
13. Fleischmann, R.D., Adams, M.D., White, O., Clayton, R.A., Kirkness, E.F., Kerlavage, A.R., Bult, C.J., Tomb, J.F., Dougherty, B.A., Merrick, J.M., *et al.*(1995) Whole-genome random sequencing and assembly of *Haemophilus influenzae* Rd, *Science* **269**, 496-512.

14.   Fraser, C.M., Gocayne, J.D., White, O., Adams, M.D., Clayton, R.A., Fleischmann, R.D., Bult, C.J., Kerlavage, A.R., Sutton, G., Kelley, J.M., *et al.* (1995)The minimal gene complement of *Mycoplasma genitalium, Science* **270**, 397-403.

15.   Kotler, L. E., Zevin-Sonkin, D., Sobolev, I. A., Beskin, A. D., and Ulanovsky, L. E. (1993) DNA sequencing: modular primers assembled from a library of hexamers or pentamers, *Proc. Natl. Acad. Sci. U. S. A.* **90**, 4241-4245.

16.   Szybalski, W. (1990) Proposal for sequencing DNA using ligation of hexamers to generate sequential elongation primers, *Gene* **90**: 177-178.

17.   Kieleczawa, J., Dunn, J. J., and Studier, F. W. (1992) DNA sequencing by primer walking with strings of contiguous hexamers, *Science* **258**, 1787-1791.

18.   Venter, J. C., Smith, H.O., and Hood, L. (1996) A new strategy for genome sequencing, *Nature* **381**, 364-366.

19.   Woolley, A. T. and Mathies R. A. (1994) Ultra-high-speed DNA fragment separation using microfabricated capillary array electrophoresis chips, *Proc. Natl. Acad. Sci. U. S. A.* **91**, 11348-11352.

20.   Cantor, C.R., Tinoco Jr., I., and Peller, L. 1964. Exoenzyme Kinetics with Applications to the Determinations of Nucleotide Sequences. Biopolymers 2: 51.

21.   Southern, E.M., Maskos, U., and Elder J.K. (1992) Analyzing and comparing nucleic acid sequences by hybridization to arrays of oligonucleotides: Evaluation using experimental models, *Genomics* **13**, 1008-1017.

22.   Drmanac, R., Labat, I., Brukner, I., and Crkvenjakov, R. (1989) Sequencing of megabase plus DNA by hybridization: theory of the method,*Genomics* 4, 114-128.

23.   Bains, W., and Smith, G.C. (1988) A novel method for nucleic acid sequence determination,*J.Theor. Biol.* **135**, 303-307.

24.   Khrapko, K. R., Lysov, Y. P., Khorlin, A. A., Ivanov, I. B., Yershov, G. M., Vasilenko, S. K., Florentiev, V. L. & Mirzabekov, A. D. (1991) A method for DNA sequencing by hybridization with oligonucleotide matrix, *J. DNA Sequencing Mapping* 1, 375-388.

25.   Abrams, E.S., Murdaugh, S.E., Lerman, L.S., (1995) Intramolecular DNA melting between stable helical segments: melting theory and metastable states, *Nucl. Acids. Res.* **23**, 2775-83.

26.   Soto, D., and Sukamar, S. (1992) Improved detection of mutations in the p53 gene in human tumors as single-stranded conformation polymorphs and double-stranded heteroduplex DNA, *PCR Methods and Applications* **2**, 96-98.

27.   Youil, R., Kemper, B.W., and Cotton, R.G.H. (1995) Screening for mutations by enzyme mismatch cleavage with T4 endonuclease VII, *Proc. Nat. Acad. Sci. USA* **92**, 87-91.

28.   Jacobson, K.B., Arlinghaus, H.F., Schmitt, H.W., Sachleben, R.A., Brown, G.M., Thonnard, N., Sloop, F.V., Foote, R.S., Larimer, F.W., Woychik, R.P., *et al.*(1991) An approach to the use of stable isotopes for DNA sequencing, *Genomics* **9**, 51-59.

29.   Tang, K., Taranenko, N.L., Allman, S.L., Chang, L.Y., and Chen, C.H. (1994) Detection of 500-nucleotide DNA by laser desorption mass spectrometry,*Rapid Commun. Mass spectrom.* **8**, 727-730.

30.   Wunschel, D.S., Fox, K.F., Fox, A., Bruce, J.E., Muddiman, D.C., and Smith, R.D. (1996) Analysis of double-stranded polymerase chain reaction products from the Bacillus Cereus group by electrospray ionization Fourier transform ion cyclotron resonance mass spectrometry, *Rapid Comm. Mass. Spec.* **10**, 29-35.

31.   Little, D.P., Thannhauser, T.W., McLafferty, F.W. (1995) Verification of 50- to 100-mer DNA and RNA sequences with high-resolution mass spectrometry, *Proc. Natl. Acad. Sci. USA* **92**, 2318-22.

32.   Jurinke, C, Zöllner, B., Feucht, H.-H., Kirchhübel, J., Lüchow, A., van den Boom, D., Laufs, R. and Köster, H. (1996) Detection of hepatitis B virus DNA in serum samples via nested PCR and MALDI-TOF mass spectrometry, *Genet. Anal. (Biomol. Eng.)* [in press]

33.   Yaar, R., Szafranski, P., Cantor, C.R., and Smith, C.L. (1996)*In situ* detection of tandem repeat length: differential enzymatic labeling, *Genet. Anal. (Biomol.Eng.)* [submitted].

34.   Unrau, P., and Deugau, K.V. (1994) Non-cloning amplification of specific DNA fragments from whole genomic DNA digests using DNA 'indexers,' *Gene* **145**, 163-169.

35.   Fu, D.J., Broude, N.E., Köster, H, Smith, C.L., and Cantor C.R. (1995) A DNA sequencing strategy which requires only five bases of known terminal sequence for priming, *Proc. Nat. Acad. Sci. USA* **92**, 10162-10166.

36.   Fu, D. J., Broude, N.E., Köster, H., Smith, C.L., and Cantor, C.R. (1995) Efficient preparation of short DNA sequence ladders potentially suitable for MALDI-TOF sequencing, *Genet. Anal. (Biomol. Eng.)* **12**: 137-142.

37.   Broude, N.E., Sano, T., Smith, C.L., and Cantor, C.R. (1994) Enhanced DNA sequencing by hybridization, *Proc. Nat. Acad. Sci. USA* **91**, 3072-3076.

38.   Kallioniemi, A., Kallioniemi, O.P., Sudar, D., Rutovitz, D., Gray, J.W., Waldman, F., and Pinkel, D. (1992) Comparative genomic hybridization for molecular cytogenetic analysis of solid tumors, *Science* **258**, 818-821.

39.   Schena, M., Shalon, D., Davis, R.W., Brown, P.O., (1995) Quantitative monitoring of gene expression patterns with a complementary DNA microarray, *Science* **270**, 467-470.

40.   Veculescu, V.E., Zhang, L., Vogelstein, B., and Kinzler, K.W. (1995) Serial Analysis of Gene Expression, *Science* **270**, 484-487.

41.	Harding, J.D., Keller, R.A. (1992) Single-molecule detection as an approach to rapid DNA sequencing, *TIBTECH* **10**, 55-57.
42.	Grothues, D., Cantor, C.R., and Smith, C.L. (1994) Top-down construction of an ordered Schizosaccharomyces pombe cosmid library, *Proc. Nat. Acad. Sci. USA* **91**, 4461-4465.

LARGE-SCALE CHROMATIN STRUCTURE

A.S. BELMONT
Department of Cell and Structural Biology
University of Illinois, Urbana-Champaign
B107 Chemical and Life Sciences Building
601 S. Goodwin Ave.
Urbana, IL 61801 USA

1. Introduction

Analysis of cellular ultrastructure currently is seriously handicapped by the lack of structural tools capable of visualizing macromolecular assemblies and organelles which extend over large subcellular volumes. This problem is particularly severe with regard to structural dissection of chromosome architecture in which a hierarchy of folding motifs give rise to linear compaction ratios tens of thousands to one for mitotic chromosomes and hundreds or thousands to one for interphase chromatids. Whereas analysis of the highest levels of chromosome folding requires tracing individual chromatids over at least several um distances, analysis of the lowest levels of chromatin folding requires a resolution of at least several nm. In practice, because of the extremely high DNA packing ratios, these different levels of chromosome folding are not easily isolated for separate, sequential structural determination. This is particularly true for mitotic chromosomes, in which the several nm spatial separation between larger units of chromatin folding is comparable to the spacing between lower levels of chromatin folding contained within these larger domains. Therefore direct identification within native chromosomes of structural motifs underlying the highest level of chromosome folding may first require the unambiguous tracing of the lower levels of chromatin folding.

This limitation of available structural tools for investigation of nuclear ultrastructure has been complicated by previous difficulties in staining DNA specifically for transmission electron microscopy (TEM). An added problem is the difficulty associated with localizing specific DNA sequences or chromosome regions without perturbing higher order chromosome ultrastructure. The only general methodology for this purpose, in situ hybridization, requires denaturation of DNA under relatively harsh conditions.

In this chapter we will briefly review previous structural investigations of large-scale chromatin structure which have largely focused on the structure of isolated mitotic chromosomes, for which the 3-dimensional range can be reduced by spreading chromosomes on a 2-dimensional surface and for which selective DNA stains are not required. We will then outline the general experimental approach our laboratory has adopted which instead is focused on identifying intermediates in the pathway of chromosome condensation or decondensation by examining changes in interphase

C. Nicolini (ed.), Genome Structure and Function, 261–278.
© 1997 *Kluwer Academic Publishers. Printed in the Netherlands.*

chromosome structure during cell cycle progression. We will summarize results obtained using this approach in which we have identified large-scale chromatin fibers, roughly 0.1 μm in diameter, which can be traced over several micron distances within interphase nuclei and which appear to irregularly coil and uncoil during chromosome condensation and decondensation. Finally we will describe our development of a new method for in situ localization of specific chromosomal loci, based on lac operator / repressor recognition, which allows direct visualization of these large-scale chromatin fibers in living cells. Collectively, our results have led to a "chromonema" model of mitotic and interphase chromosome structure in which the irregular folding, kinking, and supercoiling of these large-scale chromatin, or chromonema, fibers gives rise to chromosome condensation.

2. Models of Mitotic Chromosome Structure

As introduced above, most investigations of higher order chromosome structure have focused predominately on mitotic chromosomes, and in particular metaphase chromosomes. Paradoxically it is during metaphase that chromosomes are maximally condensed and reveal the least substructure. Generally speaking, most investigations in the past have dealt with this structural complexity by choosing approaches aimed at reducing this complexity in some way. In practice, this has meant focusing on particular features of chromosome structure, such as surface topology [1-6] or recognized size regularities as seen in certain views [7-10], or employing various extraction methods designed to progressively unfold the chromosome [3, 9, 11-21].

Early electron microscopic investigations by DuPraw suggested a complex and irregular association of uniformly sized, 30 nm diameter fibers within the metaphase chromosome, leading to a folded fiber model of chromosome structure [22]. In contrast, a combination of light microscopy observations, scanning and transmission electron microscopy, and methods for uncoiling isolated metaphase chromosomes, led instead to a model in which the last stage of chromosome condensation corresponded to a helical coiling of a underlying structure of ~0.2-0.4 μm diameter. This was inferred from visualization of surface features of metaphase chromosomes from SEM [2, 16, 23], by visualization of apparent uncoiled gyres by light microscopy after treatment of metaphase chromosomes with specific buffers [2, 11, 14, 16, 17, 23], by visualizing chromosomes in specific cell lines which appeared to be defective in final stages of chromosome condensation [23], and by banding patterns seen in TEM thin sections and suggestions of gyres in whole mount metaphase chromosomes viewed as stereo-pairs [7] .

A very different conceptual model of mitotic chromosome structure emerged from experiments employing extraction methods which removed the majority of chromosomal proteins. Earlier work in prokaryotic cells (and with bacteriophage) had demonstrated extraction conditions which led to the removal of most DNA associated proteins but which retained topologically restrained DNA domains [24, 25]; examination by transmission electron microscopy (TEM) revealed loops of DNA extending from the remnants of the cell membrane [26]. Application of comparable methods to eukaryotic nuclei by Worcel [27] led to similar observations of negatively supercoiled DNA domains after removal of histones and most nonhistone proteins. Likewise, by applying comparable extraction conditions to isolated metaphase chromosomes and directly visualizing the extracted chromosomes using electron microscopy, Paulson and Laemmli obtained striking images of DNA apparently looping out from a residual chromosome "scaffold" structure [19]. In these images the DNA was too dense to actually trace the path of an individual strand. However, by using alternative DNA spreading techniques for TEM visualization, partially spread chromosomes were obtained which showed a lower density of DNA loops protruding from a thickened, higher density chromosome "scaffold"; in these preparations, individual loops could be traced as emanating from closely spaced regions of the residual chromosome structure [19].

These results led to the formulation of the radial loop model in which large loops of DNA, ~50-100 kbp in mammalian chromosomes, are organized radially around a chromosome scaffold organized by nonhistone proteins. Subsequent TEM visualization of isolated metaphase chromosomes, with and without histone H1 extraction, gave results consistent with the radial loop model; specifically what appeared to be loops of chromatin fibers either 30 nm in width, or 10 nm after histone H1 extraction, were seen protruding from a dense, closely packed chromosome core in which fibers could not be traced [3, 19, 21]. An attractive aspect of the radial loop model was its suggestion for a mechanism underlying chromosome condensation based on self-assembly of a chromosome scaffolding and anchoring chromosomal loops. Later experiments revealed just two major components of the residual chromosome scaffolding [28] as well as specific DNA sequences (SAR/MAR) associated with both chromosomal and nuclear scaffold preparations [29, 30]. These two major protein components of the chromosome scaffold have been identified. SCI, or scaffold associated protein 1, was demonstrated as being topoisomerase II, while SCII, recently identified in several organisms as a member of a protein family represented in both prokaryotes and eukaryotes, also has features suggestive of enzymatic or motor activities [31-33]. These features include a NTP binding site at one end of the molecule separated by a long coiled coil domain from another globular region with a "DA" sequence motif which may represent a second NTP binding site [32]. Preliminary data suggests the possibility that SCII interacts with SCI, topoisomerase 2 [32].

For several reasons, however, the radial loop model of chromosome structure has remained controversial. A key experimental concern has been the degree to which artifactual protein cross-linking during the extraction procedure influences the final scaffold structure. In fact, isolation of chromosome and nuclear scaffolds is critically dependent on "stabilization" steps involving incubations at 37 degrees C in the presence of calcium ions, or incubation with Cu^{++}, and elimination of reducing agents from the extraction buffers [28, 34]. Similarly, although specific DNA sequences are retained on the scaffold after extraction and then restriction digest, apparently these scaffold associated regions are not resistant to a second round of extraction. This raises obvious questions regarding the mechanisms underlying the coisolation of these DNA fragments with the nuclear or chromosome scaffold, as discussed elsewhere [30].

Additional experiments as well raise questions regarding assumptions or predictions of the radial loop model. Recent in vitro studies of chromosome condensation using Xenopus extracts have shown that topoisomerase 2, a major scaffold protein, can be easily extracted in physiological buffers with no obvious effect on mitotic chromosome structure [35]. Previous ultrastructural localization of topoisomerase 2 suggested a diffuse distribution throughout the chromosome without an obvious central core of staining [36]. In vivo visualization of topoisomerase 2 distribution in early Drosophila embryos by light microscopy also did not reveal a central core of staining [37]; moreover, levels of chromosome associated topoisomerase 2 dropped sharply from a maximum in prophase to a minimum in anaphase without significant changes in chromosome condensation. To date although both topoisomerase 2 and SCII proteins have been demonstrated as required for normal mitotic chromosome condensation [33, 38, 39], distinguishing between an enzymatic role in this process from a structural or combined enzymatic and structural role has not been possible. Finally, electron microscopy (EM) tomography of embryonic Drosophila chromosomes [40] failed to reveal an obvious, radial looping substructure, as did similar studies on isolated mammalian chromosomes [41].

More recently, new evidence [23, 42] supporting the last stage of metaphase chromosome compaction as corresponding to a coiling of a prophase-like chromatid has revived helical models of chromosome organization. Based on these experiments, the radial loop model has been combined with the helical model to produce a radial loop-helical model of mitotic chromosome organization in which radial loops are now postulated as being organized around the core of a prophase-like chromatid which then coils to form the final metaphase chromosome [23, 42]. This requires a reduction of the predicted length of the DNA loops by a factor of 2, which may contradict earlier estimates of DNA loop sizes [19].

3. Transitions into and out of Mitosis and Interphase Chromosome Structure

Conceptually, DNA compaction during mitosis might occur by either of two mechanisms. First, it may occur during late G2-prophase through a pathway essentially independent of hierarchical chromatin folding patterns; in this case mitotic chromosome condensation would occur through similar pathways in different types of cells, for instance embryonic versus somatic, which might have very different higher order chromatin structure. In fact, comparison of early embryonic versus somatic Drosophila chromosomes suggested significant differences in large-scale chromatin folding [40], and recently experiments using Xenopus extracts for in vitro chromosome condensation have revealed chromosome condensation is not dependent on the presence of histone H1 [43]. Second, it might occur through successive levels of chromatin compaction in a hierarchical model in which the final mitotic chromosome condensation represents the highest levels of chromatin compaction. In both cases, late interphase G2 and early prophase nuclei and late telophase and early G1 nuclei should contain intermediates of mitotic chromosome condensation which should be highly informative with regard to models of chromosome organization.

Cytological examination of stages of mitosis over the last hundred years have in fact indicated a continual process of condensation during early stages of mitosis. Although most of these studies have been conducted at the light microscopy level with limiting resolution, the general impression is that there is a progressive transition from late G2 to early prophase nuclei in which a diffuse "thready" appearance at the earliest stages of condensation then progresses to yield a progressive thickening of the condensing chromosome axis (see for example, cover illustration, Mol. Biol. Cell, vol. 6, 1995). Analysis of premature chromosome condensation (PCC) suggests a similar transition of progressive increases in chromosome condensation but in this case the continuum of condensation is observed through much of interphase [44].

An additional problem of radial loop models of chromosome organization is therefore reconciling this apparently continual process of chromosome condensation with the idea of fixed DNA loop sizes. Analysis of SAR/MAR sites suggests that sequences associated with the metaphase scaffold are also associated with the nuclear scaffold [29]; this observation has led to the suggestion that loops are maintained during interphase and do not change in size. While it is easy to envision chromosome condensation as occurring through a process which organizes the base of the DNA loops with a linear chromosome scaffold, it is much harder to understand how such loops of fixed size can be arranged to give a progression of linear chromatid structures of increasing diameter.

Unfortunately, previous direct visualization of chromatin organization within interphase nuclei has not provided significant insight into large-scale chromatin organization and intermediates of chromosome condensation.

Traditional approaches to analysis of interphase DNA distribution transmission electron microscopy (TEM) have led to the textbook picture of heterochromatin and euchromatin, where heterochromatin is interpreted as the electron dense masses dispersed throughout the nucleus but concentrated adjacent to the nuclear envelope and nucleolus, and euchromatin is interpreted as the fine fibrillar material interspersed between the heterochromatin, nucleoli, and interchromatin granule clusters. This picture of interphase DNA distribution is problematic because it does not connect readily with models of chromatin organization or chromosome structure. For instance, the underlying fibrillar nature of chromatin is not apparent, with neither 30 nor 10 nm chromatin fibers easily visualized. Moreover, this image of isolated masses of heterochromatin dispersed in a sea of euchromatin does not connect readily with the linearity of interphase chromosomes or with our knowledge of the compaction of interphase chromosomes within chromosome domains, as defined by chromosome painting using in situ hybridization [45-47].

Previous studies in the literature, however, seriously question this textbook image of interphase chromosome organization. It is important to keep in mind that this textbook interpretation is based on traditional TEM staining using uranyl and lead salts which are known to be nonspecific for DNA or even for nucleic acids. In experiments performed by Skaer and Whytock [48, 49], polytene cells from Drosophila melanogaster salivary glands or rat liver were fixed using different combinations of formaldehyde, glutaraldehyde, and acrolein. All of these aldehyde cross-linking agents caused the normally clear nucleoplasm to become optically opaque, indicating aggregation; moreover the appearance and characteristic sizes of fibrillar-granular features in the "euchromatin" compartment of the nuclei varied according to the aldehyde combination used for fixation. Centrifugation of chromosomes to the bottom of nuclei prior to fixation did not change the fibrillar-granular appearance of the nucleoplasm, demonstrating that decondensed chromatin was not a major component of these nucleoplasmic features.

Similarly, correlative light and electron microscopy revealed that the DNA distribution within semi-thick, conventional Epon sections of CHO nuclei, as revealed by light microscopy DAPI staining, showed little correlation with the "euchromatin" compartment revealed by traditional TEM staining [50]; most DAPI staining appeared confined to concentrated spots or apparent fiber segments, as well as to the "heterochromatic" regions identified by TEM staining.

4. Alternative Approach to Analysis of Large-scale Chromatin Organization

In the Introduction above, we have outlined the technical difficulties associated with investigation into mitotic and interphase chromosome organization. To address these problems, we have focused on an alternative experimental approach combining three themes. First, to avoid questions of potential artifacts inherent especially in approaches relying on physical unraveling of chromosome structure through removal of chromosomal proteins, we are emphasizing preservation of in vivo chromosome structure. Second, to deal with the high density and complexity of chromosome structure, we are emphasizing the analysis of chromosome folding intermediates during the physiological transition into and out of mitosis, rather than direct analysis of metaphase chromosomes. More recently, this is being combined with analysis of large-scale chromatin decondensation associated with initiation of DNA replication or transcriptional activation. Third, even with our emphasis on interphase chromosome structure, the observed structures are still highly complex and densely packed. Therefore we are applying three-dimensional reconstruction and visualization techniques with the goal of computationally "unraveling" the chromosome structure as a means of recognizing the underlying folding motifs giving rise to this level of chromosome architecture.

Our work to date has represented cycles of technique development and application. Most of our results so far have been obtained through a first pass through such a cycle in which we have examined cell-cycle modulated changes in bulk large-scale chromatin organization. More recently, we have developed new technology for visualizing specific chromosomal regions and for carrying out EM tomography and serial section reconstructions. A second cycle is therefore just beginning in which we will apply this type of approach to the analysis of specific chromosomal segments and specific genes.

5. Preservation of in vivo Chromatin Packing Allows Visualization of Large-scale Chromatin Domains

Conventional TEM staining methods using uranyl and lead salts are not specific for DNA, let alone nucleic acids. One way to circumvent this problem is to work with isolated nuclei or detergent permeabilized cells in which the soluble nucleoplasmic protein has been extracted. However, as a polyelectrolyte, chromatin at all levels of organization is very sensitive to relatively small variations in ionic conditions. This has been a great source of confusion in the literature in that mitotic chromosome structure can change dramatically as a function of isolation buffers, and especially the concentrations of divalent and polyvalent cations. In an early approach to this problem, Sedat and Manulidas chose a polyamine containing isolation buffer which preserved chromatin organization as assayed by maintenance of birefringence within polytene interphase chromosomes [8]. Using this buffer with permeabilized nuclei revealed large-scale chromatin domains, well above 30 nm in diameter, within diploid nuclei [8]. Similarly, Drosophila chromosomes were demonstrated as also containing such large-scale chromatin domains, particularly when chromosomes were isolated after the early, precellular blastoderm rapid division period or from tissue culture cells [40, 50].

To examine this issue more carefully, subsequent experiments set out to compare isolation buffers based on the criteria of preservation of in vivo chromosome structure [50]. This was assayed in two ways. The first was to verify whether the particular isolation procedures used at least preserved in vivo chromosome structure to the resolution provided by light microscopy; comparisons could then be made with the appearance of mitotic and interphase chromosomes stained in living cells with a DNA specific fluorescent dye. The second method was to assume that the overall large-scale chromatin morphology was preserved using conventional TEM methods and to compare the appearance of mitotic chromosomes and interphase nuclei in cells fixed directly in glutaraldehyde versus in cells which had been permeabilized in different isolation buffers prior to fixation.

The results of these experiments demonstrated that physiological buffers such as PBS in the absence of divalent cations produced an obvious perturbation in mitotic and interphase chromosome structure relative to that observed in vivo. The margins of mitotic chromosomes became diffuse and lost all substructure, and interphase nuclei showed a loss of discretely stained regions. At the TEM level of resolution, metaphase chromosomes isolated in these buffers showed again the loss of sharp, distinct borders, and the absence of obvious substructure larger than the 30 nm chromatin fiber. The appearance of these chromosomes was consistent with radial loop models in showing protruding, looped ends of 30 nm fibers emanating from a densely packed core, while interphase nuclei gave the appearance of a random coiling of 10-30 nm chromatin fibers.

In contrast, using isolation buffers containing either polyamines or divalent cations it was possible to preserve the sharply defined chromosome margins and the banding substructure observed in vivo for mitotic chromosomes and to preserve the discrete staining observed in vivo for interphase nuclei. By TEM, using these isolation buffers preserved the appearance of ~100-130 nm diameter chromatin domains which were observed within metaphase chromosomes in cells fixed directly in glutaraldehyde.

Significantly, application of these same isolation buffers to interphase nuclei now revealed a striking pattern of similarly sized, 80-130 nm diameter, large-scale chromatin domains now visualized as spatially distinct entities. Stereopairs of semi-thick Epon sections suggested that these domains in fact corresponded to actual fibers which could be traced for short distances. Moreover, the majority of chromatin packing within these nuclei appeared to be within these large-scale domains. Close examination of nuclei in cells fixed directly in glutaraldehyde revealed similar domains which could be visualized within the nucleoplasmic background ("euchromatin") background staining. These could be seen more easily in telophase nuclei. Quantitative measurements demonstrated a similar histogram of fiber widths in telophase nuclei from intact cells fixed directly with glutaraldehyde as compared to cells which had been permeabilized in a polyamine buffer prior to fixation.

These results led to the suggestion that interphase chromosome organization, at least in these cycling cells, was due to the folding and kinking of such large-scale domains, which appeared to be discrete fibers, while most of the traditional "euchromatin" compartment did not correspond to decondensed chromatin [50].. Conversely, this also implied that a large fraction of chromatin was contained within the traditional "heterochromatin" compartment, as defined by electron dense, chromatin masses above 30 nm in diameter. A serious concern with these studies was that they depended on extraction in isolation buffers; therefore, despite attention to preserving in vivo structure, as assayed by light microscopy, the large-scale chromatin domains observed could be artifactually condensed by exposure to these polycation buffers.

For these reasons, the more recent development of three new, DNA specific staining protocols for TEM is of great interest [50-55]. Significantly, preliminary investigations using all three of these staining protocols have yielded pictures of intranuclear DNA distribution very similar to that obtained by prior detergent permeabilization in appropriate isolation buffers [50] - namely the absence of DNA stain from the "euchromatic" compartment, and the packaging of most DNA within condensed structures well above the 30 nm chromatin fiber. More impressively, both the osmium ammine and the chemical pretreatment method (NAMA-Ur) have the sensitivity to detect mitochondrial and viral DNA as well as highly decondensed nucleolar DNA, presumably representing the rDNA loci. This means that the failure to see DNA stained structures within the euchromatic compartment reflects the true absence of decondensed chromatin rather than a sensitivity limitation of these staining methods.

As developed below, the large-scale chromatin domains described above appear to represent cross-sections through more continuous fibers, which we have labeled "chromonema" fibers [50, 56], formed by the folding of 10 and 30 nm chromatin fibers. These chromonema fibers therefore provide a bridge between lower, better characterized levels of chromatin folding, and the expected linear structure of more highly compacted mitotic and interphase chromosomes.

6. Cell Cycle Modulation in Chromonema Fiber Structure and Folding

To analyze the significance of these large-scale chromatin domains with regard to mitotic and interphase chromosome architecture, we have carried out a careful cell cycle analysis. These studies were combined with 3-dimensional reconstruction techniques in order to visualize these large-scale domains more clearly and, in the case of interphase nuclei, to demonstrate that these large-scale chromatin domains in fact correspond to distinct fibers.

To date we have examined cell cycle stages from anaphase through early S phase and from late S phase through G2 and prophase. These studies have been mostly in CHO cells, although mitotic and early G1 stages have also been examined in Hela cells as well. Similar large-scale domains have been observed in Drosophila Kc cells.

Metaphase chromosomes in Drosophila, Hela, and CHO tissue culture cells all show the presence of large-scale chromatin domains [40, 50]. These can be appreciated in TEM thin sections or stereopairs of semi-thick sections, although the tight packing within these chromosomes makes them difficult to distinguish. In Fig. 1 (top) we show results from EM serial thin sections of Hela metaphase chromosomes. In the serial section data of Hela metaphase chromosomes, the presence of large-scale chromatin domains can be seen especially in the grazing sections. In this case the section thickness was ~80 nm, which is large relative to the size of the domains and, in particular, to the separation between adjacent domains. The greater ease in visualizing these domains in the peripheral regions may reflect problems related to the close packing of these domains. Recent SEM imaging of isolated chromosomes has shown surface features consistent with these large-scale chromatin domains [6], as has preliminary AFM imaging of isolated metaphase CHO chromosomes [Yang, G., G. Li, C. Bustamante, A.S. Belmont, unpublished data].

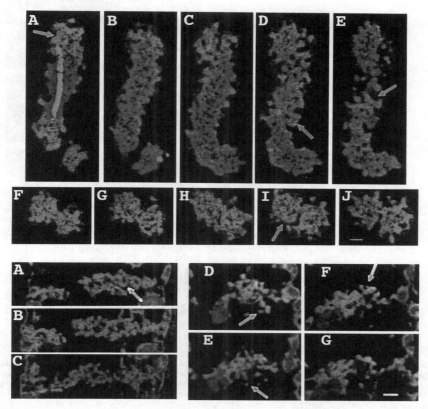

Figure 1: (top) Serial sections (80 nm thick) of a longitudinally oriented metaphase chromosome (A-E) and of a transversely oriented metaphase chromosome (F-J) from Hela cells (orientation relative to sectioning plane). Arrows mark examples of large-scale domains, roughly 100-130 nm in width. In (A-E) the arrangement of these large-scale domains is suggestive of irregular coiling, or loops formed by these structures. Scale bar= 270 nm. Negatives were digitized and a local contrast algorithm applied (40). Pixel size was 9 nm. (bottom) Serial sections (80 nm) (A-C) and (D-G) through longitudinally oriented chromosomes from Hela telophase nuclei. Fiber-like, large-scale chromatin domains are less tightly packed than in metaphase chromosomes and therefore more easily visualized. Arrows point to apparent loops or coils of these structures within the chromosomes. Scale bar=250 nm.

Examination of mitotic chromosomes from early prophase [57] and telophase [50, 56, 57] reveals similar ~100 nm diameter large-scale domains; in these cases, however, these features are more easily visualized in these less condensed chromosomes where the separation between adjacent domains increases. Importantly, these domains are evident at all stages of mitotic chromosome condensation and the observed images are consistent with mitotic chromosome condensation/decondensation occurring with an irregular folding of these large-scale chromatin domain. Fig. 1(bottom) shows serial thin sections of decondensing chromosomes from telophase Hela nuclei. Fig. 2 shows an example of EM tomography carried out on the tip of a telophase chromosome within a 0.5 μm thick Epon section.

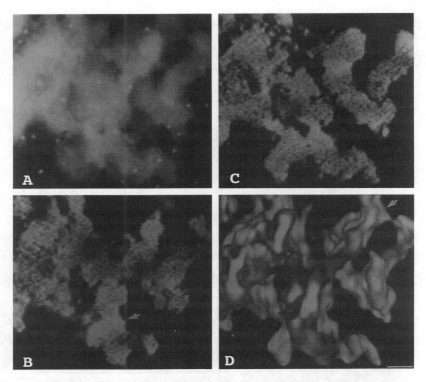

Figure 2: EM tomography reconstruction of oblique, 0.5 um thick section through decondensing chromatid arm within Hela telophase nucleus. (A) corresponds to raw image, or projection, at zero degree tilt, after mass normalization and digital local contrast enhancement. (B-C) are computational slices at different planes, parallel to the physical section, through the reconstruction. Under the buffer conditions used (2x normal polyamine concentrations), 30 nm chromatin fibers are very tightly packed within the larger ~130 nm structures, with very little contrast between fibers, and are not resolved in this reconstruction. However, large-scale domains are easily visualized. The chromatid appears to be decondensing through an unfolding of these large-scale domains. (D) shows a solid model, viewed from the side of the chromatid contained entirely within the physical section; to emphasize the large-scale structure, the reconstruction was blurred using a box convolution prior to calculating the solid model. Arrow in (D) points to region formed by folded coil of large-scale fiber. Arrow in (B) points to region suggestive of a twisting or supercoiling of large-scale fiber. Scale bar = 92 nm (A-D) [65].

Although these large-scale domains become progressively easier to recognize as distinct features as earlier stages in prophase or later stages in anaphase and telophase are examined, the close packing in these condensed chromosomes and regions of overlap between adjacent domains still make it impossible to determine unambiguously whether they correspond to actual, spatially discrete fibers. Examination of G1 [56] and G2 [57] nuclei does make this possible. EM tomography of selected regions of G1 nuclei showed short fiber segments, as shown in Fig. 3; however as only small subvolumes of nuclei could be reconstructed for technical reasons [58], these fibers could not be traced over significant distances without fiber segments leaving the confines of the reconstruction volume. Instead, serial thin sections, 30-40 nm in thick, in conjunction with new algorithms for serial section alignment [59], allowed individual chromonema fibers to be traced unambiguously as isolated fibers in both G1 and G2 nuclei, through much larger nuclear subvolumes. In particular it was possible to trace individual chromonema fibers for 1-2 μm as distinct fibers; extended fibers could be traced over even longer distances- in these cases, though, regions of overlap between adjacent fibers still left questions of continuity throughout the extent of the fibers.

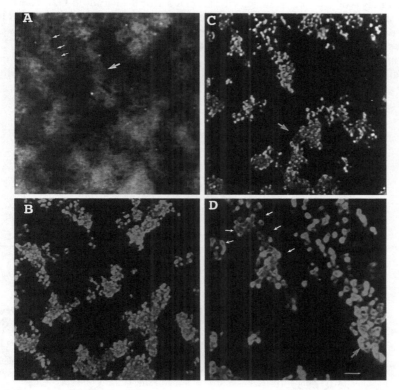

Figure 3: EM tomography of Hela early G1 nucleus shows chromonema fibers. (A) shows 0 degree projection of 0.5 um thick section. (B-C) represent computational "slices", calculated from the tomographic reconstruction. These slices are parallel to the original physical section but from two different depths within the section. The reconstruction slices have been image enhanced using an operation involving the 3-dimensional gradient. Arrow in (C) points to a ~100-130 nm chromonema fiber which can be followed through slice. Large arrow in (A) points to similar ~100-130 nm fiber recognized in original projection, but with orientation such that only smaller segments are shown in a given slice as in (B) or (C). (D) is a larger magnification view of the upper left corner of (C). The larger arrow points to an ~30 nm fiber folded within segment of larger, chromonema fiber. Small arrows in (A) and (D) indicate region suggestive of chromonema fiber decondensing into smaller units. Note the internal folding of 30 nm fibers (dark lines through 30 nm fiber axis is result of the gradient operation). Scale bars= 120 nm (A-C), 60 nm (D). [65]

To place these observations in perspective, a rough calculation shows that to stretch a 30 nm fiber across and back the width of a ~120 nm fiber would correspond to ~20 kb of DNA. Therefore the extended fibers visualized in these reconstruction examples must correspond to the organized folding within these large-scale chromonema fibers of many hundreds to thousands of kbp. Fig. 3D shows results from EM tomography of an early G1 Hela nucleus in which irregular folding of 30 nm chromatin fibers within a chromonema fiber is visualized.

272

A summary of the results of extensive survey and serial section reconstructions from cell populations progressing through G1 and G2 is provided in Fig. 4. G1 progression is accompanied by a progressive straightening and unfolding of chromonema fibers simultaneous with a decondensation of the fibers to yield predominately ~60-80 nm diameter chromonema fibers in middle G1. These fibers initially are mostly peripherally located near the nuclear envelope in early G1 but become dispersed throughout the nuclear interior with G1 progression. Although certain examples suggest that the 100-130 nm diameter fibers observed in early G1 may correspond to a coiling or torsional winding of an underlying ~80 nm fiber, our current data is still inconclusive. A second possibility which we can not exclude is that there is a more continuous variation in chromonema fiber diameter with cell cycle progression.

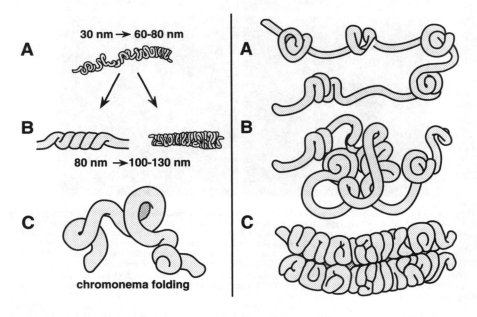

Figure 4: Working model of interphase and mitotic chromosome structure. Left- (A) Chromonema fiber is formed by irregular folding of 10 and 30 nm chromatin fibers into ~60-80 nm fiber. This is predominant fiber width in CHO cells in middle to late G1 and G2. (B) In early G1 and late G2, thicker chromonema fibers, ~100-130 nm diameter, are seen. These may represent a torsional twisting or coiling of 60-80 nm fibers into thicker fiber. Alternatively, there may be a continual transition in fiber diameters to explain this increased diameter. (C) Within interphase nuclei, these chromonema fibers are irregularly folded, with twisting, kinking, and supercoiling as prominent features. Right- (A) Chromosome condensation in late G2 nuclei begins by local regions of folding of chromonema fibers separated by extended fibers. (B) A collapse of chromatin towards nuclear envelope and into locally condensed masses occurs in late G2-early prophase. (C) Continued folding of chromonema fibers is accompanied by resolution of condensed masses into extended, linear chromosome with obvious axis.

In late S phase and early G2 nuclei, chromonema fibers are predominately ~60-80 nm in width and are quite extended as linear fibers and still distributed throughout the nuclear interior. With G2 progression, there is a collapse of these fibers towards the nuclear envelope in condensing chromosome masses. These then resolve into ~0.2-0.3 μm diameter, linear prophase chromatids; at all stages of this condensation, the substructure of these large-scale chromatin fibers is visible, with the ~0.2-0.3 μm diameter prophase chromosomes appearing to be formed by the folding of these fibers. In both early G1 and late G2 nuclei, the tight folding of these chromonema fibers appears to be accompanied by supercoiling and kinking.

7. In vivo Confirmation of Large-scale Chromatin Fibers

The general impression of interphase and mitotic chromosome structure being organized through the tight folding of these large-scale chromatin fibers is intellectually satisfying in terms of providing a continuity between the better characterized lower levels of chromatin folding and chromosome architecture. It is also quite obviously different from standard textbook models. One source of concern with our work has been the degree to which the results have depended on examination of detergent permeabilized cells in which the buffers used might induce changes in chromatin organization. Although we have attempted to choose buffer conditions which preserved large-scale chromatin organization, as described earlier, questions still can be raised concerning the accuracy of our results.

In our minds, a more serious problem has to do with the limitations in trying to deduce an overall pathway of chromosome condensation and decondensation through the cell cycle by examining patterns of condensation in bulk chromatin throughout individual nuclei. In fact our goal of attempting to identify distinct intermediates in the pathway of chromosome condensation and decondensation has been seriously hindered by the highly asynchronous conformational changes seen along a given chromosome arm during nearly all cell cycle stages in somatic cells.

We were therefore very interested in being able to develop a system which would allow us to examine a specific chromosomal region at the resolution needed to dissect chromosome ultrastructure. At the same time, we wanted a system which would allow us to directly observe the temporal sequence of cell cycle modulations in chromosome structure within individual living cells. Moreover, for analysis of chromosome folding ideally we wanted to be able to examine a chromosome region which would fold and unfold relatively synchronously. Finally, in the long term we wanted to be able to extend these structural studies to the analysis of the large-scale chromatin structure surrounding individual gene loci.

Our approach to this problem has been to develop an in situ chromosome tag based on specific protein-DNA recognition [60-62]. As a model system we chose the lac operator and repressor. A construct containing a 256 copy direct repeat of the lac operator was engineered. This was introduced into an expression vector for DHFR (dihydrofolate reductase) which was transformed into DHFR minus CHO cells. Stable transformants were selected. A further amplification of the lac operator target was carried out by applying gene amplification of the DHFR gene using selection in methotrexate (MTX). In CHO cells this typically produces stable cell lines containing multiple repeats of the DHFR locus. These are usually present within amplified chromosome regions which have been labeled homogeneously staining regions (HSRs), because of their uniform staining with chromosomal banding techniques. These HSRs typically contain many copies of a repeat, or amplicon, containing the selectable marker flanked by hundreds to thousands of kb of coamplified chromosomal DNA.

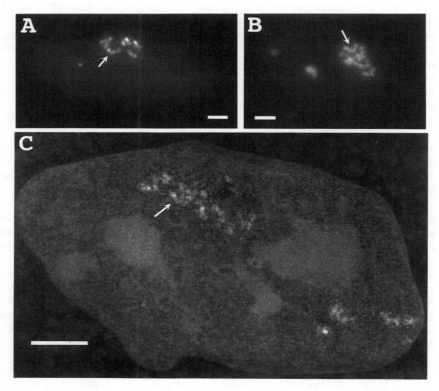

Figure 5: Visualization of amplified chromosome regions (HSRs) containing lac operator repeats using lac repressor localization (60,62). (A-B) In vivo visualization of HSRs using endogenous expression of GFP- lac repressor- NLS fusion protein. Arrows point to HSRs containing obvious fiber segments 1-2 um in length. HSR staining suggests the irregular folding of these fibers within local domains gives rise to the observed staining patterns. (C) Immunogold staining of HSR in same cells, using formaldehyde fixation of intact cells, followed by immunostaining. Visualization of 0.5 μm thick section through interphase nucleus shows HSR containing short segments of ~130 nm chromonema fibers stained by lac repressor. Again the observed staining pattern supports the irregular but tight folding of chromonema fibers within HSR. These results are significant in demonstrating the existence of large-scale chromatin fibers within living cells by light microscopy and showing that these fibers correspond to ~100-130 nm chromonema fibers, as visualized within cells fixed intact, prior to detergent permeabilization. Scale bars= 2um (A-C).

To visualize these HSRs we used three different methods of lac repressor staining- staining of fixed cells with purified lac repressor followed by a immunostaining against lac repressor, in vivo expression of a lac repressor - NLS (nuclear localization signal) fusion protein followed by fixation and immunostaining against the repressor, and finally in vivo expression of a green fluorescent protein (GFP)- lac repressor - NLS fusion protein. All three methods gave indistinguishable results at light microscopy resolution.

Examples of this staining are shown in Fig. 5 - panels (a-b) show in vivo imaging of fibers within HSRs by light microscopy while panel (c) shows fiber segments visualized within a 0.5 μm thick Epon section by TEM. Using this methodology extended large-scale chromatin fibers can be traced for distances over 5 um in length [62]. A comparison of the 0.5 μm or larger length of these HSRs in metaphase chromosomes with the apparent lengths within interphase nuclei suggests a compaction of hundreds to thousands to one for these chromonema fibers.

More recent studies have demonstrated the ability to visualize single vector copy insertions containing the 256 copy lac operator repeat using GFP-repressor expression in CHO cells [62]. In yeast, this methodology has been combined with homologous recombination to introduce this lac operator array adjacent to a specific centromere and allow in vivo visualization of sister chromatid dynamics [63, 64]. We have recently introduced the 256 lac operator repeat into Drosophila fly lines using P element transformation [G. Sudlow, C. Tolerico, A.S. Belmont, unpublished]; therefore we expect this approach will be applicable to a wide range of biological systems.

8. Summary and Future Directions

In this chapter we have briefly reviewed models of mitotic and interphase chromosome structure, and have introduced a new model for large-scale chromatin organization based on folding of ~100 nm diameter chromonema fibers. Confirmation of this model, particularly with regard to folding transitions occurring during mitosis, should come from detailed 3-dimensional analysis of folding intermediates observed in early or late stages of mitosis. The development of a new in situ localization method based on lac operator - repressor recognition should greatly enhance such analysis, by allowing us to focus on specific chromosome regions, and compute actual compaction ratios for these intermediates. By varying the size of these chromosome regions it should be possible to probe different levels of folding.

Preliminary work in this direction has focused on analysis of a late replicating, heterochromatin HSR [61]. We have observed a precise, reproducible choreography of changes in large-scale chromatin organization and positioning within the nucleus as a function of cell cycle progression for this HSR which has been verified by direct in vivo imaging. Interestingly, a dramatic decondensation and movement away from the nuclear envelope to the nuclear interior appears to precede initiation of replication in this system. We have also recently begun exploring changes in large-scale chromatin organization associated with transcriptional activation. Targeting of high concentrations of the acidic activation domain of VP-16 using a lac repressor fusion protein to chromosomal HSRs led to a dramatic unfolding of large-scale chromatin fibers within the heterochromatic HSR, and also movement into the nuclear interior, and preliminary results are suggestive of a propagation of large-scale chromatin decondensation over very large DNA distances. Future extensions of this work are aimed at visualizing the large-scale chromatin structure and dynamics of specific genes.

Most of our work to date has focused on cycling, somatic cells. In earlier work, examining chromosomes from precellular blastoderm embryos, the chromonema fibers described in this chapter were not evident. Similarly, we have not examined the large-scale chromatin organization in terminally differentiated tissues. Future work will need to address changes in this large-scale chromatin organization associated in noncycling cells as well as rapidly dividing embryos.

9. Acknowledgements

This work was supported by National Institutes of Health grant GM42516, NSF grant DIR-8907921, and a Whitaker Foundation grant to A.S. Belmont.

10. References:

1. Daskal, Y., *et al.* (1976) Use of direct current sputtering for improved visualization of chromosome topology by scanning electron microscopy. *Exp. Cell Res.*, **100**, p. 204-212.
2. Harrison, C.J., *et al.* (1982) High-resolution scanning electron microscopy of human metaphase chromosomes. *J. Cell Sci.*, **56**, p. 409-422.
3. Marsden, M.P.F. and U.K. Laemmli. (1979) Metaphase chromosome structure: evidence for a radial loop model. *Cell*, **17**, p. 849-858.
4. Stubblefield, E. and W. Wray. (1971) Architecture of the chinese hamster metaphase chromosome. *Chromosoma*, **32**, p. 262-294.
5. Takayama, S. and H. Hiramatsu. (1993) Scanning electron microscopy of the centromeric region of L-cell chromosomes after treatment with Hoescht 33258 combined with 5-bromodeoxyuridine. *Chromosoma*, **102**, p. 227-232.
6. Rizzoli, R., *et al.* (1994) High.resolution detection of uncoated metaphase chromosomes by means of field emission scanning electron microscopy. *Chromosoma*, **103**, p. 393-400.
7. Ris, H. (1981) Stereoscopic electron microscopy of chromosomes. *Methods Cell Biol.*, **22**, p. 77-96.
8. Sedat, J. and L. Manuelidis. (1977) A direct approach to the structure of eukaryotic chromosomes. *Cold Spring Harbor Symp. Quant. Biol.*, **42**, p. 331-350.
9. Zatsepina, O.V., V.Y. Polyakow, and Y.S. Chentsov. (1983) Chromonema and chromomere-structural units of mitotic and interphase chromosomes. *Chromosoma*, **88**, p. 91-97.
10. Hao, S., M. Jiao, and B. Huang. (1990) Chromosome organization revealed upon the decondensation of telophase chromosomes in Allium. *Chromosoma*, **99**, p. 371-378.
11. Utsami, K.R. and T. Tanaka. (1975) Studies on the structure of chromosomes. 1. The uncoiling of chromosomes revealed by treatment with hypotonic solution. *Cell Struct. Funct.*, **1**, p. 93-99.
12. Bak, A.L. and J. Zeuthen. (1978) Higher order structure of mitotic chromosomes. *Cold Spring Harbor Symp. Quant. Biol.*, **42**, p. 367-377.
13. Earnshaw, W.C. and U.K. Laemmli. (1983) Architecture of metaphase chromosomes and chromosome scaffolds. *J. Cell Biol.*, **96**, p. 84-93.
14. Iino, A. (1971) Observations on human somatic chromosomes treated with hyaluronidase. *Cytogenetics*, **10**, p. 286-294.
15. Mullinger, A.M. and R.T. Johnson. (1980) Packing DNA into chromosomes. *J. Cell Sci.*, **46**, p. 61-86.
16. Mullinger, A.M. and R.T. Johnson. (1983) Units of chromosome replication and packing. *J. Cell Sci*, **64**, p. 179-193.
17. Ohnuki, Y. (1968) Structure of chromosomes: 1. Morphological studies of the spiral structure of human somatic chromosomes. *Chromosoma*, **25**, p. 402-428.
18. Okada, T.A. and D.E. Comings. (1979) Higher order structure of chromosomes. *Chromosoma*, **72**, p. 1-14.
19. Paulson, J.R. and U.K. Laemmli. (1977) The structure of histone depleted chromosomes. *Cell*, **12**, p. 817-828.
20. Paulson, J.R. (1989) Scaffold morphology in histone-depleted Hela metaphase chromosomes. *Chromosoma*, **97**, p. 289-295.
21. Adolph, K.W. (1980) Organization of chromosomes in mitotic Hela cells. *Exp. Cell Res.*, **125**, p. 95-103.
22. DuPraw, E.J. (1965) Macromolecular organization of nuclei and chromosomes: a folded fibre model based on whole mount electron microscopy. *Nature*, **206**, p. 338-343.
23. Rattner, J.B. and C.C. Lin. (1985) Radial loops and helical coils coexist in metaphase chromosomes. *Cell*, **42**, p. 291-296.
24. Stonington, O.G. and D.E. Pettijohn. (1971) The folded genome of Escherichia coli isolated in a protein-DNA-RNA complex. *Proc. Nat. Acad. Sci. USA*, **68**, p. 6-9.
25. Worcel, A. and E. Burgi. (1972) On the structure of the folded chromosome of Escherichia coli. *J. Mol. Biol.*, **71**, p. 127-147.
26. Kavenoff, R. and O.A. Ryder. (1976) Electron microscopy of membrane-free folded chromosomes from Escherichia coli. *Chromosoma*, **55**, p. 13-26.

27. Benyajati, C. and A. Worcel. (1976) Isolation, characterization, and structure of the folded interphase genome of Drosophila melanogaster. *Cell*, **9**, p. 393-407.
28. Lewis, C.D. and U.K. Laemmli. (1982) Higher order metaphase chromosome structure: evidence for metalloprotein interactions. *Cell*, **29**, p. 171-181.
29. Mirkovich, J., S.M. Gasser, and U.K. Laemmli. (1988) Scaffold attachment of DNA loops on metaphase chromosomes. *J. Mol. Biol.*, **200**, p. 101-109.
30. Razin, S.V. (1996) Functional architecture of chromosomal DNA domains. *Crit. Rev. in Eukaryotic Gene Expression*, **6**, p. 247-269.
31. Strunnikov, A.V., V.L. Larionov, and D. Koshland. (1993) SMC1: an essential yeast gene encoding a putative head-rod-tail protein is required for nuclear division and defines a ubiquitous family. *J. Cell Biol.*, **123**, p. 1635-1648.
32. Saitoh, N., *et al.* (1994) ScII: An abundant chromosome scaffold protein is a member of a family of putative ATPases with an unusual predicted tertiary structure. *J. Cell Biol.*, **127**, p. 303-318.
33. Hirano, T. and T.J. Mitchison. (1994) A heterodimeric coiled-coil protein required for mitotic chromosome condensation in vitro. *Cell*, **79**, p. 449-458.
34. Jeppesen, P. and H. Morten. (1985) Effects of sulphydryl reagents on the structure of dehistonized metaphase chromosomes. *J. Cell Sci.*, **73**, p. 245-260.
35. Hirano, T. and T.J. Mitchison. (1993) Topoisomerase II does not play a scaffolding role in the organization of mitotic chromosomes assembled in Xenopus egg extracts. *J. Cell Biol.*, **120**, p. 601-612.
36. Earnshaw, W.C. and M.M.S. Heck. (1985) Localization of topoisomerase 2 in mitotic chromosomes. *J. Cell Biol.*, **100**, p. 1716-1725.
37. Swedlow, J.R., J.W. Sedat, and D.A. Agard. (1993) Multiple Chromosomal Populations of topoisomerase 2 detected in vivo by time-lapse, three-dimensional wide-field microscopy. *Cell*, **73**, p. 97-108.
38. Uemura, T., *et al.* (1987) DNA topoisomerase II is required for condensation and separation of mitotic chromosomes in S. pombe. *Cell*, **50**, p. 917-925.
39. Adachi, Y., M. Luke, and U.K. Laemmli. (1991) Chromosome assembly in vitro: topoisomerase II is required for condensation. *Cell*, **64**, p. 137-148.
40. Belmont, A.S., J.W. Sedat, and D.A. Agard. (1987) A three-dimensional approach to mitotic chromosome structure: evidence for a complex hierarchical organization. *J. Cell Biol.*, **105**, p. 77-92.
41. Harauz, G., *et al.* (1987) Three-dimensional reconstruction of a human metaphase chromosome from electron micrographs. *Chromosoma*, **95**, p. 366-374.
42. Boy de la Tour, E. and U.K. Laemmli. (1988) The metaphase scaffold is helically folded: sister chromatids have predominately opposite helical handedness. *Cell*, **55**, p. 937-944.
43. Ohsumi, K., C. Katagiri, and T. Kishimoto. (1993) Chromosome condensation in Xenopus mitotic extracts without histone H1. *Science*, **262**, p. 2033-2035.
44. Alberts, B., *et al.* (1994) *Molecular Biology of the Cell* 3rd ed. New York: Garland Publishing, Inc. 870-873.
45. Manuelidas, L. (1985) Individual interphase chromosome domains revealed by in situ hybridization. *Hum. Genet.*, **71**, p. 288-.
46. Cremer, T., *et al.* (1993) Role of chromosome territories in the functional compartmentalization of the cell nucleus, in *Cold Spring Harbor Symposia on Quantitative Biology*, Cold Spring Harbor Laboratory Press: Cold Spring Harbor. p. 777-792.
47. Bischoff, A., *et al.* (1993) Differences of size and shape of active and inactive X-chromosome domains in human amniotic fluid cell nuclei. *Microscopy Res. Tech.*, **25**, p. 68-77.
48. Skaer, R.J. and S. Whytock. (1977) Chromatin-like artifacts from nuclear sap. *J. Cell Sci.*, **26**, p. 301-310.
49. Skaer, R.J. and S. Whytock. (1976) The fixation of nuclei and chromosomes. *J. Cell Sci.*, **20**, p. 221-231.
50. Belmont, A.S., *et al.* (1989) Large-scale chromatin structural domains within mitotic and interphase chromosomes in vivo and in vitro. *Chromosoma*, **98**, p. 129-143.
51. Bohrmann, B. and E. Kellenberger. (1994) Immunostaining of DNA in electron microscopy: an amplification and staining procedure for thin sections as alternative to gold labeling. *J. Histochem. Cytochem.*, **42**, p. 635-643.
52. Olins, A.L., *et al.* (1989) Synthesis of a more stable osmium ammine electron-dense DNA stain. *J. Histochem. Cytochem.*, **37**, p. 395-398.
53. Mikhaylova, V.T. and D.V. Markov. (1994) An alternative method for preparation of Schiff-like reagent from osmium-ammine complex for selective staining of DNA on thin Lowicryl sections. *J. Histochem. Cytochem.*, **42**, p. 1643-1649.
54. Vazqueznin, G.H., M. Biggiogera, and O.M. Echeverria. (1995) Activation of osmium ammine by SO2-generating chemicals for EM Feulgen-type staining of DNA. *Eur. J. Histochem.*, **39**, p. 101-106.
55. Robards, A.W. and A.J. Wilson, ed. (1993) *Procedures in Electron Microscopy*. John Wiley @ Sons: New York.
56. Belmont, A.S. and K. Bruce. (1994) Visualization of G1 chromosomes- a folded, twisted, supercoiled chromonema model of interphase chromatid structure. *J. Cell Biol.*, **127**, p. 287-302.
57. Li, G., K. Bruce, and A.S. Belmont. (1995) Visualization of early stages of chromosome condensation: a folded chromonema model of mitotic chromosome structure. *J. Cell Biol*, **submitted**.
58. Delaney, A. and A.S. Belmont. (1994) Deblurring of high tilt projections for EM tomography. *Ultramicroscopy*, **56**, p. 319-335.

59. Basu, S. and A.S. Belmont. (1996) Serial section alignment with correction for nonlinear distortions. *in preparation*, .
60. Robinett, C., *et al.* (1994) Lac-op repressor detection of chromosome HSRs: visualization of extended chromonema fibers. *Mol. Biol. Cell*, **5 (suppl.)**, p. 3a.
61. Li, G., *et al.* (1995) Interphase dynamics of a late replicating, heterochromatic chromosome HSR. *Mol. Biol. Cell.*, **6 (supplement)**, p. 71a.
62. Robinett, C.C., *et al.* (1996) In vivo localization of DNA sequences and visualization of large-scale chromatin organization using lac operator/repressor recognition. *J. Cell Biol.*, **in press, Dec.**
63. Minshull, J., *et al.* (1996) Protein phosphatase 2A regulates MPF activity and sister chromatid cohesion in budding yeast. *Current Biol.*, **in press, Dec.**
64. Straight, A.F., *et al.* (1996) GFP tagging of budding yeast chromosomes reveals that protein-protein interactions can mediate sister chromatid cohesion. *Current Biol.*, **in press, Dec.**
65. Belmont, A.S., J.W. Sedat, D.A. Agard, unpublished data

PLANT GENE TECHNOLOGY

I.L. BAGYAN[1], I.V. GULINA[1], A.S. KRAEV[1], V.N. MIRONOV[1], L.V. PADEGIMAS[1], M.M. POOGGIN[1], E.V. REVENKOVA[1], A.V. SHCHENNIKOVA[1], O.A. SHOULGA[1], M.A. SOKOLOVA[1], J. VICENTE-CARBAJOSA[1], G.A. YAKOVLEVA[2], K.G. SKRYABIN[1]

[1]Centre of Bioengineering of Russian Academy of Sciences, prospect 60-letiya Oktyabrya, 7/1, 117312, Moscow, Russia

[2]Potato Research Institute of Agricultural Academy of Byelorus, Samochvalovichy, 223013 Minsk, Byelorus

In the past 7-8 years considerable effort has been expended to produce transgenic plants that resist virus infection, insects, herbicides and disease development. The crop improvement and the desired phenotype is conferred by transformation, usually with a single foreign gene. Selection of transformants demands the use of selectable markers, which usually encode enzymes which inactivate either a herbicide or an antibiotic. Also we use certain foreign genes known as reporter genes (for review see [7]).

Agrobacterium-mediated transformation is now used to introduce foreign genes into the nucleus of many dicots. For some species such as potato, tomato and tobacco, specific varieties have been found to be extremely readily transformed by Agrobacterium, that is, transgenic calli are produced efficiently and are regenerated with high frequency into viable, fertile plants. However, for other dicotyledonous plants Agrobacterium-mediated transformation may proceed but regeneration may be infrequent, or transformation itself may prove a stumbling block. For this reason, tobacco, tomato and potato feature prominently in the list of plants engineered for crop improvement [7].

Some of the monogenic traits, that are already widely used for the improvement of different properties of agronomically important plants by the transfer of appropriate genes, are listed in the Table below.

The progress with plant molecular biology and gene technology critically depends on the discovery of new genes, further study of genome organization, gene functions and patterns of their regulations, development of novel and improved methods of delivery of foreign genetic material to the genomes of agronomically important plants.

C. Nicolini (ed.), Genome Structure and Function, 279–318.
© 1997 Kluwer Academic Publishers. Printed in the Netherlands.

TABLE I. Genes of use for crop improvement.

Transformation protocols	Antibiotic resistance:
	neomycin phosphotransferase (B)
	hygromycin phosphotransferase (B)
	Herbicide resistance:
	phosphinothrycin acetyl transferase (B)
	EPSP synthase (B, P)
	Reporter genes:
	chloramphenicol acetyl transferase (B)
	betta-glucuronidase (B)
	luciferase (A)
Pest & Disease Resistance	Virus:
	coat protein (V)
	RNA-binding protein (V, antisense)
	Satellite RNA (V)
	AL1 virus replication gene (V, antisense)
	Fungus:
	Chitinase (P)
	Ribosome inactivating protein (P)
	Insect:
	Bt Cry 1A(b) (B, S)
	Bt Cry 1A(c) (B, S)
	Cowpea trypsin inhibitor (P)
	Potato inhibitor II (P)
Abiotic stress	Heavy metal:
	Metallothionen II (A)
	Salt:
	Mannitol-1-phosphate dehydrogenase (B)
	Freezing:
	Fish antifreeze protein (A)
	Oxidation:
	Mn-Superoxide dismutase (P)
Modified Product Quality	Fatty acid composition:
	ACP thioesterase (P)
	Amino acid composition:
	Methionine-rich protein (P)
	Dihydropicolinate synthase (B)
	Delayed fruit ripening:
	Polygalacturonase (P, antisense)
	ACC oxidase (P, antisense)
	ACC synthase (P, antisense)
	ACC deaminase (B)
	Flower pigments:
	Chalcone synthase (P, antisense)
Other Properties	Polyhydroxybutyrate formation:
	Acetoacetyl CoA reductase (B)
	PHB synthase (B)
	Male sterility:
	Barnase (B) and Barstar (B)
	Antibodies:
	Heavy chain (A)
	Light chain (A)

Source of genes: A = animal, B = bacterium, P = plant, S = synthetic, V = virus.

In this review I'll give several examples, based on the results of the research conducted at my Laboratory of plant Genetic Engineering, Centre of Bioengineering of Russian Academy of Sciences during last several years, illustrating the principles of study of structure and expression of plant genes and production of transgenic plants with agronomically important traits.

1. Transformation of Cotton (*Gossipium hirsutum* l.) with a Supervirulent Strain of *Agrobacterium tumefaciens* A281 [64]. Construction of a Disarmed Derivative of the Supervirulent Ti Plasmid ptibo542

The work by Revenkova and coworkers is devoted to development of more efficient and more convenient vector system for plant transformation, useful in broadening the list of plant species that can be stably transformed with different *Agrobacterium tumefaciens* systems.

Agrobacterium tumefaciens strain A281 can induce fast growing tumors with higher frequency end on a wider range of plants than do other *A. tumefaciens* strains. A281 is virulent on *Solanaceae*, *Umbelliferae*, *Fabaceae*, *Asteraceae*, *Crassulaceae*, *Malvaceae* species, and also on *Picea abies* (*Gymnospermae*) [52, 85].

Transgenic cotton plants (*Gossipium hirsutum* L.) were first obtained in 1987. Both binary and integrative vector systems were used for transformation with *Agrobacterium tumefaciens*, and the efficiency of both systems was relatively low. Increasing efficiency of transformation may be necessary, for example, in experiments in which transformed plants or tissues are used for studying the nature of gene expression. In these cases, it is desirable that the number of independently obtained transformants be adequate for statistically reliable evaluation of the average level of expression.

Using strains of *A. tumefaciens* that have an increased virulence with respect to the given plant species is one way of increasing the frequency of transformation. Strain A281 has a broader circle of hosts than a number of other strains [15]. It carries plasmid pTiBo542, termed "supervirulent" since it was demonstrated that when it is used as a plasmid helper in a binary system, the transformation frequency increased by more than 10 times [19]. A segment of the region *vir* pTiBo542 that has the supervirulent phenotype was identified [41].

We compared the efficiency of transformation of *Gossipium hirsutum* L. by the vectors listed in the Table below. Since genes of phytohormone biosynthesis are found in T-DNA pTiBo542, selection of transformed tissues was done on media without phytohormones. When pBI101 and pBI121 carrying the *npt* II gene in the T-DNA were used, on media without phytohormones with 25 mg/l kanamycin, double transformants were selected that had obtained both T-DNA pTiBo542 and T-DNA pBI101 or pBI121.

Along the edges of explants that were cocultured with *A. tumefaciens*, tumors 0.5-1 mm in diameter developed in 10 days, while explants not treated with agrobacteria displayed only points of root growth in zones where the cut intersected vascular bundles.

The growth points on the explants were counted after 3 weeks (Table II). After coculturing with A281 and selection on medium with agarose without phytohormones and kanamycin, five or more tumors were formed on each explant, and segments of compact growth were found on some explants.

TABLE II. Average Number of Growth Points (callus or tumor) on Explant.

Agrobacteria strain	Medium without phytohormones*				Medium for callus formation*	
	without kanamycin		with kanamycin**		without kanamycin	with canamycin**
	bactoagar	agarose	bactoagar	agarose		
Without agrobacteria	0 (20)	0 (20)	0 (20)	0 (20)	CG***	0 (20)
A281	3.6 (108) 4.0 (20)	8.5 (20)	0 (20)	0 (35)		
A281 (pBI101)		4.5 (20)	2.1 (42)	0.9 (75)		
A281 (pBI121)	2.3 (13)	5.8 (12)	0.6 (70)	2.0 (52)		
LBA4404					CG***	0 (20)
LBA4404 (pBI101)					CG***	0.5 (16)
LBA4404 (pBI121)					CG***	0.4 (103)

*Number of explants is indicated in parentheses.
**25 mg/l kanamycin.
***Compact growth.

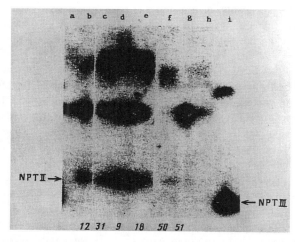

Figure 1. Determination of neomycinphosphotransferase (*npt* II) activity in cotton tumors: a-h - extracts from tumors obtained as a result of transformation with strains; a, h - A281 (control); b-e - A281 (pBI101); f,g - A281 (pBI121); i - control extract from *A. tumefaciens*, carrying te gene of resistance of kanamycin. In bacterial cells, *npt* type III is expressed which differs according to electrophoretic mobility from *npt* II. Numbers of tumors are indicated below.

The picture was similar in transformation with A281 (pBI101) and A281 (pBI121) in growth on medium without kanamycin, but on medium with 25 mg/l kanamycin, the number of tumors decreased by approximately 3-5 times and their growth was somewhat slowed.

npt II Activity was found in eight independently obtained tumors. The test confirmed the presence of this activity in all tumors obtained on medium with kanamycin while no *npt* II was found in the control variants (tumors obtained as a result of transformation with A281) (Fig. 1).

DNA was isolated from eight tumors (one of these was from the control, transformation with A281). Hybridization according to Southern's method (Fig. 2) showed that in the DNA of six tumors resistant to kanamycin, there are sequences of the *npt* II gene. In DNA of tumor 31, resistant to kanamycin, just as in the control, we found no fragments hybridizing with the probe. In DNA of tumor 50, one hybridizing fragment was found, in the others, two or more. As we expected, the lengths of the hybridizing fragments in various independent transformants differed since T-DNA pBI101 and pBI121 do not contain internal *Eco*RI fragments.

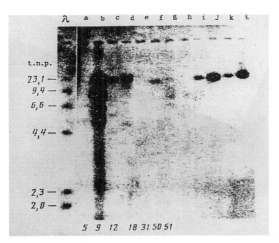

Figure 2. Hybridization according to Southern's method, DNA of cotton tumors separated with *Eco*RI. As a matrix for synthesis of the labeled probe, we used single-strand DNA of M13 phage containing the *npt* II gene. λ is DNA of phage λ separated with *Hind*III restrictase and labeled at the ends with a Klenow fragment; a-g - DNA from tumors obtained as a result of transformation with strains: a-e - A281 (pBI121); h - A281 (control); i, j - plasmid pBI101 separated with *Eco*RI; k, l - plasmid pBI121 separated with *Eco*RI. Plasmid DNA was introduced in an amount corresponding to 1 (i, k) and 5 (j. l) copies of the *npt* II gene per cell. Tumor numbers are indicated below. Fragment size is on the left.

Since the Ti plasmid of strain LBA4404 is "disarmed" when it is used as a component of a binary system, the transformed tissues grew on medium with phytohormones. As a result of the transfer of T-DNA pBI101 and pBI121, we obtained calluses that grew on medium with 25 mg/l kamamycin.

The results of counting growth points (tumors or calluses) are presented in Table II. Formation of calluses on medium with phytohormones and with kanamycin occurs quite efficiently (compact growth along the edge of the explant), but in selection on kanamycin, we obtained in average of 0.4-0.5 growth points per explant. In transformation with A281 on medium without hormone but with agarose, we obtained, on an average, no fewer than 4.5 growth points per explant, and with double selection on medium with agarose, 0.9 for pBI101 and 2.0 for pBI121.

In transformation of cotton with strain LBA4404 using other phytohormones, the average number of calluses per explant fluctuated from 1.3 to 4.1, and the authors indicate in a note that as a result of optimization of transformation conditions, 6.8 could be obtained. In our experiments, with transformation with A281 (selection on medium without phytohormones and kanamycin), the highest average number of growth points per explant was 8.5, and this number is low since extensive zones of compact growth along the edge of the explant were counted as one point.

Thus, strain A281 displays a high virulence with respect to cotyledon tissues of cotton seedlings. This system may be used for studing the promoters of cotton genes under model conditions: undifferentiated tumor tissue. Also, strain A281 may be used in the future as a component of a binary system for obtaining transgenic plants.

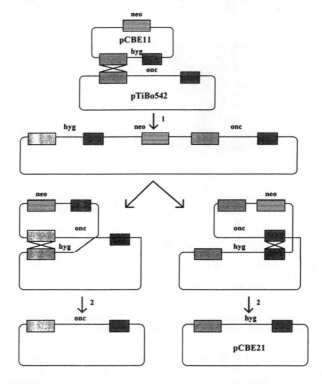

Figure 3. The scheme of homologous recombination for the obtaining of pCBE21 non-oncogenic derivative of pTiBo542 from *Agrobacterium tumefaciens* A281.

The supervirulent phenotype of A281 is conferred by its Ti plasmid pTiBo542 together with the chromosomal background. A fragment of pTiBo542 which is responsible mainly for supervirulence was identified in the *vir* region [12]. This fragment contains the 3'-end of the *vir*B operon and the *vir*G locus. It can enhance the virulence of A348, which has the same chromosomal background and cryptic plasmids as does A281. The chromosomal background of A281 is also important since this strain has an increased host range as compared to Bo542, which was the Ti plasmid donor for A281 [8].

On the basis of pTiBo542 we have constructed a helper plasmid in which a part of T$_L$-DNA carrying the genes responsible for tumor formation was exchanged for a chimaeric hygromycin resistance gene. This plasmid may prove useful as a part of a certain binary or integrative system for high efficiency plant transformation.

To construct a derivative of pTiBo542 we have used the general approach developed by Van Haute *et al.* [18], pCBE21 is a disarmed derivative of pTiBo542 in which a plant selective marker was substituted for T-DNA oncogenes by homology recombination (Fig. 3).

This is thought to be useful as a helper for a binary vector carrying T-DNA free of plant selective markers. We assume that the supervirulent strain can provide a high frequency of co-transfer of two T-DNAs. Transformants obtained in such an experiment can be initially selected on hygromycin, and subsequently be screened for the presence of the second T-DNA, for example, by the polymerase chain reaction technique. Since the two T-DNA should not be linked, they could be separated in the progeny of the transformants, and we would be able to obtain transgenic plants with a minimal portion of foreign DNA, lacking antibiotic resistance markers. Our data show that pCBE21 transfers with a high frequency both its Ti plasmid T-DNA and a binary vector T-DNA both in tobacco leaf disk transformation, and in transformation of a number of fruit plants, like apple, cherry, plum and others.

2. 5'-Regulatory Region of *Agrobacterium tumefaciens* t-DNA Gene 6b irects Organ-specific, Wound-inducible and Auxin-inducible Expression in Transgenic Tobacco [6]

The work of Irina Bagyan and coauthors concerns one of the very intersting feautures of genetic interactions between the natural plant genetic engineer - *Agrobacterium tumefaciens* and the host plant - the evolutionary obtained capability of procaryotic genes, encoded by the T-DNA of the *Agrobacterium* Ti plasmids to respond to the eukaryotic signals of gene expression when transferred and integrated in the plant genome.

The soil bacterium *Agrobacterium tumefaciens* causes crown gall tumours on many plant species. The biochemical and molecular genetic basis of this phenomenon was extensively studied (for review see [36]). After transfer of a specific DNA segment (T-DNA) of tumour-inducing (Ti) plasmid from the agrobacterial to the infected plant cell genes carried on the T-DNA are expressed that disrupt the cell metabolism. Three genes (*iaaH*, *iaaM* and *ipt*) provide for the biosynthesis of the phytohormones auxin and cytokinin, which leads to plant cell proliferation and formation of a tumor; gene 5 modulates the auxin response of tumor cells, and gene 6b remains the only one gene of the 'core' T-DNA for which the role in crown gall tumorigenesis is still poorly understood. It participates in the regulation of the tumor cell responses to auxins and cytokinins [74, 79-81]. We have previosly cloned [63] the 5'-regulatory region (826 bp from the start of translation) of gene 6b from the TL-DNA of pTiBo542 [35], fused it to the *gus*-coding region and the 3'-regulatory region of the gene coding for nopaline synthase (*nos*) and transferred this fusion gene into tobacco plants (*Nicotiana tabacum* L. cv. Petit Havana SR1) via the *Agrobacterium* binary vector system. We also produced transgenic tobacco plants carrying the *gus*-coding region under the control of the 35S-e promoter, the CaMV 35S promoter enhanced by the duplication of the enhancer region (EP 0 339009 A2) and the *nos* 3'-regulatory region. Analysis of GUS activity in roots and young leaves of the primary transformants revealed that the P6b-*gus* plants contain 20-80-fold higher GUS activity in roots than in young leaves [5]. Here we report a more detailed analysis of the activity of the 5'-regulatory region of gene 6b in different organs of the transgenic plants, as well as the effect of wounding and phytohormones on the activity of the regulatory region.

We analysed GUS activity in 40-day-old kanamycin-resistant seedlings of the F1 transgenic tobacco plants carrying the P6b-*gus* fusion (Table III). The highest GUS activity was found in roots, followed by cotiledons and hypocotyls.

TABLE III. GUS activity in seedlings of F1 transgenic plants carrying the P6b-*gus* fusion.

Number of the independent transgenic line	GUS activity, nmol MU per minute per mg protein		
	roots	hypocotyls	cotyledons
11	28.7	0.8	4.4
20	10.2	0.1	0.8
21	22.3	0.3	0.9
45	18.9	0.1	1.2
51	56.5	1.0	0.2
60	27.8	0.3	1.9

For each independent transgenic line roots, hypocotyls and cotyledons of 30 seedlings grown on MS medium [63] containing 100 mg/l of kanamycin were pooled and used for extract preparation. Explants were frozen in liquid nitrogen and stored at -70°C until analysis. Plant extracts were prepared and GUS activity was measured fluorometrically according to Jefferson *et al.* [40]. Protein content of the extracts was determined by the method of Bradford [10]. BSA was taken as a protein standard. Experimental variations in the GUS activity values in all our assays reported here did not exceed 10% of an average value. MU-4-methylumbelliferone.

We also determined GUS activity in different organs of the P6b-*gus* and P35S-e-*gus* plants grown in vitro to the 12-leaf stage (Fig. 4). The results indicated that the 6b 5'-regulatory region provided maximal GUS activity in roots followed by stems and leaves. We found a gradual increase of GUS activity from the shoot apex towards the base in stems and leaves of the P6b-*gus* plants. The basipeptal gradient of GUS activity is not simply the result of the accumulation of GUS, since P35S-e-*gus* plants did not show this pattern of expression (Fig.4). The expression profiles of P6b-*gus* and P35S-e-*gus* fusions differed considerably (Fig. 4). P6b-*gus* plants showed much lower GUS activities in roots, upper and middle leaves, but higher activities in middle and lower stem than P35S-e-*gus* plants.

GUS activity was differentially distributed within a mature leaf. In the leaves of the P6b-*gus* plants the lowest GUS activity was found in the leaf blades, while middle veins showed a higher GUS activity, followed by petioles (Table IV). We also analysed GUS activity in flowers of the P6b-*gus* primery transformants (Table V). GUS activity was differentially distributed within a flower with a maximum in the corolla and a minimum in the ovary.

nmolMU/(min x mg protein)

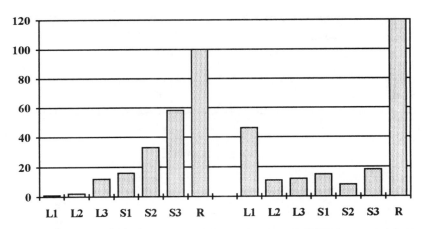

Figure 4. The average values of the GUS activity in the different organs of F1 P6b-*gus* and P35S-e-*gus* plants. For each independent transgenic line on representative F1 plant grown on MS medium containing 100 mg/l of kanamycin was taken. Extracts were prepared from upper leaf (the 1st leaf from the apex, 2.5-3.5 cm long) (L1), middle leaf (the 5th or 6th leaf from the apex, 4.5-6.5 cm long) (L2), lower leaf (the 12th leaf from the apex, 3-4 cm long) (L3), stem sections (2-3 mm thick) (S1, S2, S3) and roots (R). Leaf explants were taken avoiding veins; stem sections were taken from the internodes just below the site of attachment of the respective leaf; roots were taken as total. This figure represents the average values for eight independent P6b-*gus* transformants and six independent P35S-e-*gus* transformants, standard deviations are shown as a numbers above the respective bars. L1, L2, L3, S1, S2, S3, R (left) - data for F1 P6b-*gus* plants. L1, L2, L3, S1, S2, S3, R (right) - data for F1 P35S-e-*gus* plants.

TABLE IV. Distribution of GUS activity within a mature leaf of P6b-*gus* plants, nmol MU per minute per mg protein.

Plant No	Leaf blade	Middle vein	Petiole
10	2.4	12.2	46.3
11	2.6	15.5	22.4
20	1.3	3.5	10.7
21	0.7	6.1	14.8
38	4.2	16.8	34.3
45	0.6	4.4	9.1
51	3.6	17.8	29.4
60	0.4	3.0	5.4

We analysed GUS activity in the extracts prepared from different parts of mature leaf (5th or 6th leaf from the apex, 4.5-6.5 cm long). GUS activity was determined as described in the legend to Table III.

It is worth to be noted that we also found GUS activity in the cells of the *Agrobacterium* strain carrying the binary vector plasmid containing the P6b-*gus-nos* 3' fusion, the strain we used to produce the transgenic plants. GUS activity was not found in the cells of the original *Agrobacterium* strain B6S3-SE [25] which we used as a *vir* helper in plant transformation. Expression in *Agrobacterium* was earlier found for some other T-DNA genes [27, 28, 39].

It was shown for 6b genes from different Ti plasmids (pTiAch5, pTiC58 and pTiTm4) that they participate in the regulation of cellular responses to auxins and cytokinins [74, 79-81].

TABLE V. Distribution of GUS activity within a flower of P6b-*gus* plants, nmol MU per minute per mg protein.

Plant No	Corolla	Stamens	Ovary	Sigma and style	Sepal	Leaf
10	29.7	11.5	0.08	24.7	2.2	0.7
11	9.5	20.0	0.15	10.4	2.8	2.4
20	62.7	52.4	0.08	21.1	2.1	1.3
21	20.7	12.5	0.16	19.2	1.3	0.6
38	26.9	9.2	0.10	9.8	1.6	1.7
51	46.7	0.4	0.07	11.6	4.8	1.8
60	29.2	10.5	0.04	10.0	2.1	2.0

For each independent transgenic plant grown on soil substitution mixture we pooled different organs of three flowers the 1st-2nd day from the inforescence. GUS activity was determined as described in the legend to Table III.

relative GUS activity

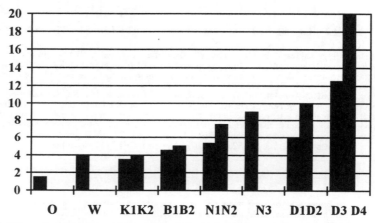

Figure 5. Effect of auxin and cytokinin on the activity of the 6b 5'-regulatory region. Mature leaves of 12-leaf stage plants grown on soil substitution mixture were cut into ca. 0.5 cm^2 sections and the slices were floated on sterile water or solutions of auxins or cytokinins for 22 h. GUS activity was measured in extracts of the leaf fragments before (O) and after incubation (W, in water; K1, in solution containing 0.1 mg/l kinetin; K2, 5.0 mg/l kinetin; B1, 0.1 mg/l BAP; B2, 5.0 mg/l BAP; N1, 0.1 mg/l NAA; N2, 2.0 mg/l NAA; N3, 5.0 mg/l NAA; D1, 0.1 mg/l 2,4-D; D2, 0.5 mg/l 2,4-D; D3, 1.0 mg/l 2,4-D; D4, 5.0 mg/l 2,4-D). The ratio between the GUS activity in leaf fragments incubated in a given solution and the GUS activity in the leaf fragments before incubation was calculated for each independent transgenic plant. The figure represents the average values for six independent P6b-*gus* transformants; standard deviations are shown as a number above the respective bars.

Confirming our previous results on the stimulation of the activity of the pTiBo542 6b 5'-regulatory region in explants of mature transgenic leaves by naphthaleneacetic acid (NAA), and benzylaminopurine (BAP) [5], we have discovered the profound effect of different auxins and cytokinins on the activity of the 6b 5'-regulatory region (Fig. 5). GUS activity in leaf explants of P6b-*gus* plants was enhanced by auxin treatment, with the effect depending upon the type and the concentration of auxin. 2,4-dichlorophenoxyacetic acid (2,4-D) had a more significant effect than NAA. The enhancement of GUS activity increased with the auxin concentration. No significant effects were observed with any of cytokinins tested, including BAP, kinetin (see Fig. 5), *trans*-zeatin and isopenthenyladenine (data not shown). GUS activity in the leaf explants incubated in sterile water was also increased (Fig. 5 and 6a) suggesting wound inducibility of the 6b 5'-regulatory region activity. Wounding of leaves of the intact P6b-*gus* plants grown *in vitro* also induced 6b-driven GUS activity, although to a lesser extent (Fig. 6b). The 35S-e promoter activity was not influenced by phytohormones or wound treatment (data not shown).

nmolMU/(min x mg protein)

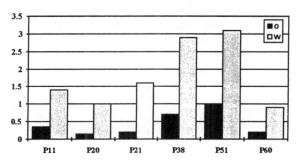

nmolMU/(min x mg protein)

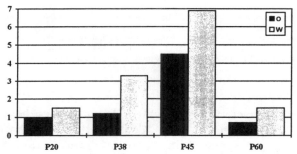

Figure 6. Wound induction of the 6b 5'-regulatory region activity on leaves of P6b-*gus* plants. a (above). GUS activity in the extracts of the leaf explants before (O) and after (W) floatation on sterile water for 22 h (see legend to Fig. 4). b (below). GUS activity in the extracts of a leaf before wounding (O) and of the same leaf left on the plant for 22 h after wounding (W); F1 plants grown *in vitro* to 12-leaf stage were used in the analysis, wounding was carried out by making two holes (5 mm diametre) in the leaf blade (avoiding veins) of a mature leaf. After 22 h, 1 mm thick rings around the holes were taken and analysed for GUS activity.

Organ-specific activity in transgenic plants was observed earlier for the promoters of many other T-DNA genes, including *nos* [3], octopine synthase gene (*ocs*) [26, 47], mannopine synthase gene (*mas*) [51, 77], agropine synthase gene (*ags*) [38], gene 5 [46], isopentenyltransferase gene (*ipt*) [20, 75]. Most of the studied T-DNA promoters are preferentially active in roots and lower parts of transgenic plants, which may be associated with the fact these are the usual sites of agrobacterial infection in mature. The expression characteristics of the 6b 5'-regulatory region resemble those of the other auxin-inducible T-DNA promoters, especially those of *nos* and *mas*: the highest activity in the roots of transgenic plants, basipetal pattern of expression in the aerial parts of the plants, wound inducibility and auxin inducibility. However, the *mas* promoter was also inducible by cytokinin (BAP), albeit to a lesser extent, than by auxin (NAA) [2, 3, 51]. It would be interesting to elucidate *cis*-regulatory elements within the 6b 5'-flanking region responsible for its differential activity and compare them with known elements [9, 30, 47, 48, 56] from the promoters of other T-DNA genes. We have found in the 6b 5'-regulatory region a sequence with partial homology to the 16 bp palindromic element of the *ocs* promoter (Fig. 7); sequences homologous to the *ocs* element have been identified previously for the other T-DNA genes and some viral genes [9]. The importance of the sequence for the 6b promoter function remains to be determined.

	ocs-like element region		distance ocs-TATA, bp
"ocs" consensus	T	CGA G T AA	
	TG ACG AA	G T ACG	
	C	GAC T C CC	
gene 6b	AA ACGCAC C A T T T	GCG CTG ATTG AA	218
ocs	AA ACGTAA	GCGCTTACGT AC	141
mas2'	TG ACGCTC	GCGGTGACGC CA	39
nos	TG AGCTAA	GCACATACGT CA	85

Figure 7. Sequence homologous to the *ocs* 16 bp palindrome found in the 6b promoter region and the homologous sequences from *ocs*, *mas 2'* and *nos* promoters. The consensus sequence for the *ocs* element, derived from the alignment of *ocs*-like elements of the promoters of seven opine synthase genes of T-DNAs from Ti and Ri plasmids and three plant viral genes, was taken from the paper by Bouchez *et al.* [9]. The 5 bp gap was introduced into the consensus sequence, *ocs*, *mas 2'* and *nos* sequences to align them with the 6b sequence.

3. Molecular Characteristics of the Chalcone Synthase Gene Family from Two Cotton Species [13]. Identification of Chalcone Synthase Genes Specifically Expressed in Petals of Two Cotton Species [14]

Several plant genes are known to be responsible for determining the colouring of flowers in plants. These genes are of potential interest to floriculture. The flower-specific pattern of expression of these genes also is intriguing with respect of understanding of the basic mechanisms determining the organ and- tissue-specific gene expression in plants.

The work of Marina Byzova and coworkers was focused on the genes, responsible for colour formation in cotton.

Chalcone synthase (CHS) is a key enzyme in the biosynthesis of all classes of flavonoids in plant tissues. It catalyzes the condensation of three malonyl-CoA molecules with one ρ-coumaroyl-CoA molecule, forming naringenin-chalcone, the major precursor in the biosynthesis of flavonols, flavones, isoflavonoids, and anthocyans [32]. The flavonoids are included in the biosynthesis of plant pigments, and in compounds that are involved in protecting plants against damage by ultraviolet light and by various pathogenes [17]. In legumes flavonoids trigger the mechanism forming root nodules, including the expression of the *nod* genes in *Rhizobium* [22].

By now *chs* genes have been cloned from several plant species [57]. Their number in the genome varies substantially from species to species. Certain plant species contain only one *chs* gene (*Petroselinum hortense, Antirrhinum majus, Arabidopsis thaliana*) [21, 62, 72] whereas in others these genes are represented by a family of 6-10 members (*Petunia hybrida, Phaseolus vulgaris, Glycine max*) [29, 44, 65]. Despite the extensive investigations, detailed data on the molecular organization of the families of *chs* genes and tissue-specific expression of several genes contained in the family have been obtained as yet only for *Petunia hybrida* (V30) [45].

The structural conservatism of the chalcone synthase gene permits its use for establishing the evolutionary relationships between taxonomic groups of plants [57]. Up to this time, however, systematic investigations on this subject have been limited by the technical difficulties of cloning and determination of the primery structure of genes. In most cases sequences accessible for study have been cloned for other purposes and belong to distant groups, chiefly from the asterid subclass (*Asteridae*).

In this work we selected two species of cotton as the objects of investigation: one of the closest to the wild species, *G. herbaceum*, and a cultivated variety widespread until recently, *G. hirsutum* 108F. *G. herbaceum* is a diploid species, which currently grows in difficulty accessible zones of tropical Africa [82]. *G. hirsutum* is a natural amphidiploid, which was formed by a cross of two diploid species of cotton: *G. herbaceum* of Afro-Asian origin and *G. raimondii* of South American origin [78].

The cultivar *G. hirsutum* 108F was produced by crossing *G. hirsutum* 17687 and *G. hirsutum* 36M$_2$ [76].

Figure 8 presents radioautographs of the blot hybridization of *G. hirsutum* 108F and *G. herbaceum* DNA, digested by restriction endonucleases *Hind*III, *Bam*HI, and *Eco*RI. Two probes, synthesized on M13*mp*10 clones, containing the 3'-region of the *chs* gene of *G.hirsutum* 108F, cloned before and differing in position within the *chs* gene, were used for hybridization (Fig. 8). A comparative analysis of the results obtained revealed that all the hybridization bands that appear in the *G. herbaceum* genome are also detected in the *G. hirsutum* genome. However, in the *G. hirsutum* 108F DNA, in addition to the bands that concade with the hybridization bands of *G. herbaceum*, supplementary ones are detected. Thus, in the case of digestion of DNA by the restriction endonuclease *Hind*III and its hybridization with probe 1 in the *G. hirsutum* 108F genome under mild conditions of washing of the filter, seven bands are detected (Fig. 8, Ia), and under rigorous conditions five are observed (Fig. 8, Ib); in the genome of *G. herbaceum*, four bands are detected under any conditions of washing. When probe 2 is used, the number of bands of *G. herbaceum* is equal to five under any conditions (Fig. 8, II), whereas for *G. hirsutum* 108F it is equal to 10 under mild conditions of washing of the filter (Fig. 8, IIa), and eight under rigorous conditions (Fig. 8, IIB). The results obtained suggest that in the genome of the wild species *G.*

292

herbaceum the *chs* gene family is represented by four to six different sequences. In the genome of the tetraploid variety *G. hirsutum* 108F, the analogous family consists of 8-10 genes, part of which are close to the genes of *G. herbaceum*. Considering the data on the origin of *G. hirsutum*, we can suggest that the *chs* gene family of this cotton species consists of two groups of genes, one of which arose from *G. herbaceum*, while the other was derived from the genome of *G. raimondii*.

To confirm these hypotheses we selected a strategy of cloning fragments of the *chs* genes, preliminary amplified by two versions of the polymerase chain reaction with primers designed based on the arrangement of consensus peptides isolated from the sequences of analogous genes of other plant species and on the basis of the nucleotide sequence of the incomplete genomic clone CGH3 that we produced [57].

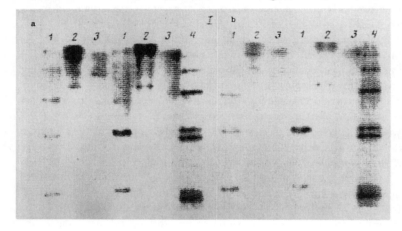

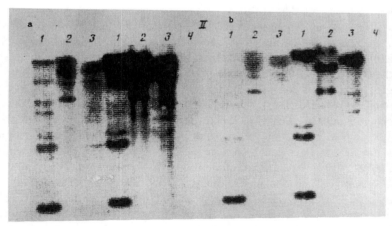

Figure 8. Blot hybridization of genomic DNA of *G. hirsutum* 108F and *G. herbaceum*. I - Probe 1 (see Fig. 9); II - probe 2 (see Fig. 9); a - mild conditions of washing: 100 mM phosphate buffer, pH 7.2, 0.1% Na dodecyl sulfate, 60°C; b - rigorous conditions of washing: 40 mM phosphate buffer, pH 7.2, 0.1% Na dodecyl sulfate, 60°C; genomic DNA digested by restriction endonuclease *Hind*III (1); *Bam*HI (2); *Eco*RI (3); phage λ DNA, digested by restriction endonuclease *Hind*III (4).

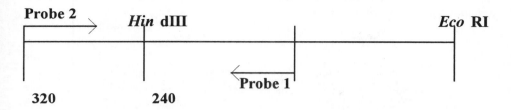

Figure 9. Mutual arrangement of probes used in blot hybridization of genomic DNA of *G. hirsutum* 108F and *G. herbaceum. Hin*dIII *Eco*RI structure of M13*mp*10, containing a fragment of the genomic clone CGH₃ of *G. hirsutum* 108F [16]: * the 3'-end of the coding sequence.

A comparative analysis of the DNA structure of the clones obtained, containing the amplified DNA species showed that the distribution of conservative portions and single substitutions in the nucleotide sequences is very similar within each gene family. This is evidence of a high degree of homology between the *chs* genes of *G. herbaceum* and *G. hirsutum* 108F. In a further comparison of the clones of one cotton species with the other it was found that two clones of *G. herbaceum* are identical with two clones of *G. hirsutum* 108F; one of these pairs (a clone of type 5 of *G. hirsutum* 108F and a clone of type 4 of *G. herbaceum*) is identical with the genomic clone CGH₃.

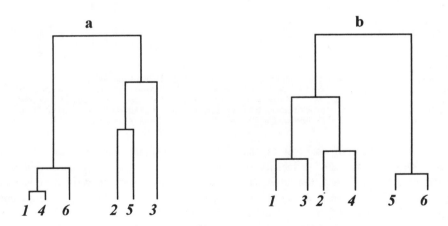

Figure 10. Dendrograms of clones containing fragments of *chs* genes of *G. hirsutum* 108F (a) and *G. herbaceum* (b), constructed on the basis of the nucleotide sequences; the numbers denote the numbers of the clones.

Figure 10 presents dendrograms reflecting the evolutionary relationship of the *chs* genes among the cotton species investigated. An analysis of the dendrograms revealed a similar development of the genes of both families, which can be represented as a series of successive duplications of the initial sequence with accumulation of point mutations in the duplicated copies.

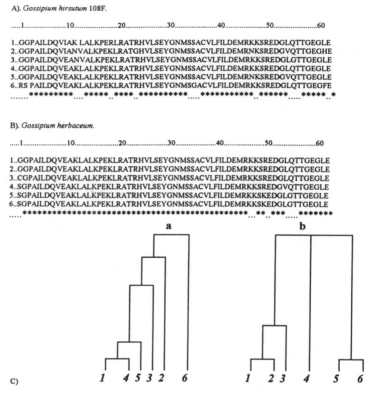

A). *Gossipium hirsutum* 108F.

```
.....1...............10................20................30................40................50................60

1..GGPAILDQVIAK LALKPERLRATRHVLSEYGNMSSACVLFILDEMRKKSREDGLQTTGEGLE
2..GGPAILDQVIANVALKPEKLRATGHVLSEYGNMSSACVLFILDEMRNKSREDGVQTTGEGHE
3..GGPAILDQVEANVALKPEKLRATRHVLSEYGNMSSACVLFILDEMRKKSREDGLGTTGEGLE
4..GGPAILDQVEAKLALKPEKLRATRHVLSEYGNMSSACVLFILDEMRKKSREDGLQTTGEGLE
5..GGPAILDQVEAKLALKPEKLRATRHVLSEYGNMSSACVLFILDEMRNKSREDGVQTTGEGLE
6..RS PAILDQVEAKLALKPEKLRATRHVLSEYGNMSGACVLFILDEMRKKSREDGLQTTGEGFE
........*********....*****..****.**********.....***********.******....*****.*
```

B). *Gossipium herbaceum.*

```
.....1...............10................20................30................40................50................60

1..GGPAILDQVEAKLALKPEKLRATRHVLSEYGNMSSACVLFILDEMRKKSREDGLQTTGEGLE
2..GGPAILDQVEAKLALKPEKLRATRHVLSEYGNMSSACVLFILDEMRKKSREDGLQTTGEGLE
3..CGPAILDQVEAKLALKPEKLRATRHVLSEYGNMSSACVLFILDEMRKKSREDGLQTTGEGLE
4..SGPAILDQVEAKLALKPEKLRATRHVLSEYGNMSSACVLFILDEMRKKSREDGVQTTGEGLE
5..SGPAILDQVEAKLALKPEKLRATRHVLSEYGNMSSACVLFILDEMRKKSKEDGLGTTGEGLE
6..SGPAILDQVEAKLALKPEKLRATRHVLSEYGNMSSACVLFILDEMRKKSKEDGLGTTGEGLE
.....**********************************************...**..***.....*******
```

C)

Figure 11. Comparison of amino acid sequences derived on the basis of the DNA structure of *chs* fragments from *G. hirsutum* 108F (A) and *G. herbaceum* (B). *Conservative amino acids; the numbers of amino acids from the beginning of the cloned fragments are indicated above the sequences; the numbers on the left indicate the numbers of the clones. (C) Dendrograms of clones containing fragments of *chs* genes of *G. hirsutum* 108F(a) and *G. herbaceum* (b), constructed on the basis of the amino acid sequences.

A comparative analysis of the CHS protein fragments obtained (data not shown) showed that the variability within each family at the protein level is significantly lower than at the DNA level. We have compared the derived consensus sequences of *G. hirsutum* 108F and *G. herbaceum* and found that these CHS fragments of two cotton species differ in only six positions (Fig. 12).

The dendrograms (Fig. 11) obtained on the basis of the investigated amino acid sequences of CHS from *G. hirsutum* 108F and *G. herbaceum* illustrate our hypotheses on the course of evolution of the investigated CHS, while the branching after the first duplication of the initial sequence of clone 6 of *G. hirsutum* 108F again indicates that its genome contain a second group of *chs* genes, evidently derived from the second parental form - *G. raimondii*.

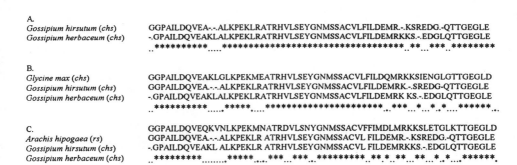

Figure 12. A) Comparison of consensus amino acid sequences of fragments of chalcone synthase (*chs*) clones of *G. hirsutum* 108F and *G. herbaceum*, derived on the basis of the clones obtained; B) comparison of the amino acid sequences of a fragment of the chalcone synthase gene *chs*1 of *Glycine max* [1] and consensus amino acid sequences of *G. hirsutum* 108F and *G. herbaceum*; C) comparison of the amino acid sequences of a fragment of the resveratrol synthase (*rs*) gene of *Arachis hypogaea* [67] and consensus sequences of *G. hirsutum* 108F and *G. herbaceum*. *Conservative amino acids.

Recently the gene for resveratrol synthase (*rs*) - a key enzyme of stilbene biosynthesis - was cloned from the peanut (*Arachis hypogaea*) genome [67]. Just like *chs*, *rs* utilizes three malonyl-CoA molecules and one coumaroyl-CoA molecule as substrates; however, the reaction products are different for these enzymes: naringenin chalcone for *chs* [32] and resveratrol for *rs* [34]. It is known that the structure of the two enzymes are very close; therefore it is believed that it will be possible to prove whether the investigated genes clone *chs* or *rs* only by isolating the protein products corresponding to them and checking their enzymatic activity [53]. Without having such data available as yet, at this stage of the investigations we compared the consensus amino acid sequences derived for *G. hirsutum* 108F and *G. herbaceum* with the analogous sequences of peanut *rs* [34] and soybean (*Glycibe max*) *chs* [1], i.e., *rs* and *chs* genes from plants of the same family were selected for comparison. As can be seen from the data presented, the sequences that we cloned are closer in number, arrangement, and nature of the amino acid substitutions to the soybean *chs* than to the peanut *rs*. These results enabled us to assert that the gene fragments that we cloned code for the enzyme CHS.

An analysis of the DNA structure and amino acid sequences of the clones obtained showed the evolutionary interrelationship of the *chs* genes and their protein products within the families investigated. The currently available data do not permit to establish the biological meaning of the existence of a family of genes for an enzyme that performs a quite specialized function. However, the absolute identity of two pairs of

296

fragments for both species investigated may be considered as an indication that precisely these sequences correspond to the actively expressed members of the family.

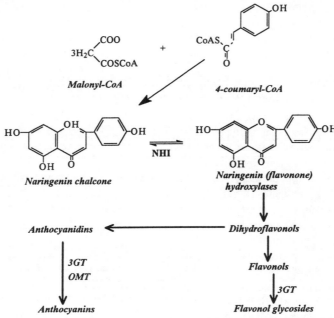

Figure 13. Biochemical pathway for the synthesis of flavonoids in plant cells. CHS - chalcone synthase.

Thus we have shown that in two cotton species (*G. hirsutum* 108F and *G. herbaceum*) CHS is encoded by gene families. These families include 8-10 members in *G. hirsutum* 108F and 6-8 members in *G. herbaceum* [13]. The next stage of our research was the identification of patterns of expression of CHS genes in petal tissues of these cotton species.

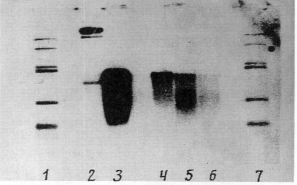

Figure 14. Blot hybridization of RNA isolated from G. hirsutum 108F and G. herbaceum tissues: 1, 7 - phage λ DNA cleaved by HindIII; 2 - DNA of M13mp10 clone containing the 3'-terminal fragment of the G. hirsutum 108F chs gene; 3 - RNA isolated from G. hirsutum 108F flower tissues; 5, 6 - RNA isolated from G. herbaceum mature leaves.

An autoradiograph of the hybridization of RNA isolated from G. hirsutum 108F and G. herbaceum petals and G. hirsutum 108F mature leaves is presented in Fig. 14. Single-stranded DNA fragment 150-200 nucleotides in length, encompassing the 3'-terminus of the translated region of the G. hirsutum 108F chs gene, was used as a probe. RNA blot hybridization revealed the presence of a single transcript, about 1200 nucleotides in length in all tissues examined, which is in agreement with the length of known CHS mRNAs from other plant species. However, the level of chs gene expression in G. hirsutum 108F petals was significantly higher than in G. herbaceum petals and G. hirsutum 108F leaves.

In studing the composition chs gene families in G. hirsutum 108F and G. herbaceum, we showed that in the 3'-transcribed region the degree of homology between members of the same family is equal to 92-97%. This means that actively transcribed genes can be identified only by cloning and sequencing.

To solve this problem we chose the reverse transcriptase-PCR method [70] using total RNA preparations from G. hirsutum 108F and G. herbaceum petals. The oligonucleotides used for PCR were the same as those used for studing the molecular organization of the chs genes in the two cotton species. The DNA fragments obtained by PCR were cloned in the M13mp10 phage vector. The clones were analyzed by sequencing, and their sequences were compared with the chs gene sequences of G. hirsutum 108F and G. herbaceum. An analysis of the data showed that the same chs gene is expressed in petal tissues of both cotton species (chs5 in G. hirsutum 108F and chs4 in G. herbaceum) [13]. It is interesting to note that this gene is also found in the genomic CGH$_3$ clone.

Thus by using the reverse transcriptase-PCR procedure with total RNA preparations isolated from G. hirsutum 108F and G. herbaceum petals, we were able to show that only one chs gene of the multigenic family is expressed in these tissues. These findings do not contradict the data on chs gene expression in other plant species. One of the charcreristics features of all chs genes is the dependence of their expression on the stage of plant development and on the tissue type [33, 65]. So in petunia plants only two of eight chs genes - chs-A and chs-J - are expressed in flowers, whereby the level of expression of the chs-J gene is only 10% of that of the chs-A gene [43]. During the early stages of flower development chs genes were found to be expressed only in anthers tissues. As flower development continues, the level of chs genes expression in anthers falls, while in petals it increases rapidly and remains at a constant level until the flower fades. Similar data were obtained for other plant species [33, 66]. It is possible that in the cotton species studied by us the expression of individual members of the chs family also depends on the flowers developmental stage. Furthermore, since G. hirsutum 108F is a natural amphidiploid obtained by cross-pollination of G. herbaceum and G. raimondii and contains, as we propose, two classes of chs genes within this family [13], other, yet unidentified genes of the G. raimondii genome are likely to be expressed in G. hirsutum 108F petals (in addition to the chs5 gene, which is identical to the chs4 gene of G. herbaceum).

4. Conserved Structure and Organization of B Hordein Genes in the *Hor2* Locus of Barley [83].

One of the promising representatives of the plant gene families, useful in the improvement of food quality of cereals are those coding for the seed storage proteins. However, before the potential of genetic manipulating with the seed strorage protein genes can be put to the service of plant breeders, substantial gaps remain to be filled.

The work by Vicente-Carbajosa and coworkers in our lab [83] presents new information on the structure and organization of B hordein genes in the *Hor2* locus of barley.

B hordeins are the major group of seed storage proteins that accumulate in the barley endosperm. Genetic studies and extensive characterization at the protein level showed that they are encoded by a multigene family tightly linked at the *Hor2* locus. Further work demonstrated the existence of 10-25 members organized in 2 subfamilies (B1 and B3) [49] that comprise at least 85 kb of DNA. Three genes belonging to the B1 subfamily [11, 15, 23] and several cDNAs of B1 and B3 types have been cloned and characterized (for a review see [69]).

We have used the Southern blot analysis to investigate the organization of the B hordein genes in the *Hor2* locus of barley. Considering data from previously characterized B hordein genes and cDNAs, a range of restriction endonucleases were selected according to the following criteria (a) enzymes with more than one site within the coding region, which would provide information on sequence conservation and possible insertion/deletion events in the genes; (b) enzymes with only one or no sites in the coding region. Two classes of enzymes were selected in this group: frequent cutters (e.g. *Sau*3A1) and rare cutters (e.g. *Bam*HI) to obtain information on the structure of the intergenic spacers at the gene boundaries and at longer distances respectively.

An example of the results obtained from such analysis is presented in Fig. 15A and B and can be summarized as follows. The first group of enzymes (e.g. *Sau*3A1: lanes 1 and 8 in Fig. 15A and B) produce a pattern upon hybridization with a B hordein probe (see Fig. 15D for a probe description) in which most of the signal is contained in one or a few fragments; indicating a conservation of their sites in most members of the gene family and hence suggesting a high degree of homology among them. In contrast when enzymes of the second group are used, both rare cutters (e.g. *Bam*HI or *Bgl*II) and frequent cutters (e.g. *Dde*I) give a picture of multiple bands of different lengths and intensities (see Fig. 15A and B). This heterogeneity is consistent with the presence of polymorphic intergenic spacers in which the occurence of these sites is variable.

Altogether these data strongly suggest that the *Hor2* locus is composed of a conserved repeated unit (here detected in the *Taq*I digestion) containing the whole coding region and about 400 bp of the promoter of the B hordein genes (Fig. 15D). This unit may be present in variable numbers of copies, separated by stretches of different length and structure accounting primarily for the polymorphism of the locus [16].

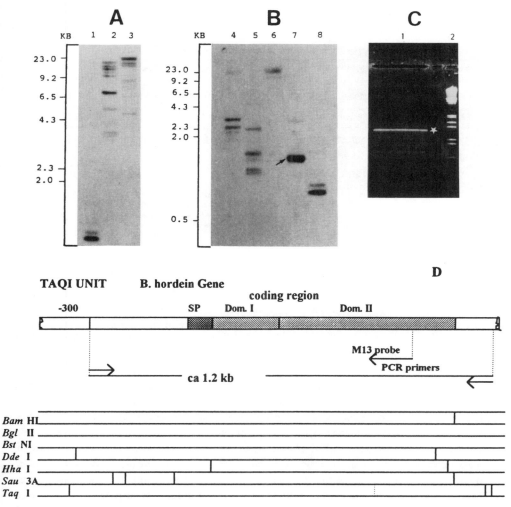

Figure 15. A and B. Southern blot analysis of genomic DNA from barley (cv. Moskovsky 3) using an M13 B hordein probe (see part D). Digestions of DNA with frequent cutter (f.c.) or rare cutter (r.c.) restriction endonucleases correspond to lanes 1 and 8: *Sau*3A1 (f.c.); 2: *Bgl*II (r.c.); 3: *Bam*HI (r.c.); 4: *Bst*NI (f.c.); 5: *Dde*I (f.c.); 6: *Hha*I (f.c.); 7: *Taq*I (f.c.). The arrow in B indicates the *Taq*I band containing the conserved unit described in the text.

C. PCR amplification of B hordein sequences. Lane 1: separation of reaction products using 200 ng of barley genomic DNA under conditions of [2 min at 92°C + 2 min at 37°C + 4 min at 72°C] x 25 cycles with primers 5'-ATGTAAAGTGAATAAGGT-3' and 5'-CTACATCGACATATACATC-3'. The star shows the 1.2 kb band from which the B hordein sequences described were obtained. Lane 2: *Eco*RI & *Hin*dIII digestion of λ DNA.

D. Proposed structure of the conserved repeated unit containing the B hordein genes detected by Southern blot analysis. Different elements and domains of the genes are indicated as well as the regions from which the B hordein probe and the PCR primers were derived. Solid triangles show the positions of the *Taq*I target sites. Other restriction sites are indicated according to the sequence of pBHR184 [6]. Note that this clone (exceptionally) contains an internal *Taq*I site (dotted line) interrupting the coding region.

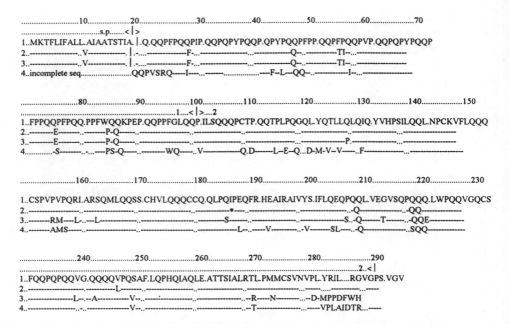

Figure 16. Deduced protein sequences of the products encoded by pcr47 (2.) and pcr31 (3.) compared to those of pBHR184 (1.) [23] and pB7 (4.) [24] representing B1- and B3-type hordeins respectively. Variable positions with respect to pBHR184, the predicted signal peptide cleavage site and limits of the two domains in the proteins are indicated. The asterik at position 188 in 2. indicates the stop codon in clone pcr47.

The detection of a repeated unit with a conserved structure by Southern blot analysis suggested that PCR amplification could be used to isolate further genomic clones as the most straightforward procedure. Two primers were designed for this purpose, one corresponding to the -300 element, a putative *cis*-acting factor highly conserved in the promoters of nearly all prolamin genes sequenced so far and presumably associated with their seed-specific expression [23]. A second primer was based on the region containing the 3' *Taq*I sites since results obtained by Southern blot analysis (see Fig. 15 and discussion above) indicated that it is also conserved in most of the hordein genes. The use of these primers under non-stringent conditions allowed the amplification of a band of the expected size that was isolated and cloned for further analysis (see Fig. 15C). Other minor bands resulting from the amplification were also isolated and analysed but did not contain B hordein-related sequences.

In the analysis and selection of representative clones from the amplified band the following stages were carried out. Four amplification experiments were performed using barley genomic DNA under conditions described in Fig. 15C. The products corresponding to the expected 1.2 kb were independently cloned and random clones selected from each individual experiment. Terminal sequences of these clones allowed a preliminary classification into two groups based on homologies at the 3' end. Subsequently, the complete nucleotide sequences of clones representing each of the two groups were determined [83]. These clones (called pcr31 and pcr47) were selected on the basis of their presence in all four amplification-cloning experiments.

The initial classification of B hordeins into B1 and B3 types was based on protein characteristics including M_r, pI, CNBr mapping patterns and organization of the repetitive domain [69]. Nevertheless, the limited information on nucleotide sequences, in particular the lack of a complete B3 type coding sequence, makes it difficult to classify any new gene simply by sequence comparisons. Figure 15 compares the deduced amino acid sequences of pcr31 and pcr47 with those of previously characterized clones pBHR184 (B1 type) [23] and pB7 (B3 type) [24]. Like all other known B hordein genes, pcr31 and pcr47 are intronless and encode proteins with a putative signal peptide followed by two well defined domains [50]. The N-terminal domains of the proteins consist of repeats and are rich in proline and glutamine. Nucleotide substitutions are present throughout the region coding for this domain in the different clones analysed. In addition some genes have differences affecting whole blocks of repeats (data not shown), supporting a possible mechanism of slippage [19] in their evolution. The 3' regions of the genes encode a non-repetitive domain with highly conserved cysteine residues. Within this domain the C-terminus is the most divergent region with differences (both in length and in sequence) between the B hordein subfamilies.

The C-terminus of the protein encoded by pcr31 differs from those of B1 and B3 hordeins so far described, but homology in the 3'-untranslated region suggest that it may be more closely related to the B3 type (see Fig. 16). Apart from punctual substitutions, pcr47 is essentially a B1 member equivalent to pBHR184. It is also probably a pseudogene since it contains an internal stop codon. Similar cases have previously been identified in other prolamin gene families are reported in maize to give rise to truncated proteins [84]. In the present case it is difficult to establish whether pcr47 is a true pseudogene due to the characteristics of the cloning strategy adopted where polymerase misincorporation can not be easily ruled out.

Altogether, results reported here provide a basis for further studies of B hordein structure and organization, and detailed examination of the Hor2 locus by cloning and analysis of DNA from the intergenic spacers between the genes. In particular, the use of amplification conditions described above can considerably accelerate the validation of positive clones in the screening of genomic libraries.

5. The Obtaining Transgenic Plants Resistant to Viruses, Herbicides, Insects

During the last fifteen years the research in our Centre "Bioengineering" have yielded an impressive collection of transgenic plants. and some of those are listed in the table below. The introduced genes confer various useful traits, like resistance to viruses, herbicides, insects. They can also be used to study some basic feautures of gene expression in plants, for instance, position effects, the phenomena of co-suppression, gene silencing and others.

TABLE VI. Transgenic plants obtained in the Centre of Bioengineering of Russian Academy of Sciences.

Plant name	Introduced gene	Methods of analysis
Nicotiana tabacum Nt-XA3'	Antisense sequence of 3'-end region of genomic RNA of PVX which contain the genes of 12 kDa, 8 kDa and coat protein (CP)	Blot hybridization analysis of plant DNA, RNA; Western blot analysis
Nicotiana tabacum Nt-X3'	Block of genes of 12 kDa, 8 kDa and CP PVX	Blot hybridization analysis of plant DNA, RNA
Nicotiana tabacum Nt-XP	Gene of CP PVX under the control of CaMV 35S promoter and terminator	Blot hybridization analysis of plant DNA; Western blot analysis
Nicotiana tabacum Nt-XA	Antisense RNA of gene CP PVX under the control of CaMV 35S promoter and terminator	Blot hybridization analysis of plant RNA
Nicotiana tabacum Nt-Ad	Gene of rat (2-B)-oligo-adenylatsynthase under the control of CaMV 35S promoter and terminator	Blot hybridization analysis of plant DNA, RNA
Nicotiana tabacum Nt-alfa14	Gene of human alfa-interferon under the control of CaMV 35S promoter and NOS terminator	Blot hybridization analysis of plant DNA, RNA; Western blot analysis
Nicotiana tabacum Nt-alfa15	Gene of human alfa-interferon under the control of CaMV 35S promoter and NOS terminator	Blot hybridization analysis of plant DNA, RNA; it was shown the synthesis of alfa-interferon
Nicotiana tabacum Nt-bar	Gene of phosphinothricin-acethyltransferase (*bar*) under the control of CaMV 35S promoter and terminator	PCR analysis; blot hybridization analysis of plant DNA, RNA
Nicotiana tabacum Nt-YP	Gene of CP PVY under the control of CaMV 35S promoter and terminator	PCR analysis; blot hybridization analysis of plant DNA, RNA; Western blot analysis
Solanum tuberosum SB-Y3	Gene of CP PVY with the 5'-lider sequence of PVX under the control of CaMV 35S promoter and terminator	PCR analysis; blot hybridization analysis of plant DNA, RNA; Western blot analysis; field performance
Solanum tuberosum SB-Y1	Gene of CP PVY under the control of CaMV 35S promoter and terminator	PCR analysis; blot hybridization analysis of plant DNA, RNA; Western blot analysis; field performance
Solanum tuberosum ST-delta1	BT gene of delta-endotoxin from *Bacillus thuringiensis* var. *tenebrionis*	Kanamycin resistant regenerants; PCR analysis; blot hybridization of plant RNA; Western blot analysis; field performance
Solanum tuberosum ST-YP	Gene of CP PVY under the control of CaMV 35S promoter with 5'-lider sequence of PVX and CaMV 35S terminator	Kanamycin resistant regenerants; PCR analysis

PVX - Potato Virus X. PVY - Potato Virus Y.

The description of the procedures used for the construction of some of these plants and the comparative analysis of their performance in laboratory and field conditions are given in more details in the sections below.

5.1. TRANSGENIC PLANTS RESISTANT TO HERBICIDE PHOSPHINOTHRICIN [59].

Transgenic mechanisms of herbicide resistance have obvious commercial significance in addition to their use as selectable markers for transformation. Considerable progress has been made in devising strategies for increasing herbicide tolerance in plants (for review see [7]). One of this strategies envisions the transformation of plants with foreign genes which detoxify herbicides. An example is provided by the herbicide Basta/phosphinothricin, which inhibits glutamin synthetase, leading to accumulation of ammonia. We obtained field resistance to Basta through use of the *bar* gene from *Streptomyces hydroscopicus*. The *bar* gene encodes an acetiltransferase which acetylates the active component of Basta, and renders it inactive as a herbicide. This strategy is reminiscent of the mechanism to which certain corn lines owe their natural tolerance of atrazine: detoxification of the herbicide through conjugations with glutathione-S-transferase. In the context of selectable marker genes, the *bar* gene is probably the most popular of the herbicide resistance genes.

The development of Basta-resistant crop plants will permit the use of this herbicide in new situations and encourage the phasing-out of the highly toxic herbicides currently in use. This strategy, of course, would eventually be rendered ineffective if there were gene transfer between the transgenic crop and wild relatives.

The aim of our work consist of obtaining and studing of transgenic plants of tobacco (*Nicotiana tabacum*) SR1 and "Prigojy-2" potato plants (*Solanum tuberosum*) which have resistance to herbicide L-phosphinothricin. We have previously cloned [68] the *bar* gene in pVGB6. The wild *bar* gene isolated from *Streptomyces hygroscopicus* has the GTG start codon. However, this codon non effective for gene expression in plants, 80-85% from effectivity of ATG start codon. We have modified the wild bar gene by PCR by alter GTG on ATG codon and cloned the resulting open reading frame under the control of CaMV 35S promoter and *nos*-terminator into pBI121 [40]. So, we have obtained the expression cassette of *bar* gene in the pBIBar (Fig. 17).

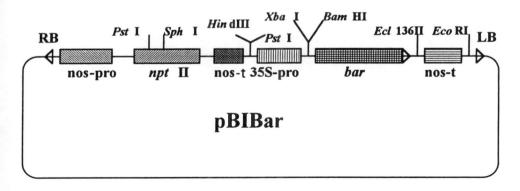

Figure 17. Cassete of expression of *bar* gene in the pBIBar.

Then we have obtained the transformed by pBIBar *Agrobacterium* strain LBA4404 and used this strain for obtaining transgenic plants.

We have obtained transgenic tobacco plants by agrobacterial transformation of leaf disks [37] and transgenic potato plants by agrobacterial transformation of the stem segments [55].

The results of analysis showed that the integration was occured in 1-3 sites of plant genome DNA.

Then we have studied the effect of transcription of transferred gene in transgenic plants by Northern hybridization. Compairing of the expression level and the number of *bar* gene copies we suggested that in most of cases the gene expression level depends on the number integrated gene copies. However, TabBar-12 plant has a higher transcription rate and PBar-5 has a lower transcription rate. We could not find the transcripts of *bar* gene in TabBar26 plant in spite of this plant has two copies of *bar* gene.

Probably the higher transcription rate was caused by integration of cassete of *bar* gene expression into actively transcribing plant genome loci. We suggest that either lower transcription or absence of transcription were caused by methylation process because *bar* gene contain GC-rich (about 68%) sequence with many sites which able to methylate.

We have carryed out the test of obtained transgenic tobacco and potato plants in the laboratory conditions and field conditions as well. After 10 days of herbicide treatment all control plants died while all transgenic plants were resistant independently of number of bar gene copies and transcription rate. Treated transgenic tobacco and potato plants have had the normal flowering and tobacco plants have produced the seeds and potato plants have produced the normal tubers. There were not any differs between the tubers of transgenic potato plants and the tubers of control nontreated potato plants.

In our growth chamber tests transgenic potato plants were shown to be resistant to treatment by herbicide phosphinotricin (100g/ha, 1000g/ha, 4000g/ha). Treated transgenic plants were true to type and gave mini-tubers that were used for field testing 1994 to evaluate their efficacy under normal field conditions.

One mounth after sprouting transgenic plants were handtreated by herbicide phosphinothricin 1000g/ha (seriesA) and 4000g/ha (seriesB). After 10 days all PAT(-) control plants were dead (seriesB) whereas the transgenic potato PAT(+) plants were resistant to this treatment. The transgenic potato plants exhibited a normal phenotype and gave tubers that were true to type. The principal morphological plant and tuber characteristics were studied. These included plant type, leaf characteristics and tuber characteristics like type and color of the skin, depth of eyes and etc. Obtained data suggest that approximately 70% of the Prigozii-2 transgenic plants were true to type.

5.2. TRANSGENIC POTATO PLANTS RESISTANT TO Y-VIRUS INFECTION [60, 61, 71].

It has been known for many years that plants can be protected against a virulent strain of certain viruses by prior infection with an attenuated atrain of the same virus or a related virus. This phenomenon is known as cross-protection (for review see [7]). In the case of positive-strand RNA viruses, a similar protective effect is observed in transgenic plants expressing the coat protein gene of the virus. The genes used in this type of transformation are double-stranded cDNAs derived from the viral RNA. Although the mechanism of cross-protection is not fully understood, it does not always appear to require high levels of expression of the coat protein itself.

The transgenic "Byelorusskii -3" potato plants that accumulated coat protein (CP) from potato virus Y (PVY) or expressed sense- or antisense CP transcripts and accumulated no CP PVY have been produced. The PVY CP is generated by cleavage (Glu/Gly) from the polyprotein precursor by a virus encoded protease. Using PCR we attached an artificial ATG codon to the PVY CP coding sequence and cloned the resulting open reading frame under the control of CaMV 35S promoter and terminator in both sense and antisense orientation. Moreover, the 5'-untranslated leader of potato virus X genome RNA has been engineered in front of the start codon to increase translational efficiency of CP RNA.

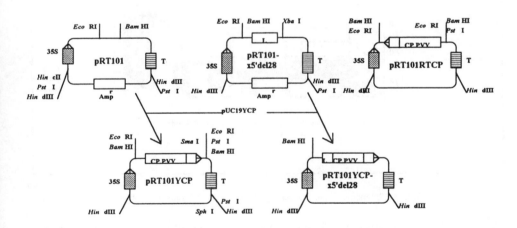

Figure 18. Cassettes of expression of CP PVY gene and asRNA of CP PVY.

The resulting constructs (Fig. 18) were transfered into potato genome. *Agrobacterium tumefaciens*-mediated transformation of the steam segments. Several

306

independent lines of transgenic potato plants for each cassettes were regenerated and analysed by Southern, Northern and Western blot hybridization methods.

The introduction the 5'-leader of PVX genomic RNA just upstream of the ATG start codon increases both mRNA steady-state levels and the coat protein accumulation. The coat protein level in plants that contained the cassett without PVX leader was below the detection threshold.

All transgenic potato plants were challenged by mechanical inoculation with PVY. Transgenic potato plants with low level of PVY CP RNA expression and undetectable CP were revealed to be protected from PVY infection better than transgenic plants accumulated PVY CP or antisense PVY RNA. The results obtained suggest that the resistance observed in the transgenic plants is principally based on the presence of PVY CP RNA rather than on the accumulation of virus coat protein.

All transgenic potato plants gave mini-tubers that were used for field testing to evaluate their efficacy under normal field conditions. A total of 10 different morphological plant and tuber characteristics were studied. These included plant type, leaf characteristics and tuber characteristics like type and color of the skin, depth of eyes and etc. Obtained data suggest that approximately 80% of the potato transgenic plant true to type.

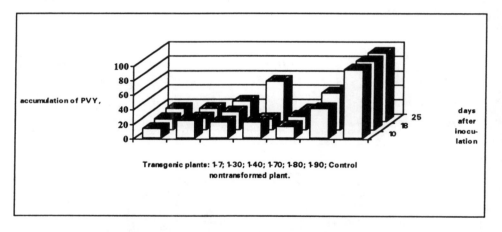

Figure 19. Diagram of accumulation of PVY in "CP " transgenic plants.

Field resistance of wild type and transgenic potato plants (cv.Byelorusskii-3) that accumulated the coat protein (CP) of potato virus Y (PVY) (lines 3/CP), or expressed plus- (lines 1/RNA) or minus-sense CP transcripts and accumulated no CP (lines 4/asRNA) was studied. Plants were subjected to mechanical inoculation with PVY. It was shown that transgenic plants of the lines 1/RNA accumulate no PVY and did not develop the disease symptoms. Besides there was no PVY in the plants 1/RNA that were grown from mini-tubers. At the same time all transgenic plants that were grown from mini-tubers of the lines 3/CP and 4/asRNA contained high levels of PVY although parental plants accumulated at least 4-fold less PVY as compared to controls. Thus low levels of CP RNA expresion appears to reduce multiplication of the PVY in the tubers (line 1/RNA) while expression of the PVY CP or CP asRNA do not prevent PVY accumulation in the tubers of the transgenic plants of the lines 3/CP and 4/asRNA. The results obtained suggest that the transgenic potato plants resistant to viruses are effective under field conditions.

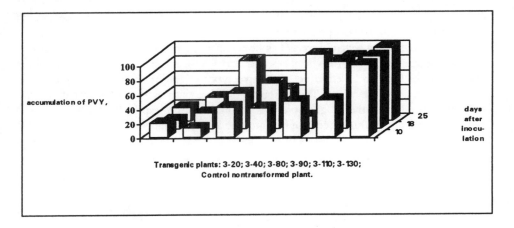

Figure 20. Diagram of accumulation of PVY in "CP⁺" transgenic potato plants.

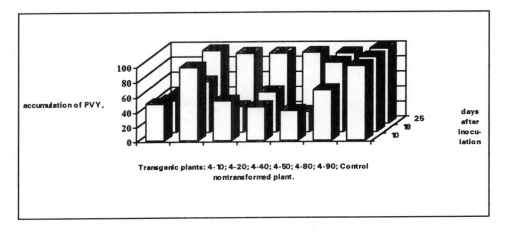

Figure 21. Diagram of accumulation of PVY in "antisense" trangenic potato plants.

5.3. TRANSGENIC POTATO PLANTS RESISTANT TO COLORADO POTATO BEETTLE [31].

Insects are the major source of yield loss, especially in the countries of the humid and sub-humid tropics. The potential attractiveness of the transgenic approach to insect control arises from five considerations: (a) the economic and human health costs of insecticide use, (b) the development of insecticide resistance in pests, (c) the counter-productive effects of insecticides on many of the natural enemies of crop pests, (d) the absence of effective host plant resistance to many insect pests, and (e) the tendency of effective host plant resistance, when it does not exist, to break down in the face of adaptive changes in the pest population.

Resistance to several insects has been enhanced through expression in plants of *Bacillus thuringiencis* (BT) toxin genes and also genes encoding proteinase inhibitors (for review see [7]). To be effective, these two types of inhibitor must be expressed in tissue consumed by the insect. The inhibitors interfere with aspects of insect digestion. BT toxins bind to epithelial glycoproteins of the intestine, especially the midgut, and cause fatal leakage of fluids between the intestine and the hemocoel.

It is known that BT genes to be poorly expressed in transgenic plants. One of the ways to solve this problem is to alter the gene base sequence, using code degeneracy.

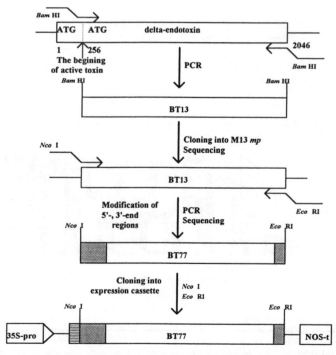

Figure 22. Scheme of modification of gene of delta-endotoxin from *Bacillus thuringiensis* var. *tenebrionis* and obtaining the cassete of expression the modified BT gene in plant genome.

The aim of our work was to obtain transgenic potato plants which have resistance to Colorado Potato Beetle. Therefore we needed to construct of modified BT gene of delta-endotoxin from *Bacillus thuringiensis* var. *tenebrionis*.

Such a modified gene was constructed (Fig. 22) and cloned into the pMON505 (Fig. 23).

This binary transformation vector have been introduced into *Agrobacterium tumefaciens* strain, containing helper disarmed Ti-plasmid CBE21 (see above). The *Agrobacterium* strain containing this vector were used to transform potato stem segments (*Solanum tuberosum*, cultivars: Desiree, Resy, Byelorussckiy-3, Temp, Granat) and selected on kanamycin containing media.

Western blot analysis revealed that transgenic plants produce the BT protein in range of 0.005-0.02% of total protein (data not shown).

After that we have examined the phenotypic variation of resistance to Colorado Potato Beetle of the obtained transgenic potatoes in the laboratory conditions and their field performance.

The leaves from 3rd-4th tiers of transgenic and nontransformed potato plants were placed into the Petry plates. 10 larvaes first age of Colorado Potato Beetle were placed on these Petry plates. Average weight of these larvaes have been about 0.6-0.7 mg. Then we have placed Petry plates into the laboratory conditions with light day 12 h and temperature 25-30°C. Every day the plant leaves have been changed. The determination of weight and number of survived larvaes have been carried out on the 6th and 21st days from the begining of nutrition (Fig. 24).

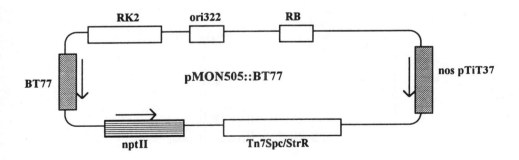

Figure 23. The binary transformation vector pMON505:BT77. RK2 - fragment of plasmid with a wide host range which contain the sites of initiation of replication and transfer; ori322 - fragment which cantain the site of initiacion of replication of pBI322; RB - the right border of T-DNA from pTiT37; *nos* pTiT37 - gene of nopalinsynthase from pTiT37; Tn7 Spc/StrR - fragment from Tn7 which contain a gene of resistance to spectinomycin/streptomycin; *npt* - chimeric gene of neomycinphosphotransferase II under the promoter and terminator of nopalinsynthase gene from T-DNA which provide with resistance to kanamycin; BT77 - partial modified gene of delta-endotoxin from *Bacillus thuringiensis* var. *tenebrionis*. The arrows indicate of direction of transcription.

310

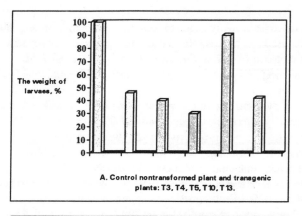

A. Control nontransformed plant and transgenic plants: T3, T4, T5, T10, T13.

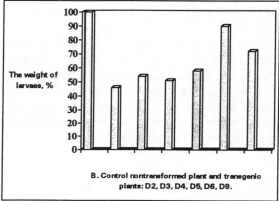

B. Control nontransformed plant and transgenic plants: D2, D3, D4, D5, D6, D9.

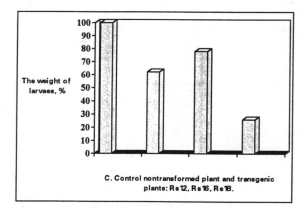

C. Control nontransformed plant and transgenic plants: Rs12, Rs16, Rs18.

Figure 24. The weight of larvaes in percents from control after 6 days of begining of nutrition by transgenic potato plant leaves. Cultivars Temp (A), Desiree (B), Resy (C).

At the same time we have carried out the visual estimation of damage of leaf surface (Table VII).

TABLE VII. Damage of leaf surface of transgenic potato plants of cultivars Temp, Desiree, Resy.

Plant number (Temp)	% of damage	Plant number (Desiree)	% of damage	Plant number (Resy)	% of damage
3	20	2	60	12	40
4	55	3	45	14	50
5	40	4	30	15	40
10	40	5	30	control	100
18	50	6	60		
control	100	9	55		
		control	100		

It is known that the Colorado Potato Beetle larvaes in the process of development have a three casting the coat. After the 2nd casting the coat the 3rd age of larvae is coming. In this period larvaes are most resistant to different influences. For the determination of oppression degree of general transgenic potato plants population on the laboratory beetle population the statistic analysis of discrimination of weight the 3rd age larvaes on the 21st day of the nutrition by transgenic plants have been carried out (Fig. 25). The weight of most number of larvaes is about 50 mg while the maximal weight of control larvaes is 116.8 mg.

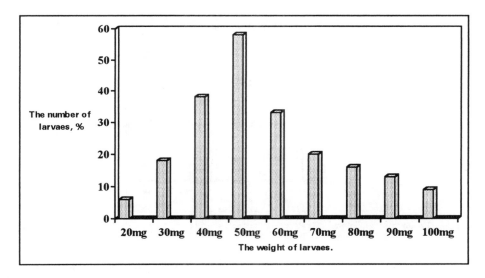

Figure 25. Independent distribution in population of transgenic potato plants of cultivar Temp which oppress the population of Colorado Potato Beetle larvaes of third age.

We have carryed out the statistic analysis of distribution of weight of third age larvaes on the 21st day of nutrition by transgenic potato plants. As shown the weight of most of larvaes is about 50 mg while the maximal weight in control is 116.8 mg.

Also we have carried out the determination of larvaes mortality which have feeded on transgenic potato plants (Fig. 26).

312

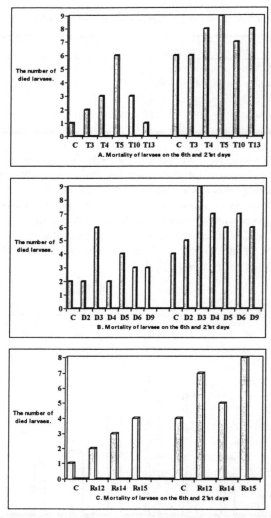

Figure 26. The mortality of Colorado Potato Beetle larvaes in conditions of nutrition by leaves of transgenic potato plants of cultuvars Temp (A), Desiree (B) and Resy (C).

Bt genes are known to be poorly expressed in transgenic plants therefore in order to increase Bt gene expression in plants we have considered that there is a necessity of modification of bacterial genes by altering the gene base sequence using code degeneracy.

The aim of one of our projects is to construct the completely modified gene of δ-endotoxin from *Bacillus thuringiensis* var. *tenebrionis* using PCR-techniques, subcloning and site-directed mutagenesis. We hope, that such modified gene will confer much better protection against CPB in transgenic potatoes.

6. CONCLUSIONS

Plant gene technology has an enormous potential in changing the current agriculture towards higher productivity, environment safety and efficeincy. Some of the transgenic varities of economically important plants with significantly improved qualities have already been approved by responsible agencies for large-scale cultivation, sale and consumption. It is becoming increasingly evident that concerns about the potential harmful consequences of transgenic plants to agricultural ecology are rather ill-substantiated as compared to the obvious hazards of various chemicals currently used in crop protection and obvious benefits of the applications of plant gene technology to modern agriculture.

In view of the urgent demands to feed the exponentially growing earth population, plant gene technology has to be considered as one of the key high technologies, capable to meet the challenges of the next century.

7. References

1. Akada, S., Kung, S.D., Dube, S.K. (1991) The nucleotide sequence of gene 1 of the soybean chalcone synthase multigene family, *Plant Mol. Biol.* 16, 751-752.
2. An, G., Costa, M.A., Ha, S-B. (1990) Nopaline synthase promoter is wound inducible aux auxin inducible, *Plant Cell* 2, 225-233.
3. An, G., Costa, M.A., Ha, S-B., Marton, L. (1988) Organ-specific and developmental regulation of the nopaline synthase promoter in transgenic tobacco plants, *Plant Physiol.* 88, 547-552.
4. An, G. Watson, B.D., Stachel, S., Gordon, M.P., Nester, E.W. (1985) New cloning vehicles for transformation of higher plants, *EMBO J.* 4, 277-284.
5. Bagyan, I.L., Revenkova, E.V., Kraev, A.S., Skryabin, K.G. (1994) Functional analysis of the 5'-flanking region of gene 6b from TL-DNA pTiBo542 in transgenic tobacco plants, *Mol. Biol.* 28, 487-492.
6. Bagyan, I.L., Revenkova, E.V., Kraev, A.S., Skryabin, K.G. (1995) 5'-Regulatory region of *Agrobacterium tumefaciens* T-DNA gene 6b directs organ-specific, wound-inducible expression in transgenic tobacco, *Plant Mol. Biol.* 29, 1299-1304.
7. Bennett, J. (1993) Genes for crop improvement, *Genetic Engineering* 15, 165-189.
8. Blochlinger, K., Diggelmann, H. (1984) Hygromycin B phosphotransferase is a selective marker for DNA transfer experiments with higher eukariotic cells, *Mol. Cell Biol.* 4, 2929-2931.
9. Bouchez, D., Tokuhisa, J., Llewellyn, D., Dennis, E., Ellis, J. (1989) The *ocs*-element is component of the promoters of several T-DNA and plant viral genes, *EMBO J.* 8, 4197-4204.
10. Bradford, M.M. (1976) A rapid and sensitive method for the quantitation of microgram quantities of protein utilizing the principle of protein dye binding, *Anal. Biochem.* 72, 248-254.

314

11. Brandt, A., Montembault, A., Cameron-Mills, V., Rasmussen, S.K. (1985) Primery structure of a B1 hordein gene from barley, *Carlsberg Res. Commun.* 50, 333-345.

12. Byrne, M.C., McDonell, R.E., Wright, M., Carnes, M.G. (1987) Strain and cultivar specificity in the *Agrobacterium*-sobean interaction, *Plant Cell* 8, 3-15.

13. Byzova, M.V., Kraev, A.S., Pozmogova, G.E., Skryabin, K.G. (1992) Molecular characteristics of the chalcone synthase gene family from two cotton species using the polymerase chain reaction, *Mol. Biol.* 26, 335-342.

14. Byzova, M.V., Kraev, A.S., Pozmogova, G.E., Skryabin, K.G. (1992) Identification of Chalcone Synthase genes specially expressed in petals of two cotton species, *Mol. Biol.* 26, 614-616.

15. Chernyshev, A.I., Davletova, S.H., Bashkirov, V.I., Chakhmanov, N.V., Mekhedov, S.L., Ananiev, E.V. (1989) Nucleotide sequence of a B1 hordein gene of barley *Hordeum vulgare* L., *Genetika (USSR)* 8, 1349-1355.

16. Chernyshev, A.I., Khadirzhanova, D.K., Pomortsev, A.A., Ananiev, E.V., Sozinov, A.A. (1988) Restriction fragment length polymorphism of the B hordein genes of barley, *Genetika (USSR)* 10, 1841-1849.

17. Cramer, C.I., Ryder, T.B., Bell, J.N., Lamb, C.J. (1985) Rapid switching of plant gene expression induced by fungal elicitor, *Science* 227, 1240-1243.

18. Datla, R.S.S., Hammerlindl, J.K., Pelcher, L.E., Crosby, W.L., Selvaraj, G. (1991)A bifunctional fusion between b-glucuronidase and neomycin phosphotransferase: A broad-spectrum marker enzyme for plants, *Gene* 101, 239-246.

19. Dover, G. (1989) Slips, strings and evolution, *Trends Genet.* 5, 100-102.

20. Dymock, D., Risiott, R., de Pater, S., Lancaster, J., Tillson, P., Ooms, G. (1991) Regulation of *Agrobacterium tumefaciens* T- *cyt* gene expression in leaves of transgenic potato (*Solanum tuberosum* L. cv. Desiree) is strongly influenced by plant culture conditions, *Plant Mol. Biol.* 17, 711-725.

21. Feinbaum, R.L., Ausubel, F.M. (1988) Transcriptional regulation of the *Arabidopsis thaliana* chalcone synthase gene, *Mol. Cell Biol.* 8, 1985-1992.

22. Firmin, J.L., Wilson, K.E., Rossen, L., Johnston, A.W.B. (1986) Flavonoid activation of nodulation genes in *Rhizobium* reversed by other components present in plants, *Nature* 324, 90-92.

23. Forde, B.G., Heyworth, A., Pywell, J., Kreis, M. (1985) Nucleotide sequence of a B1 hordein gene and the identification of possible upstream regulatory elements in endosperm storage protein genes from barley, wheat and maize, *Nucl. Acids Res.* 13, 7327-7339.

24. Forde, B.G., Kreis, M.S., Williamson, R.P., Fry, J., Pywell, P.R., Shewry, P.R., Bunce, N., Miflin, B.J. (1985) Short tandem repeats shared by B- and C-hordein cDNAs suggest a common origin for two groups of cereal storage protein genes, *EMBO J.* 4, 9-15.

25. Fraley, R.T., Rogers, S.G., Horsch, R.B., Eichholtz, D.A., Flick, J.S., Fink, C.L., Hoffmann, N.L., Sanders, P.R. (1985) The SEV system: a new disarmed Ti plasmid vector system for plant transformation, *Bio/technology* 3, 629-635.

26. Fromm, H., Katagiri, F., Chua, N.H. (1989) An octopine synthase enhancer element directs tissue-specific expression and binds ASF-1, a factor from tobacco nuclear extracts, *Plant Cell* 1, 977-984.

27. Gelvin, S.B., Gordon, M.P., Nester, E.W., Aronson, A.I. (1981) Transcription of the *Agrobacterium* Ti plasmid in the bacterium and in crown gall tumors, *Plasmid* 6, 17-29.

28. Gelvin, S.B., Karcher, S.J., Goldsbrough, P.B. (1985) Use of a TR-DNA promoter to express genes in plants and bacteria, *Mol. Gen. Genet.* 199, 240-248.

29. Grab, D., Loyal, R., Ebel, J. (1985) Elicitor-induced phytoalexin synthesis in soybean cells: Changes in the activity of chalcone synthase mRNA and total population of translatable mRNA, *Arch. Biochem. Biophys.* 243, 423-529.

30. Guevara-Garcia, A., Mosqueda-Cano, G., Arguello-Agtorga, G., Simpson, J., Herrera-Estrella, L. (1993) Tissue-specific and wound-inducible pattern of expression of the mannopine synthase promoter is determined by the interaction between positive and negative *cis*-regulatory elements, *Plant J.* 4, 495-505.

31. Gulina, I.V., Shulga, O.A., Mironov, V.N., Revenkova, E.V., Kraev, A.S., Pozmogova, G.E., Yakovleva, G.A., Skryabin, K.G. (1994) Expression of particular modified gene of Ω-endotoxin from *Bacillus thuringiensis* var. *tenebrionis* in transgenic potato plants, *Mol. Biol.* 28, 1166-1175.

32. Hahlbrock, K. (1981) Flavonoids, *Biochemistry of Plants*, Academic Press, New York, 7, 425-456.

33. Harker, C.L., Ellis, E.S., Coen, E.S. (1990) Identification and genetic regulation of the chalcone synthase multigene family in pea, *Plant Cell* 2, 185-194.

34. Higgins, D.G., Sharp, P.M. (1988) CLUSTALL: a package for performing multiple sequence alignment on a microcomputer, *Gene* 73, 237-244.

35. Hood, E.E., Jen, G., Kayes, L., Kramer, J., Fraley, R.T., Chilton, M.-D. (1984) Restriction endonuclease map of pTiBo542, a potential Ti plasmid vector for genetic engineering of plants, *Bio/technology* 2, 702-709.

36. Hooykaas, P.J., Schilperoort, R.A. (1992) *Agrobacterium* and plant genetic engineering, *Plant Mol. Biol.* 19, 15-38.

37. Horsch, R.B., Fry, J.E., Hoffmann, N.L., Eichholtz, D., Rogers, S.G., Fraley, R.T. (1985) A simple and general method for transferring genes into plants, *Science* 227, 1229-1231.

38. Inoguchi, M., Kamada, H., Harada, H. (1990) β-Glucuronidase gene expression by the Ti-agropine synthase gene promoter is preferential to callus tissue, *J. Plant Physiol.* 136, 685-689.

39. Janssens, A., Engler, G., Zambryski, P., Van Montagu, M. (1984) The nopaline C58 T-DNA region is transcribed in *Agrobacterium tumefaciens*, *Mol. Gen. Genet.* 195, 341-350.

40. Jefferson, R.A., Kavanagh, T.A., Bevan, M.W. (1987) GUS fusions: β-glucuronidase as a sensitive and versatile gene fusion marker in higher plants, *EMBO J.* 6, 3901-3907.

316

41. Jin, S., Komari, T., Gordon, M.P., Nester, E.W. (1987) Genes responsible for the supervirulent phenotype of *Agrobacterium tumefaciens* A281, *J. Bacteriol.* 169, 4417-4425.
42. Kim, Y., Buckley, K., Costa, M.A., An, G. (1994) A 20 nucleotide upstream element is essential for the nopaline synthase (*nos*) promoter activity, *Plant Mol. Biol.* 24, 105-117.
43. Koes, R.E., Spelt, C.E., Mol, J.N.M. (1989) The chalcone synthase multigene family of *Petunia hybrida* (V30): differential, light-regulated expression during flower development and UV light induction, *Plant Mol. Biol.* 12, 213-225.
44. Koes, R.E., Spelt, C.E., van den Elzen, P.J.M., Mol, J.N.M. (1989) Cloning and molecular characterization of the chalcone synthase multigene family of the chalcone synthase multigene family of *Petunia hybrida*, *Gene* 81, 245-257.
45. Koes, R.E., Spelt, C.E., Mol, J.N.M., Gerats, A.G.M. (1987) The chalcone synthase multigene family of *Petunia hybrida* (V30): sequence homology, chromosomal localization and evolutionary aspects, *Plant Mol. Biol.* 10, 159-169.
46. Koncz, C., Schell, J. (1986) The promoter of TL-DNA gene 5 comtrols the tissue-specific expression of chimaeric genes carried by a novel type of *Agrobacterium* binary vector, *Mol. Gen. Genet.* 204, 383-396.
47. Kononowicz, H., Wang, Y.E., Habeck, L.L., Gelvin, S.B. (1992) Subdomains of the octopine synthase upstream activating element direct cell-specific expression in transgenic tobacco plants, *Plant Cell* 4, 17-27.
48. Korber, H., Strizhov, N., Staiger, D., Feldwisch, J., Olsson, O., Sandberg, G., Palme, K., Schell, J., Koncz, C. (1991) T-DNA gene 5 of *Agrobacterium* modulates auxin response by auto-regulated synthesis of a grown hormone antagonist in plants, *EMBO J.* 10, 3983-3991.
49. Kreis, M., Rahman, S., Forde, B.G., Pywell, P.R., Shewry, P.R., Miflin, B.J. (1983) Sub-families of hordein mRNA encoded at the *Hor2* locus of barley, *Mol. Gen. Genet.* 191, 194-200.
50. Kreis, M., Shewry, P.R. (1989) Unusual features of cereal seed protein structure and evolution, *BioEssays* 10, 201-207.
51. Langridge, W.H.R., Fitzgerald, K.J., Koncz, C., Schell, J., Szalay, A.A. (1989) Dual promoter of *Agrobacterium tumefaciens* mannopine synthase genes is regulated by plant growth hormones, *Proc. Natl. Acad. Sci. USA* 86, 3219-3223.
52. Lichtenstein, K., Draper, J. (1987) *DNA Cloning*, Oxford Univ. Press.
53. Melchior, F., Kindl, H. (1990) Grapevine stilbene synthase cDNA only slightly differing from chalcone synthase cDNA is expressed in *E. coli* into a catalytically active enzyme, *FEBS letters* 268, 17-20.
54. Murashige, T., Skoog, F. (1962) A revised medium for rapid growth and bioassays with tobacco tissue cultures, *Plant Physiol.* 15, 473-497.
55. Newell, C.A., Rozman, R., Hinchee, M.A., Lawson, E.C., Haley, L., Sandersn, P., Kaniewski, W., Tumer, N.E., Horsch, R.B., Fraley, R.T. (1991) *Agrobacterium*-mediated transformation of *Solanum tuberosum* L. cv. "Russet Burbank", *Plant Cell Reports* 10, 30-34.

56. Neuteboom, S.T., Hulleman, E., Schilperoort, R.A., Hoge. J.H. (1993) In planta analysis of the *Agrobacterium tumefaciens* T-*cyt* gene promoter: identification of an upstream region essential for promoter activity in leaf, stem and root cells of transgenic tobacco, *Plant Mol. Biol.* 22, 923-929.

57. Niesbach-Klosgen, U., Barzen, E., Benhardt, J., Rohde, W., Schwarz-Sommer, Z., Reif, H.J., Wienand, U., Saedler, H. (1987) Chalcone synthase genes in plants: A tool to study evolutionary relationships, *J. Mol. Evol.* 26, 213-225.

58. Padegimas, L.S., Shulga, O.A., Skryabin, K.G. (1993) The analysis of transgenic plants by polymerase chain reaction, *Mol. Biol.* 27, 947-951.

59. Padegimas, L.S., Shulga, O.A., Skryabin, K.G. (1994) The obtaining of transgenic plants of *Nicotiana tabacum* and *Solanum tuberosum* resistant to herbicide phosphinotrycin, *Mol. Biol.* 28, 437-443.

60. Pooggin, M.M., Skryabin, K.G. (1992) The 5'-untranslated leader sequence of potato virus X RNA enhances the expression of a heterologous gene *in vivo*, *Mol. Gen. Genet.* 234, 329-331.

61. Pugin, M.M., Sokolova, M.A., Shulga, O.A., Skryabin, K.G. (1994) The effect of 5'-leader of Potato X-virus on expression of coat protein gene of Potato Y-virus in transgenic plants *Solanum tuberosum*, *Mol. Biol.* 28, 752-760.

62. Reimold, U., Kroger, M., Kreuzaler, F., Hahlbrock, K. (1983) Coding and 3' noncoding nucleotide sequence of chalcone synthase mRNA and assignment of amino acid sequence of the enzyme, *EMBO J.* 2, 1801-1805.

63. Revenkova, E.V., Bagyan, I.L., Kraev, A.S., Skryabin, K.G. (1993) Primary structure of pTiBo542 T-DNA, *Mol. Biol.* 27, 28-32.

64. Revenkova, E.V., Kraev, A.S., Skryabin, K.G. (1991) Transformation of coton (*Gossipium hirsutum* L.) with a supervirulent strain of *Agrobacterium tumefaciens* A281, *Mol. Biol.* 24, 820-825.

65. Ryder, T.B., Hedrick, S.L., Bell, J.N., Liang, X., Clouse, S.D., Lamb, C.J. (1987) Organization and differential activation of a gene family uncoding the plant defense enzyme chalcone synthase in *Phaseolus vulgaris*, *Mol. Gen. Genet.* 210, 210-233.

66. Schmelzer, E., Jahnen, W., Hahlbrock, K. (1988) *In situ* localization of light-induced chalcone synthase mRNA, chalcone synthase, and flavonoid and products in epidermal cells of parsley leaves, *Proc. Natl. Acad. Sci. USA* 85, 2989-2993.

67. Schroder, G., Brown, J.W.S., Schroder, J. (1988) Molecular analysis of resveratrol synthase cDNA, genomic clones and relationship with chalcone synthase, *Eur. J. Biochem.* 172, 161-169.

68. Sezonov, G.V, Tabacov, Yu.V., Kudryashova, E.A. (1990) The expression of *bar* gene in the *Streptomyces* strains, *Anthibiotics and Chemiotherapia* (Russian) 35, 24-26.

69. Shewry, P.R., Tatham, A.S. (1990) The prolamin storage protein of cereal seeds: structure and evolution, *Biochem J.* 267, 1-12.

70. Singer-sam, J., Robinson, M.O., Bellve, A.K., Simon, M.I., Riggs, A.D. (1990) Measurement by quantitative PCR of changes in HPRT, PGK-1, PGK-2, APRT,

318

MTase, and Zfy gene transcripts during mouse spermagenesis, *Nucl. Acids Res.* 18, 1255-1259.

71. Sokolova, M.A., Pugin, M.M., Shulga, O.A., Skryabin, K.G. (1994) The obtaining of transgenic plants of *Solanum tuberosum* resistant to Potato Y-virus, *Mol. Biol.* 28, 1002-1008.

72. Sommer, H., Saedler, H. (1986) Structure of the chalcone synthase gene of *Antirrhinum majus*, *Mol. Gen. Genet.* 202, 429-434.

73. Southern, E.M. (1975) Detection of specific sequences among DNA fragments separated by gel electrophoresis, *J. Mol. Biol.* 98, 503-517.

74. Spanier, K., Schell, J., Schreier, P.H. (1989) A functional analysis of T-DNA gene 6b: The fine tuning of cytokinin effects on shoot development, *Mo;. Gen. Genet.* 219, 209-216.

75. Strabala, T.J., Crowell, D.N., Amasino, R.M. (1993) Levels and location of expression of the *Agrobacterium tumefaciens* pTiA6 *ipt* gene promoter in transgenic tobacco, *Plant Mol. Biol.* 21, 1011-1021.

76. Straumal, B.P. (1977) *Cultivar 108F* [in Russian], FAN.

77. Teeri, T.H., Lehvaslaiho, H., Franck, M., Uotila, J., Heino, P., Palva, E.T., Van Montagu, M., Herrera-Estrella, L. (1989) Gene fusions to *lacZ* reveal new expression patterns of chimaeric genes in transgenic plants, *EMBO J.* 8, 343-350.

78. Ter-Avapesyan, D.V. (1973) *The Cotton Plant* [in Russian], Kolos, Leningrad.

79. Tinland, B., Fournier, P., Heckel, T., Otten, L. (1992) Expression of a chimaeric heat-shock-inducible *Agrobacterium* 6b oncogene in *Nicotiana rustica*, *Plant Mol. Biol.* 18, 921-930.

80. Tinland, B., Huss, B., Paulus, F., Bonnard, G., Otten, L. (1989) *Agrobacterium tumefaciens* 6b genes are strain-specific and affect the activity of auxin as well as cytokinin genes, *Mol. Gen. Genet.* 219, 217-224.

81. Tinland, B., Rohfritsch, O., Michler, P., Otten, L. (1990) *Agrobacterium tumefaciens* T-DNA gene 6b stimulates *rol*-induced root formation, permits growth at high auxin concentrations and increases root size, *Mol. Gen. Genet.* 223, 1-10.

82. Turcotte, E.L., Konel, R.J. (1985) Genetics, cytology and evolution of *Gossipium*, *Adv. Genet.* 23, 271-375.

83. Vicente-Carbajosa, J., Beritashvili, D.R., Kraev, A.S., Skryabin, K.G. (1992) Conserved structure and organization of B hordein genes in the *Hor2* locus of barley, *Plant Mol. Biol.* 18, 453-458.

84. Viotti, A., Cairo, G., Vitale, Sala, E. (1985) Each zein gene class can produce polypeptides of different sizes, *EMBO J.* 4, 1103-1111.

85. Zambryski, P. (1982) *The use of Ti plasmids as cloning vectors for genetic engineering in plants*, EMBO Course, Lab. Genetics, Rijkuniversiteit, Ghent, 24-25.

THE 5'-UNTRANSLATED LEADERS OF BSMV RNAgamma AND PVX COAT PROTEIN mRNA AS TRANSLATIONAL ENHANCERS IN TOBACCO PROTOPLASTS

I.L. BAGYAN[1], A.S. KRAEV[2], G.E. POZMOGOVA, K.G. SKRYABIN
Centre of Bioengineering, Academy of Sciences of Russia, Moscow, Russia.
[1] *Current address: Department of Biochemistry, University of Connecticut Health Center, Farmington CT 06032, USA.*
[2] *Current address: Institute of Biochemistry, Swiss Federal Insitute of Technology (ETH) , 8092 Zurich, Switzerland.*

Heterologous expression of proteins can be dramatically improved as a result of optimized mRNA translation. 5'-Untranslated mRNA sequences (leaders of translation) are one of the possible targets for such an optimization. "Translational enhancers" for foreign protein expression in plants could be derived from 5'-leaders of genomic RNAs of positive-sense RNA viruses. Genomic RNA of these viruses functions as messenger RNA in infected plant cell and apparently efficiently competes with host mRNAs for the components of translation machinery. Alternatively, subgenomic RNAs for viral coat proteins, featuring their own translation initiation sites should be efficiently translated, since coat proteins are synthesized in a substantial amount during viral reproduction.

In our search for new efficient leaders of translation we have tested two leaders of viral origin: leader of BSMV (Russian strain) RNAγ_b and leader of PVX coat protein mRNA. We tested the ability of these leader sequences to provide efficient translation of a *gus* reporter gene in a transient expression system in tobacco protoplasts and compared of their efficiency with well known translational enhancers - 5'-leaders of AlMV RNA 4 and BMV RNA 3.

1. Materials and Methods

Double stranded DNA fragments containing the leader sequences to be tested were obtained by PCR using chemically synthesized single stranded oligonucleotides as templates. PCR products were treated with T4 polynucleotide kinase and *Nco*1 and cloned into pMon755 (EP 0 339009 A2) vector, cut with *Stu*1 and *Nco*1 , between promoter CaMV 35Se and the GUS ORF. Reference plasmid, containing *cat* reporter

C. Nicolini (ed.), Genome Structure and Function, 319–323.
© 1997 *Kluwer Academic Publishers. Printed in the Netherlands.*

gene, was obtained by replacement of the GUS ORF in pMon755 with the cat ORF from the mammalian vector pSV2cat.

A rapidly growing suspension culture of *Nicotiana tabacum* cv. *Wisconsin* 38 was used for protoplast isolation. Plasmid preparations used for protoplast transformation were purified by RNAse treatment, phenol extraction and PEG-precipitation. Protoplasts were transfected with DNA by electroporation using a Cellject device (Eurogentec SA), essentially as recommended by the manufacturer. After incubation of electroporated protoplasts in incubation medium for 36-42 hours they were harvested and extracts were prepared. GUS activity was assayed in freeze-thaw extracts of the protoplasts according to [5], using 4-MUG as a substrate; fluorescence values were converted to pmoles MU produced per minute.

CAT activity in the extracts was analyzed as described [1] using 14C-chloramphenicol (Amersham plc) as a substrate. Radioactivity in the chromatographically separated acetylated derivatives was quantitated with a scintillation counter. CAT activity was calculated in relative units - as a ratio of radioactivity associated with mono- and diacetylchloramphenicol and total radioactivity per assay, and expressed as a fraction of 1. These values were used to correct the GUS activity values for the differences in transfotmation efficiency.

2. Results and Discussion

A number of plant viral leader sequences were previously shown to be efficient enhancers of heterologous mRNA translation, such as 5'-leader of genomic RNA of tobacco mosaic virus (TMV) [3, 14], alfalfa mosaic virus (AlMV) RNA 4 [4, 6], brom mosaic virus (BMV) RNA3 [4], tobacco etch virus (TEV) RNA [2], potato virus X (PVX) genomic RNA [12, 16], pea seedborne mosaic virus (PSbMV) RNA [10], and 5'-leader of potato virus S (PVS) coat potein mRNA [15]. In the course of our search for new efficient leaders of translation we chose for testing two leaders of viral origin, the 5'-leader of genomic RNAγc of barley stripe mosaic virus (BSMV) Russian strain [8] (fig.1), and the region 81 nt upstream of PVX coat protein open reading frame [13], herein referred to as PVX cp mRNA leader. Both BSMV RNAγb and PVX coat protein mRNA were known to be efficiently translated *in vitro* [9, 11]. To provide a negative control (presumably an inefficient leader of translation) we took a leader of RNAγb of BSMV strain Type which is known to be translated unefficiently [11]. This leader contains a short open reading frame (sORF) due to single base pair substitutions in the positions 57 and 73 from the 5'end. This sORF was shown to be responsible for an inefficient translation of the viral RNA *in vitro* [11]. We also tested a coincidential deletion variant of the BSMV RNAγc leader (designated as BSMV RNAγb(del27)) lacking 27 nt from the central part of the leader (see fig. 1).

The leaders, chosen for this work, were compared with the two previously described translational enhancers - 5' leaders of AlMV RNA 4 [4] and BMV RNA 3 [4] for their ability to provide efficient expression of *gus* reporter gene in a transient expression system in tobacco protoplasts. DNA fragments corresponding to the leader sequences were cloned between CaMV 35Se promoter and the GUS ORF in the plasmid pMon755 (EP 0 339009 A2). The resulting plasmids were introduced into protoplasts of *Nicotiana tabacum* cv. *Wisconsin* 38 via electroporation. After electroporation, protoplasts were incubated in recovery medium for 36-42 hours to allow expression of the introduced *gus* gene. GUS activity was determined in the crude extracts of the protoplasts. To take into account variations in electroporation efficiency which affect level of transient expression, in each transformation we introduced two plasmids to the protoplasts - one containing a leader-*gus* fusion and the other plasmid (the same in all transformations) containing *cat* reporter gene. CAT activity was measured in the same extracts as the GUS activity and taken as a measure of elecroporation efficiency. Then we divided GUS activity value by the CAT activity value to correct the GUS activity values for different electroporation efficiencies, the resultant value was taken as a measure of the leader efficiency. The values thus obtained were further normalized so that the value corresponding to the AlMV RNA4 leader was taken to be 1.0. Table I summarizes the results. It represents the data obtained from three independent experiments. Each experiment included two transformations for each leader. For each leader, the results from independent experiments actually differed by no more than 10 per cent of the average value.

TABLE I.

Leader	Relative efficiency
AlMV RNA4	1
BSMV RNA γ_b (Russian strain)	1.6
BSMV RNA γ_b (strain Type)	0.2
BMV RNA3	1
BSMV RNA γ_b (del27)	1.6
PVX cp mRNA	1

The results show that the PVX coat protein mRNA leader proved to be equally efficient and BSMV RNA γ_b (Russian strain) leader almost two fold more efficient than AlMV RNA4 and BMV RNA3 leaders. Our test system was clearly capable of detecting inefficient leaders - the leader of RNA γ_b of BSMV strain Type, containing a short open reading frame (sORF), was 8 times less efficient than the leader of the RNA γ_c of BSMV Russian strain, which does not contain any sORF due to single base pair substitutions. Our data are thus in compliance with the ribosome scanning model of translation [7]. Deletion of 27 nucleotides from the central portion of the BSMV RNA γ_b leader does not affect its efficiency as a translational enhancer. One can argue that the closeness of the figures for BSMV RNA γ_b (Russian strain) and BSMV RNA γ_b (del27) leaders could be due to a saturation of the system capacity. To test this possibility the increase of GUS activity in protoplasts transformed with different BSMV leader-*gus* fusions was monitored during the first day after electroporation. GUS activity was increasing with the same rate for both BSMV RNA γ_b (Russian strain) and BSMV RNA γ_b (del27) leaders, but with considerably lower rate for the leader of RNA γ_b of BSMV strain Type, indicating that the former two leaders have really equal efficiencies in tobacco protoplasts. Thus, 5'-leaders of BSMV RNA γ_b (Russian strain)

322

and PVX coat protein mRNA can be used to optimize expression of heterologous genes in plant cells.

BSMV (Russian strain)	ACACGTTATAGCTTGAGCATTACCGTCGTGT
BSMV (strain Type)	ACACGTTATAGCTTGAGCATTACCGTCGTGT
BSMV (del27)	ACACGTTATAGCTTGAGCATTACCGTCGTGT
BSMV(Russian strain)	AATTGCAACACTTGGCTTGCCAAATAACGCT
BSMV (strain Type)	AATTGCAACACTTGGCTTGCCAAATGACGCT
BSMV (del27)	AATTGCAACACTT[
BSMV(Russian strain)	AAAGCGTTCACGAAACAAACAACAATTCGG
BSMV (strain Type)	AAAGCGTTCTTGAAACAAACAACAATTCGG
BSMV (del27)	]TTGAAACAAACAACAATTCGG

Figure 1. Sequences of the three variants of the BSMV RNAγ leader, tested in this work. Nucleotide substitutions present in the leader of RNA γ of BSMV (strain Type) are underlined. Region absent from BSMV RNA γ(del27) leader is indicated with brackets.

3. Acknowledgments

This work was done in frame work of the joint project Monsanto company-Russian Academy Sciences Laboratory
The Authors are grateful to A. Shchennikova for help in the manuscript preparation.

4. References

1. Current Protocols in Molecular Biology, unit 9.7A, John Willey and Sons, Inc., 1995.
2. Carrington, J.C., Freed, D.D. (1990) Cap-independent translation by a plant potyvirus 5'-nontranslated region, *J. Virol.* 64, 1590-1597.
3. Gallie, D.R., Sleat, D.E., Watts, J.W., Turner, P.C., Wilson, T.M.A. (1987) The 5'-leader sequence of tobacco mosaic virus RNA enhances the expression of foreign gene transcripts *in vitro* and *in vivo*, *Nucl. Acids Res.* 15, 3257 -3273.
4. Gallie, D.R., Sleat, D.E., Watts, J.W., Turner, P.C., Wilson, T.M.A. (1987) A comparison of eukaryotic viral 5'-leader sequences as enhancers of mRNA expression *in vivo*, *Nucl. Acids Res.* 15, 8693-8711.
5. Jefferson, R.A., Kavanagh, T.A., Bevan, M.W. (1987) GUS fusions: b-glucuronidase as a sensitive and versatile gene fusion marker in higher plants, *EMBO J.* 6, 3901-3907.
6. Jobling, S.A., Gehrke, L. (1987) Enhanced translation of chimaeric messenger RNA containing a plant viral untranslated leader sequence, *Nature* 325, 622-625.
7. Kozak, M. (1987) Effects of intercistronic length on the efficiency of reinitiation by eucaryotic ribosomes, *Mol. Cell. Biol.* 7, 3438-3445.

8. Kozlov, Iu.V., Afanas'ev, B.N., Rupasov, W., Golova, Iu.B., Kulaeva, O.I., Dolia, W., Atabekov, I.G., Baev, A.A. (1989) Primary structure of RNA3 of barley stripe mosaic virus and its variability, *Mol. Biol.* 23, 1080-1090.

9. Morozov, S.Yu., Miroshnichenko, N.A., Solovyev, A.G., Fedorkin, O.N., Zelenina, D.A., Lukasheva, L.I., Karasev, A.V., Dolja, V.V., Atabekov, J.G. (1991) Expression strategy of the potato virus X triple gene block, *J. Gen. Virology* 72, 2039-2042.

10. Nicolaisen, M., Johansen, E., Poulsen, G.B., Borkhardt, B. (1992) The 5'untranslated region from pea seedborne mosaic potyvirus RNA as a translational enhancer in pea and tobacco protoplasts, *FEBS Lett.* 303, 169-172.

11. Petty, I.T.D., Edwards, M.C., Jackson, A.O. (1990) Systemic movement of an RNA plant virus determined by a point substitution in a 5'-leader sequence, *Proc.Natl.Acad.Sci USA* 87, 8894-8897.

12. Pooggin, M.M., Skryabin, K.G. (1992) The 5'-untranslated leader sequence of potato virus X RNA enhancers the expression of a heterologous gene *in vivo*, *Mol. Gen. Genet.* 234, 329-331.

13. Skryabin, K.G., Kraev, A.S., Morozov, S.Yu., Rozanov, M.N., Chernov, B.K., Lukasheva, L.I., Atabekov, J.G. (1988) The nucleotide sequence of potato virus X RNA, *Nucl. Acids Res.* 16, 10929-10930.

14. Sleat, D.E., Gallie, D.R., Jefferson, R.A., Bevan, M.W., Turner, P.V., Wilson, T.M.A. (1987) Characterization of 5'-leader sequence of tobacco mosaic virus RNA as a general enhancer of translation *in vitro*, *Gene* 217, 217-225.

15. Turner, R., Bate, N., Twell, D., Foster, G.D. (1994) *In vivo* characterization of a translational enhancer upstream from the coat protein open reading frame of potato virus S, *Arch. Virol.* 137, 123-132.

16. Zelenina, D.A., Kulaeva, O.I., Smirnyagina, E.V., Solovyev, A.G., Miroshnichenko, N.A., Fedorkin, O.N., Rodionova, N.P., Morozov, S.Yu., Atabekov, J.G. (1992) Translation enhancing properties of the 5'-leader of potato virus X genomic RNA, *FEBS Lett.* 296, 267-270.

SUBJECT INDEX